Equations of State in Engineering and Research

Equations of State in Engineering and Research

Kwang Chu Chao, Editor
Purdue University

Robert L. Robinson, Jr., Editor
Oklahoma State University

Based on a symposium sponsored by the Division of Industrial and Engineering Chemistry at the 176th Meeting of the American Chemical Society, Miami Beach, Florida, September 11–14, 1978.

ADVANCES IN CHEMISTRY SERIES **182**

AMERICAN CHEMICAL SOCIETY

WASHINGTON, D. C. 1979

Library of Congress CIP Data

Equations of the state of engineering and research.

(Advances in chemistry series; 182 ISSN 0065-2393)

"Based on a symposium sponsored by the Division of Industrial and Engineering Chemistry at the 176th Meeting of the American Chemical Society, Miami Beach, Florida, September 11–14, 1978."

Includes bibliographies and index.

1. Equations of state. 2. Engineering mathematics. I. Chao, Kwang-chu, 1925– . II. Robinson, Robert Louis, 1937– . III. American Chemical Society. Division of Industrial and Engineering Chemistry. IV. Series.

QD1.A355 no. 182 [QC173.3] 540'.8s [541'.042]
ISBN 0-8412-0500-0 79-23696 ADCSAJ 182
1-466 1979

PRINTED IN THE UNITED STATES OF AMERICA

FOREWORD

ADVANCES IN CHEMISTRY SERIES was founded in 1949 by the American Chemical Society as an outlet for symposia and collections of data in special areas of topical interest that could not be accommodated in the Society's journals. It provides a medium for symposia that would otherwise be fragmented, their papers distributed among several journals or not published at all. Papers are reviewed critically according to ACS editorial standards and receive the careful attention and processing characteristic of ACS publications. Volumes in the ADVANCES IN CHEMISTRY SERIES maintain the integrity of the symposia on which they are based; however, verbatim reproductions of previously published papers are not accepted. Papers may include reports of research as well as reviews since symposia may embrace both types of presentation.

CONTENTS

PREFACE

Since the pioneering work of Benedict and co-workers in the early 1940's, equations of state have assumed an expanding role in engineering in the representation of fluid volumetric, thermodynamic, and phase-equilibrium behavior. The past decade has seen the equation-of-state method assume center stage in property predictions, particularly in fluid-phase equilibria, with an attendant rapid growth of literature on the subject. Several forces have converged to promote this growth. High-speed computing capabilities have continued to improve, permitting new equations to be developed and tested in much more detail and much more critically than was previously possible. New data have been appearing at an accelerated rate, much of it designed specifically to be useful for developing and testing models for systems of interest. Significant progress in the theoretical understanding of dense fluids has been made through computer-generated data from molecular dynamics and Monte Carlo calculations and from perturbation and other statistical mechanical theories. Last, but not least, computer-implemented process design systems have been gaining ground, calling for properties' prediction packages for which equations of state are in great demand.

The present volume brings together a wide spectrum of investigations of equations of state in research and engineering. The extensive experiences of recent years in using cubic equations of the Redlich–Kwong type are described; the strengths and limitations of these equations in a variety of engineering applications are discussed. New directions are explored, including new experimental data and equation-of-state analyses of: the critical region; gases dissolved in liquids; using bifurcation and instability in equation development; the representation of polar mixture behavior; and of new equations of great accuracy and complexity. The problem of mixing rules is addressed and new methods are suggested.

This volume shares some common interest with the state-of-the-art ACS Symposium Series No. 60 entitled "Phase Equilibria and Fluid Properties in the Chemical Industry." However, the present volume focuses more sharply on equations of state and is not exclusively a collection of reviews; the two volumes complement each other.

The editors appreciate the encouragement and support of the Division of Industrial and Engineering Chemistry of the American Chemical Society for organizing this symposium. Special thanks go to those who prepared the manuscripts, participated as speakers at the symposium, and reviewed the papers for this volume. Finally we gratefully acknowledge Purdue and Oklahoma State Universities for their generous support.

KWANG CHU CHAO
Purdue University

ROBERT L. ROBINSON, JR.
Oklahoma State University

May, 1979

1

Practical Calculations of the Equation of State of Fluids and Fluid Mixtures Using Perturbation Theory and Related Theories

DOUGLAS HENDERSON

IBM Research Laboratory, San Jose, CA 95193

The properties of a fluid are determined largely by short-range repulsive forces. The long-range attractive forces can be considered to be perturbations. Using these concepts, a perturbation theory of fluids is developed. In addition, the relationship of empirical equations of state to the perturbation theory is examined. The major weakness of most empirical equations is the use of the van der Waals free-volume term, $(V-Nb)^{-1}$*, to represent the contributions of the repulsive forces. Replacement of this term by more satisfactory expressions results in better agreement with experiment.*

The requirements of an equation of state from a theoretical chemist and a chemical engineer are somewhat different. The theoretical chemist desires to understand the origin of the properties of the fluid he is studying and often is less interested in obtaining highly accurate agreement with experimental data. On the other hand, the chemical engineer wants a simple, empirical equation of state which is in close agreement with experimental data. The question of whether this empirical equation of state has any theoretical basis is less interesting.

Because of this apparent divergence of interests and needs, there has been little interaction between theoretical chemists and chemical engineers working on the equation of state of fluids. This is unfortunate because the requirements of the two groups are compatible. No theoretician would claim to understand fully some phenomena if he could not obtain reasonable quantitative agreement with experiment. On the other hand, an empirical equation of state with a weak or even faulty theoretical

0-8412-0500-0/79/33-182-001$07.50/1

basis is no more than an interpolation scheme and is quite useless for extrapolation to thermodynamic states for which experimental data are not available. Presumably chemical engineers would prefer to have some predictive capability and if the results of the theoretician can be expressed in some useful form which is convenient for quick calculation, chemical engineers would be interested.

I assume that this gap between theoretical chemists and chemical engineers exists because, until very recently, theoreticians had little to offer concerning the theory of fluids. However, during the past decade rapid progress has been made in this area and an attempt to bridge this gap now seems appropriate. This is an ambitious task and given the deadlines which are an unavoidable part of any conference, it is not a task that I would claim to accomplish fully here. However, I hope that this chapter will contribute to the bridging of this gap.

I shall attempt to survey recent progress in the theory of dense fluids. I will provide references to all of the major techniques. However, I will emphasize perturbation theory because I feel that this is the technique which is most interesting to chemical engineers. Further, I will show that the perturbation theory can be used in part to justify common empirical equations of state. Many of these equations are well founded in theory. However, we shall see that there is one term which seems to appear in all empirical equations of state. This term has absolutely no theoretical basis and it should be discarded and replaced by a more satisfactory expression.

Some General Considerations

The basic result in statistical mechanics is that the probability of a system being in a state specified by an energy E_i is proportional to the Boltzmann factor $\exp\{-\beta E_i\}$, where $\beta = 1/kT$. With this result, thermodynamic properties may be specified. For example, the thermodynamic energy is

$$U = \frac{\sum_i E_i \exp\{-\beta E_i\}}{\sum_i \exp\{-\beta E_i\}}$$

$$= -\frac{\partial \ln Z_N}{\partial \beta}, \tag{1}$$

where

$$Z_N = \exp\{-\beta A\}$$
$$= \sum_i \exp\{-\beta E_i\} \tag{2}$$

is called the partition function and A is the Helmholtz free energy (i.e., $A = U - TS$). With exception of a few fluids, such as helium and hydrogen, the energy levels E_i form a continuum and the sum in Equation 2 can be replaced by an integral. Thus, for a system of N molecules

$$Z_N = \frac{1}{h^s N!} \int \exp\{-\beta \mathcal{H}\}\, dp_s\, dq_s \tag{3}$$

where the p's and q's are the generalized momenta and coordinates, s is the number of degrees of freedom, and $\mathcal{H}$ is the Hamiltonian of the system. The factor h^s arises from the volume associated with quantum states in phase space and the factor $N!$ appears because the molecules in a fluid are indistinguishable.

Generally the molecules of the fluid will have internal degrees of freedom. If these internal degrees of freedom are independent of the density of the fluid (as is often the case) they make no contribution to the equation of state and can be ignored. Since I am interested only in presenting general principles, I will ignore the contribution of internal degrees of freedom and assume that

$$\mathcal{H} = \sum_{i=1}^{3N} \frac{p_i^2}{2m} + \Phi(r_1, \ldots r_N) \tag{4}$$

where Φ is the potential energy of the molecules and depends only upon the positions of the center of mass. The theory of fluids in which internal degrees of freedom contribute to the equation of state is still under development. Using Equation 4, the partition function becomes

$$Z_N - \frac{\lambda^{-3N}}{N!} \int \exp\{-\beta\Phi\} d\mathbf{r}_1 \ldots d\mathbf{r}_N \tag{5}$$

where $\lambda = h/(2\pi mkT)^{1/2}$.

To make further progress the form of Φ must be specified. Generally the potential energy will contain terms involving the coordinates of pairs, triplets, quadruplets, etc., of molecules. A few calculations of the equation of state of a fluid with such a general form for Φ have been made. However, the common practice is to assume pair-wise additivity:

$$\Phi(\mathbf{r}_1, \ldots, \mathbf{r}_N) = \sum_{i<j} u(r_{ij}) \tag{6}$$

In this case, $u(r)$ is not the correct pair interaction but some effective pair interaction which simulates the multibody terms.

Often

$$u(r) = \epsilon\varphi(r/\sigma) \tag{7}$$

where φ is some universal function which is applicable to a wide class of substances. Substitution of Equation 7 into Equation 5 leads to the law of corresponding states which states that for such substances the thermodynamic functions are also universal functions that are scaled by appropriate combinations of ϵ and σ. Thus,

$$P^* = \mathrm{f}(\rho^*, T^*) \tag{8}$$

where $P^* = P\sigma^3/\epsilon$, $\rho^* = \rho\sigma^3$, and $T^* = kT/\epsilon$ are the reduced pressure, density, and temperature.

The above form of the law of corresponding states may seem unsuitable for a chemical engineer since it seems to require the determination of $u(r)$ for the fluid of interest. However, if the temperature and density of the fluid at some fundamental state (e.g. the critical point or the triple point) are known, then ϵ and σ may be determined using the law of corresponding states. For example,

$$\begin{aligned} \epsilon &= c_1 T_c \\ \sigma^3 &= c_2 V_c \end{aligned} \tag{9}$$

where T_c and V_c are the critical temperature and volume and c_1 and c_2 are universal constants. Thus, an equivalent statement of the law of corresponding states is

$$P/P_c = \mathrm{g}(V/V_c, T/T_c) \tag{10}$$

where P_c is the critical pressure.

One useful test of the law of corresponding states is the invariance of certain dimensionless terms: e.g., for many fluids

$$z_c = P_c V_c / NkT_c \simeq 0.291 \tag{11}$$

As we would expect, there are deviations from Equation 11 for fluids with internal degrees of freedom.

Computer Simulations and Integral Equations

I would like to turn to the question of the calculation of the partition function. There are three methods of obtaining Z_N and the thermodynamic properties. I will consider two of these methods in this section and the third in the next section. The first method is computer simulation. In this method a set of about 100 molecules in a box with periodic boundary conditions (to minimize surface effects) is considered and either the time or statistical evolution of the system followed. A detailed review (*1*) is available so I will not consider this method in detail. The method involves large computations and will not become a routine tool in chemical engineering. However, it is completely general: to date it is the only technique generally applicable to molecules with internal degrees of freedom. Further, it provides (in principle) complete information about the system, including the h-body distribution functions,

$$g(\mathbf{r}_1, \ldots, \mathbf{r}_h) = V^h \frac{\int \exp\{-\beta\Phi\} d\mathbf{r}_{h+1} \ldots d\mathbf{r}_N}{\int \exp\{-\beta\Phi\} d\mathbf{r}_1 \ldots d\mathbf{r}_N} \tag{12}$$

as well as the thermodynamic functions. Computer simulations do not give Z_N directly. However, derivatives of Z_N are obtained and Z_N can be obtained by integration over a series of states. The only limits on the computer simulation method are our ingenuity in programming the computer and our ability to cope with the numerical data. The later limitation is nontrivial and prevents detailed consideration of four- and higher-body distribution functions. Finally, we may regard the computer simulation methods as either an experimental or a theoretical tool. Generally speaking it is used as an experimental tool for providing data with which theoretical calculations may be compared. Apart from statistical problems and the question of whether the box contains a sufficiently large number of molecules, the method is exact. The results of some computer simulations of hard spheres are given in Figure 1.

The second method is the integral equation method. In this approach some approximate integral equation for the radial distribution function, $g(r)$, is formulated and solved. For the simple molecules without internal degrees of freedom which I am considering, the radial distribution function (RDF) is the pair ($h = 2$) distribution function (PDF) defined by Equation 12. The integral equation method involves much less computer usage than do the computer simulations but still involves enough to make its use as a routine chemical-engineering tool unlikely.

However, in certain cases where these integral equations yield analytical solutions, the method will be interesting to chemical engineers. For instance, Wertheim (*2,3*) and Thiele (*4*) have solved the Percus–Yevick (PY) equation for the hard-sphere potential, where

$$u(r) = \begin{cases} \infty, r < d \\ 0, r > d \end{cases} \tag{13}$$

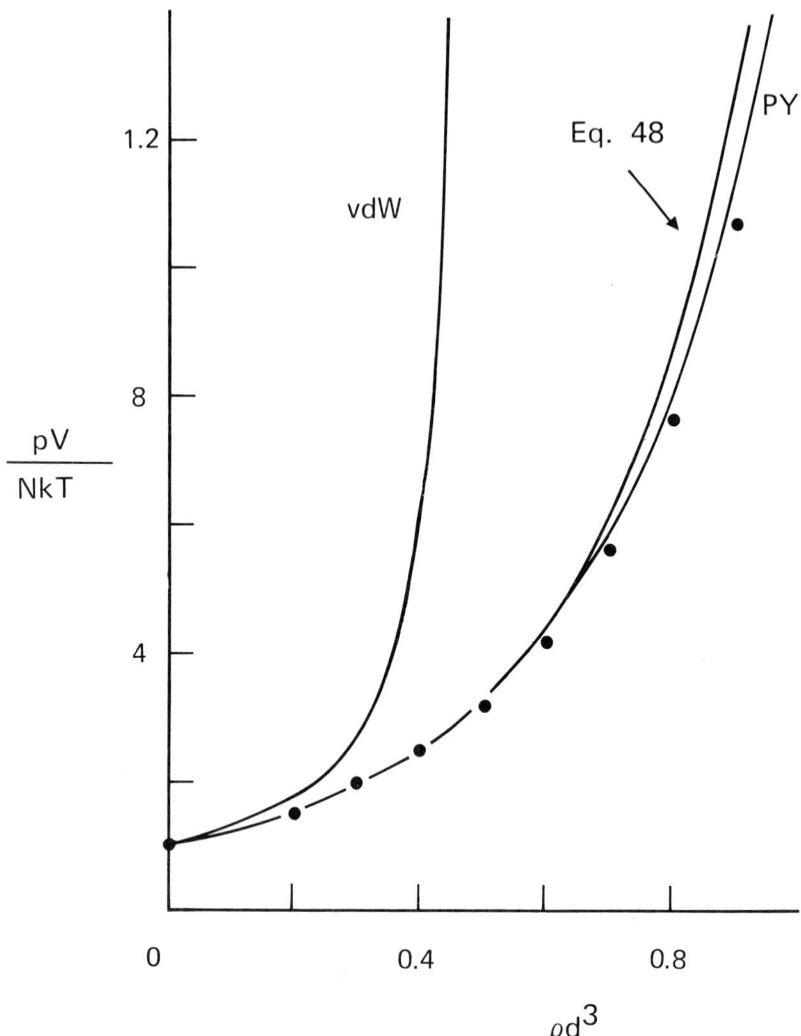

Figure 1. Equation of state of the hard-sphere fluid. The points give computer simulation results and curves give the results of three approximations.

They showed that

$$A/NkT = 3 \ln \lambda - 1 + \ln \rho - \ln (1 - \eta) + \frac{3}{2} \eta \frac{2 - \eta}{(1 - \eta)^2} \quad (14)$$

and

$$P/\rho kT = \frac{1 + \eta + \eta^2}{(1 - \eta)^3} \quad (15)$$

where $\eta = \pi \rho d^3/6$. Equation 15 is compared with computer simulations in Figure 1. They showed further that the Laplace transform of $g(r)$ is given by

$$\begin{aligned} G(s) &= \int_1^\infty xg(x) \exp(-sx) dx \\ &= \frac{sL(s)}{12\eta[L(s) + \exp(s)S(s)]} \end{aligned} \quad (16)$$

where $x = r/\sigma$,

$$L(s) = 12\eta[1 + 2\eta + (1 + \eta/2)s] \quad (17)$$

and

$$S(s) = -12\eta(1 + 2\eta) + 18\eta^2 s + 6\eta(1 - \eta)s^2 + (1 - \eta)^2 s^3 \quad (18)$$

Smith and Henderson (*5*) have inverted $G(s)$ analytically for $1 \leq x \leq 5$. For the hard-sphere potential, these PY results for *A*, *P*, and $g(r)$ are in good agreement with computer simulations.

A second case of potential interest to chemical engineers is Waisman's solution (*6*) of the mean sperical approximation (MSA) integral equation for the case where

$$u(r) = \begin{cases} \infty, r < \sigma \\ -\epsilon \exp\{-z(x - 1)\}/x, r > \sigma \end{cases} \quad (19)$$

and $x = r/\sigma$. Waisman's solution is implicit and involves the solution of six nonlinear equations. However, Henderson (*7*) has shown recently that a good parametrization of Waisman's solution is given by

$$\begin{aligned} A/NkT = A_o/NkT - \beta\epsilon v_1/2 - (\beta\epsilon)^2 v_2/4 - \\ (\beta\epsilon)^3 v_3/6 - (\beta\epsilon)^4 v_4/8 \end{aligned} \quad (20)$$

where A_0 is given by Equation 14,

$$v_1 = \frac{2z\mathrm{L}(z)}{e^{-z}\mathrm{L}(z) + \mathrm{S}(z)} \tag{21}$$

$$v_2 \simeq \frac{12}{z}\eta(1-\eta)^{12c} \tag{22}$$

$$v_3 \simeq \frac{144(2-3e^{-z}/2)}{z^3}\eta^2(1-\eta)^{19c} \tag{23}$$

$$v_4 \simeq \frac{1728(11/2-8e^{-z}+3e^{-2z})}{z^5}\eta^3(1-\eta)^{26.3c} \tag{24}$$

and

$$c = \frac{(1.25-0.35\,e^{-z^3/24})}{z^3}\left[-4+2z^2+4(1+z)e^{-z}\right] \tag{25}$$

Equation 21 gives the exact result for v_1 in the MSA. The expressions for v_2, v_3, and v_4 are parametrizations. Equation 21 is analytical and quite easy to use. For most cases, v_2, v_3, and v_4 are small when compared with v_1 so that it is best to use Equation 21 for v_1 rather than some approximation. However, if one wishes, simplifications can be obtained from an expansion of Equation 21 in powers of ρ and z. One possibility is

$$v_1 \simeq 24\eta\,\frac{1+z}{z^2}\left[1+\frac{77z^2}{10(1+z)(7+2z)}\,\eta\right] \tag{26}$$

Except for $z \to \infty$, most of the contribution to v_1 comes from the first term. The limit $z \to \infty$ is mainly of mathematical interest. In most situations of physical interest z is small.

We refer to the literature (*1*) for a discussion of the derivation of the various integral equations and for details regarding their solution (usually numerical) for other cases.

Perturbation Theory

Perturbation theory is the oldest of the three methods. We will see that it dates back to van der Waals. However, its utility was not appreciated by theorists until the last decade.

In perturbation theory we assume that we have full knowledge of some reference system (or unperturbed system) whose properties we will denote by a subscript 0. We may have obtained this knowledge by means of some computer simulations or from the solution of some integral equation. Usually, this reference system is taken to be the hard-sphere fluid, where

$$u_0(r) = \begin{cases} \infty, & r < d \\ 0, & r > d \end{cases} \quad (27)$$

We assume that, to some approximation, the pair potential is $u_0(r)$ and a small perturbation. The simplest case is

$$u(r) = u_0(r) + \epsilon w(r) \quad (28)$$

If the free energy is expanded in powers of ϵ, we have

$$\frac{A - A_0}{NkT} = \sum_{n=1}^{\infty} (\beta\epsilon)^n A_n/NkT \quad (29)$$

where

$$A_1/NkT = \frac{1}{2}\rho \int w(r) g_0(r)\, d\mathbf{r} \quad (30)$$

and $g_0(r)$ is the RDF of the reference system. The higher-order A_n involves integrals over higher-order distribution functions.

If the reference system is the hard-sphere system, A_0 may be calculated from Equation 14. Carnahan and Starling (*8*) have proposed a slight modification of Equation 14 which is slightly more accurate. Using the Carnahan and Starling expression is certainly recommended; however, I wish to give the following warning. The analogue of the Carnahan and Starling equation of state becomes very inaccurate for a mixture of hard spheres when one of the components in the mixture is very large whereas the analogue of Equation 14 remains accurate. In as much as the Carnahan and Starling-type expression is in better agreement with computer simulations for hard-sphere mixtures for diameter ratios at least as large as 3:1, it is probable that this deficiency probably is irrelevant to any practical calculation. However, one should be wary not only to avoid application of the Carnahan and Starling-type expression for extremely large diameter ratios but to examine carefully any predictions based on the use of this expression in situations in which it has not been studied in detail. For a hard-sphere reference fluid, $g_0(r)$ and thus A_1, can be obtained either from the PY hard-sphere results (*5*) or from computer simulations (*9, 10*). For most cases, A_1 must be obtained by numerical integration. However, if

$$w(r) = -\epsilon \exp\{-z(x-1)\}/x \quad (31)$$

where $x = r/\sigma$ and the PY $g_0(r)$ are used, Equation 16 may be used to yield

$$A_1NkT = -\frac{z\mathrm{L}(z)}{e^{-z}\mathrm{L}(z) + \mathrm{S}(z)} \tag{32}$$

The similarity to Equation 21 is not accidental.

If a hard-sphere reference fluid is used, the second-order term has the form

$$A_2/NkT = -\pi\rho d^3 \int_1^\infty r^2[w(rd)]^2 g_0(r)\,dr + \int_1^\infty \int_1^\infty w(r_1d)w(r_2d)\ F_0(r_1, r_2)\,dr_1\,dr_2 \tag{33}$$

Barker and Henderson (*10*) have given a convenient parametrization of $F_0(r_1, r_2)$ for the hard-sphere reference fluid.

So far all we have is formalism. One could argue that there is no reason to believe that Equation 29 is useful except at high temperatures, where $\beta\epsilon$ is small. The utility of perturbation theory even at temperatures as low as the triple point, was first pointed out a decade ago by Barker and Henderson (*11*) who argued that the relevant parameter in determining the convergence of Equation 29 was not the smallness of $\beta\epsilon$ but the smallness of the effect of the perturbation on the structure of the fluid. They noted that A_1 gives the effect of $w(r)$ on the thermodynamic properties in the absence of any changes in structure and that A_2 gives the effect of changes in structure. At high densities such changes in structure are suppressed because the molecules are packed tightly. Therefore, A_2 is small compared with A_1 (as is observed by direct calculation of A_1 and A_2). The higher-order A_n are even smaller and, as Barker and Henderson suggested, can be neglected in most calculations.

At lower densities, particularly near the critical point, changes in structure are easier and the convergence of the perturbation expansion is slower. But even there, second-order perturbation theory gives quite reasonable results.

All of this makes sense as long as $u(r)$ is the hard-sphere potential plus $\epsilon w(r)$. Unfortunately, the potentials which occur in real applications are not of this form. Nevertheless, following Barker and Henderson (*12*), we can write

$$u(r) = u_0(r) + \epsilon w(r) \tag{34}$$

where $u_0(r)$ and $w(r)$ are the positive and negative parts of $u(r)$. Thus, we can use Equations 29 and 30. However, we now have an unfamiliar reference fluid. Fortunately, for most applications $u_0(r)$ is very steep and can be replaced by the hard-sphere potential if the diameter d is chosen judiciously. By means of a formal expansion in power of an inverse steepness parameter, Barker and Henderson showed that

$$d = \int_0^\sigma [1 - \exp\{-\beta u_0(z)\}]dz \tag{35}$$

where σ is the point at which $u(r)$ changes sign. With the above choice of d, a useful second-order perturbation theory is obtained from Equations 14, 29, 30, and 32.

The results of this second-order perturbation theory for a fluid whose pair potential is the Lennard–Jones 6:12 potential,

$$u(r) = 4\epsilon\left\{\left(\frac{\sigma}{r}\right)^{12} - \left(\frac{\sigma}{r}\right)^{6}\right. \tag{36}$$

are given in Figure 2. The agreement with computer simulation is excellent.

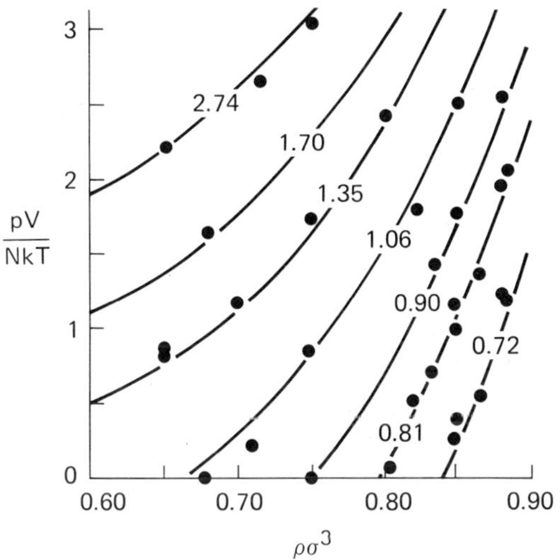

Figure 2. Equation of state of the 6:12 fluid. The points and curves give computer simulation and the second-order perturbation theory results for seven isotherms that are labelled with the appropriate values of kT/ε. For the Lennard–Jones fluid the triple-point reduced temperature and density are about 0.7 and 0.85, respectively, and the critical-point reduced temperature and density are about 1.30 and 0.30, respectively.

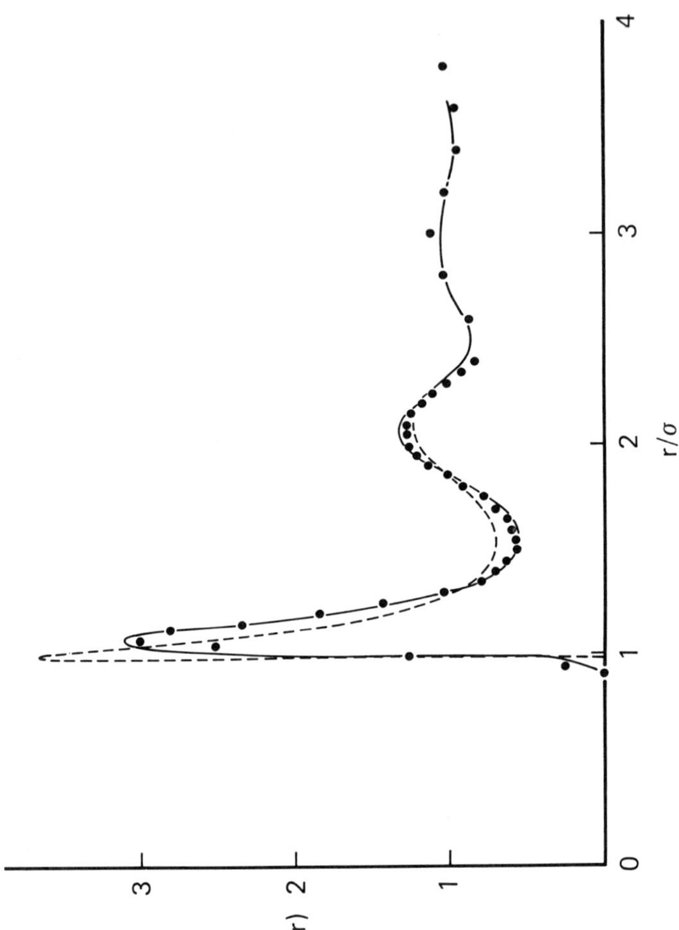

Accounts of Chemical Research

Figure 3. RDF of the 6:12 fluid near its triple point. The points give the simulation results and the broken and solid curves give the results for zero- and first-order perturbation theories. $T^* = 0.72$; $\rho^* = 0.85$.

Perturbation theory also may be applied to calculate the RDF of a fluid. The results of such a calculation of the RDF of a 6:12 fluid are given in Figure 3.

The physical content of perturbation theory can be summarized in the statement that the structure and thermodynamic properties of a fluid are determined mainly by the short-range hard (or quasi-hard) cores of the molecules. The long-range part of the pair potential is a perturbation. The fluid behaves as a hard-sphere fluid moving in a uniform background energy provided by the long-range potentials.

The perturbation theory outlined here is a practical calculational tool in chemical engineering. All that is required is some choice of a pair potential and a determination of the potential parameter. For many cases, the 6:12 potential is a convenient choice because parameters ϵ and σ are tabulated (*13*) for many substances. However, it is somewhat inconvenient because d, A_1, and A_2 must be determined numerically and, as a result, the pressure must be determined by numerical differentiation. Clearly, a simple analytical equation of state is more convenient as an engineering tool. Nonetheless, perturbation theory can be a useful guide to the selection of an empirical equation of state. This question will be considered in the next section.

Empirical Equations of State

The earliest empirical equation of state is that of van der Waals. His equation of state can be obtained from first-order perturbation theory with a hard-sphere reference system. Thus

$$A/NkT = A_0/NkT + \beta\epsilon\, A_1/NkT \tag{37}$$

In van der Waals' time (or for that matter until 20 years ago) the equation of state of a hard-sphere fluid was not known. Thus, van der Waals assumed that A_0 was equal to the free energy of a perfect gas with the volume reduced to account for the fact that the molecules occupy space. Hence, in the van der Waals (VDW) equation

$$A_0/NkT = 3 \ln \lambda - 1 + \ln \{N/(V - Nb)\} \tag{38}$$

where $b = 2\pi d^3/3 = 4\eta$. Differentiation of Equation 38 gives

$$\begin{aligned} P_0/\rho kT &= \frac{1}{1-\rho b} \\ &= \frac{1}{1-4\eta} \end{aligned} \tag{39}$$

where $\eta = \pi\rho d^3/6$. Expanding Equation 39 to the power of ρb yields

$$P_0/\rho kT = 1 + \rho b + (\rho b)^2 + (\rho b)^3 + \ldots \tag{40}$$

which is in poor agreement with the correct expansion:

$$P_0/\rho kT = 1 + \rho b + \frac{5}{8}(\rho b)^2 + 0.2869(\rho b)^3 + \ldots \tag{41}$$

Clearly, as shown in Figure 1, using Equation 39 leads to a serious overestimation of the pressure.

The VDW treatment of A_1/NkT has a better foundation. This term is not sensitive to the precise form of the pair potential. Thus, Equation 32 is satisfactory for illustrating the form of A_1. Combining Equations 21, 26, and 32 shows that, except for potentials of an unphysically short range (i.e., $z \to \infty$), A_1/NkT is independent of T and nearly linear in the density. Thus

$$\beta\epsilon\, A_1/NkT = \beta\rho a \tag{42}$$

Equations 37, 38, and 42 give the VDW equation of state

$$P = \frac{NkT}{V - Nb} - \frac{N^2 a}{V^2} \tag{43}$$

The relation between parameters a and b and the critical constants P_c, V_c, and T_c are given in Table I. One serious deficiency of the VDW equation of state is the large value of $z_c = P_cV_c/NkT_c$. This means that a and b cannot be chosen to fit P_c, V_c, and T_c simultaneously.

Carnahan and Starling (*14*) have used the VDW equation to calculate the equation of state of methyl chloride, whose critical constants (*15*) are listed in Table II. Their results are given in Table III. As expected, the VDW equation grossly overestimates the pressure.

Table I. Relation Between Empirical Parameters and Critical Point Constants[a]

	VDW	*RK*	*LHW*	*MRK*
Nb/V_c	1/3	0.260	0.522	0.333
a/bkT_c	27/8	$4.934\sqrt{T_c}$	2.646	$4.999\sqrt{T_c}$
P_cV_c/NkT_c	3/8	0.333	0.359	0.316

[a] From Carnahan and Starling (*14*).

Table II. Critical Point Constants of Methyl Chloride[a]

T_c	P_c	V_c
143.1°C	65.919 atm	2.755 cm³/gm

[a] From Hsu and McKetta (*15*).

Table III. Comparison of Experimental and Calculated Pressures of Methyl Chloride for Some Emperical Equations of State

T (°C)	V (cm³/gm)	P (Expt)[a] (atm)	P (VDW)[b] (atm)	P (RK)[b] (atm)	P (LHW)[b] (atm)	P (MRK)[b] (atm)
125	88.266	6.975	7.063	7.02	7.051	7.006
	27.774	20.049	20.629	20.555	20.523	20.132
	16.333	30.664	32.017	31.109	31.749	30.814
	11.070	39.817	42.215	40.630	41.728	40.124
	7.204	50.015	52.848	50.407	52.039	49.649
143.7	67.069	9.567	9.632	9.571	9.616	9.549
	18.099	30.398	31.288	30.670	31.123	30.449
	8.321	51.077	53.818	52.291	53.421	51.731
	1.939	69.954	251.92	103.16	145.77	83.291
	1.610	101.510	936.55	200.86	328.59	125.46
	1.400	206.85	4265.0	398.17	666.61	196.88

[a] Hsu and McKetta (*15*).
[b] Calculated by Carnahan and Starling (*14*).

The origin of these serious errors is the poor approximation to A_0 given by Equation 38. Yet despite this, the universal assumption made by workers using empirical equations of state seems to be to retain the functional form of the VDW equation and replace V by $V + Nc$. Thus

$$p = \frac{NkT}{V + Nc - Nb} - \frac{N^2 a}{(V + Nc)^2} \tag{44}$$

Equation 44 can be written in a form closer to VDW by replacing $V + Nc - Nb$ by $V - Nb'$. Clearly this approach, which dates back to Clausius, will result in a reduction of the pressure. However, it is a poor substitute for using the correct hard-sphere equation of state.

In principle, a, b, and c can be chosen to give z_c correctly. Thus, Equation 44 could be a distinct improvement over the VDW equation near the critical point. However, if the parameters are chosen so that $z_c \simeq 092.1$, poor results are obtained at high densities. It is conventional to choose c/b so that $z_c \simeq 1/3$.

Redlich and Kwong (*16*) go one step further; they introduce a weak temperature dependence into the second term, and obtain the well-known Redlich–Kwong (RK) equation:

$$P = \frac{NkT}{V - Nb} - \frac{N^2a}{T^{1/2}V(V + Nb)} \tag{45}$$

The density dependence of Equation 45 is not significantly different from that of Equation 44 with the conventional choice for *c;* e.g., as seen from the values of the critical constants listed in Table I, $z_c \simeq 1/3$. Carnahan and Starling (*14*) have used the RK equation to compute the equation of state of methyl chloride. Their results are given in Table III and are a significant improvement over the VDW equation.

The theoretical basis of the RK equation is inadequate. It retains the completely unsatisfactory VDW hard-sphere isotherm for which there is no justification. The second term in Equation 45 has more merit. Equation 26 indicates that a small addition term in the density could be included appropriately. However, Equation 26 suggests the opposite sign to that in Equation 45. This, plus the $T^{1/2}$ term, clearly suggests that the second term in Equation 45 cannot be justified as an approximation to A_1. However, it might be justified as an approximation to A_1 and higher-order terms. Barker and Henderson (*1*) have suggested that there is some evidence that A_3 and some of the higher terms may be positive at high densities. This may provide some justification for the sign in the denominator of the second term in the RK equation. Likewise, the weak temperature of the second term, although not justifiable from the form of A_1, may simulate the higher-order terms.

Despite the improved agreement, the RK equation is lacking; e.g., the RK heat capacity is

$$C = \frac{3}{2}NkT + \frac{3Na}{4T^{3/2}b} \ln \left\{ \frac{V + Nb}{V} \right\} \tag{46}$$

The heat capacity obtained from Equation 46 increases monotonically with the density. Experimentally it is known that at high densities the heat capacity decreases with increasing density.

A much more rational approach is that of Longuet–Higgins and Widom (LHW) (*17*) who suggested retaining Equations 37 and 42 but replacing Equation 38 by the correct hard-sphere equation of state. This leads to

$$P = P_0 - \frac{N^2a}{V^2} \tag{47}$$

where P_0 is the correct hard-sphere pressure. Equation 47 can be obtained from Equation 20 by taking the limit $z \to 0$. Very likely Equation 47 has been considered before the work of Longuet–Higgins and Widom since the deficiencies of Equation 38 were well known to van der Waals and his contemporaries. The quite reliable expression:

$$P_0/\rho kT = \frac{1}{(1-\eta)^4}$$

$$= 1 + \rho b + \frac{5}{8}(\rho b)^2 + \frac{5}{16}(\rho b)^3 + \ldots \qquad (48)$$

was known at least as early as 1896 (*18*). Equation 48 is compared with computer simulations in Figure 1. The results of Equation 47 are not sensitive to the precise expression used for P_0 (*19, 20*). In any case, as seen from Tables I and III, if a good expression for P_0 is used, the LHW equation of state (Equation 47) yields results which are nearly as good as those obtained from the more empirical RK equation. Since a reliable expression for P_0 has been in existence for nearly a century, it is puzzling that the obsolete and inaccurate free-volume term $(V - Nb)^{-1}$ still finds such wide usage.

Carnahan and Starling (*14*) have suggested using the correct hard-sphere equation of state together with the RK expression for the contribution of the attractive forces. Thus

$$P = P_0 - \frac{N^2 a}{T^{1/2} V (V + Nb)} \qquad (49)$$

I shall refer to Equation 49 as the modified Redlich–Kwong (MRK) equation. As seen from Table 1, the MRK equation yields a very accurate value for z_c. Furthermore, the MRK equation of state of methyl chloride, listed in Table III, is in excellent agreement with experiment. Clearly, using an accurate expression for P_0 in place of the $(V - Nb)^{-1}$ term leads to a dramatic improvement in the equation of state.

It should be noted that Equation 49 leads to Equation 46 so that the MRK and RK equations have precisely the same difficulties with the heat capacity. Unfortunately, the LHW equation-of-state heat capacity is no better.

It may be possible to do better than Equation 49. There is nothing sacred about the form of the RK expression for the attractive-force contributions. To some degree, the terms $T^{1/2}$ and $V + Nb$ were chosen to compensate for the error in the VDW hard-sphere equation of state. It is not clear that they are the best choices if some other expression for P_0 is used. A somewhat better equation of state might be obtained with a slightly different form for the second term in Equation 49.

The RK and MRK equations of state both lead to heat capacities which increase with increasing density. This is not in agreement with the experimental fact that $\partial C/\partial \rho$ is positive at low densities but negative at high densities. The VDW and LHW equations are no improvement in this regard since they both give heat capacities which are independent of density. Equation 20 is a three-parameter equation of state which may be useful in this regard since the heat capacity obtained from Equation 20 first increases with density and then decreases.

In each of the above empirical equations of state the diameter of the hard-phase fluid is assumed to be temperature independent. Some improvement might be obtained if a temperature dependence, obtained perhaps from Equation 35, were introduced.

I would like to make a few quick comments about some other empirical equations of state which are used from time to time. One is the Dieterici equation of state:

$$P = P_0 \exp \{-Na/VkT\} \tag{50}$$

This equation's main claim to fame is its low value of z_c ($z_c = 0.271$) which is in good agreement with experiment (although no better than that of the MRK equation). However, at high densities the Dieterici equation is a disaster. We can see why by expanding Equation 50 to obtain

$$P = P_0 - \frac{N^2a}{V}\left(\frac{P_0V}{NkT}\right) + \frac{1}{2kT}\,\frac{N^3a}{V^3}\left(\frac{P_0V}{NkT}\right) + \ldots \tag{51}$$

The first two terms are similar to Equation 47 and fairly reasonable. The hard-sphere equation of state in the second term overestimates the effect of this term. However, the really serious difficulties with Equations 50 and 51 are the result of the higher-order terms in β which are unboundedly large at high densities, whereas, according to the analysis of Barker and Henderson (*11*), these terms should be small at high densities. As a result, the Dieterici equation is doomed to fail at high densities independently of what expression is used for P_0.

Of more interest is the Berthelot equation:

$$P = P_0 - \frac{N^2a}{V^2T} \tag{52}$$

The only justification of Equation 52 which comes to my mind is to develop a perturbation theory for a fluid of non-spherical molecules using a hard-sphere reference fluid. The diameter of the hard spheres could be chosen to cause A_1 to vanish. In such a case, the first nonvanishing term would be A_2. If A_2 were linear in the density and the higher-order terms were neglected, Equation 52 would result.

As a result, the Berthelot equation may be of interest for fluids of nonspherical molecules. However, it can apply only to molecules which are nearly spherical. For molecules which are not close to spherical, it would be better to follow Rigby (*21*) and use Equation 47 or 49 with P_0 being the equation of state of some appropriate fluid of nonspherical molecules. Analogues of Equation 15 are known for a wide class of fluids of nonspherical molecules (*22*). For example, for a fluid of hard prolate spherocylinder

$$P_0/\rho kT = \frac{3 + (2\alpha - 6)\eta + (\alpha^2 - 3\alpha + 3)\eta^2}{(1 - \eta)^3} \tag{53}$$

where

$$\alpha = 3(1 + x)(1 + x/2)/(1 + 3x/2) \tag{54}$$

x is the length of the spherocylinder divided by its diameter, and $\eta = \pi\rho d^3(1 + 3x/2)/6$. Using Equations 47 and 53, Rigby has shown that z_c decreases as x increases in agreement with experiment.

Lastly I would like to consider the Beattie–Bridgeman equation and its offspring the Benedict, Webb, and Rubin equation which start with the VDW equation and replace $1/(V - Nb)$ by $(V + Nb)/V^2$. Several additional parameters are introduced. However, the above replacement is so meaningless at high densities that I cannot bring myself to comment further on this family of empirical equations.

Mixtures

The concept of an ideal mixture is central to the theory of liquid mixtures. This concept can be introduced by considering the partition function of a mixture of two ideal gases

$$Z_N = \lambda_1^{-3N_1}\lambda_2^{-3N_2}V^N/N! \tag{55}$$

where $\lambda_i = (2\pi m_i kT)^{1/2}$, N_i is the number of molecules of mass m_i, and $N = N_1 + N_2$. We can put Equation 55 in the form

$$Z_N = \frac{N!}{N_1!N_2!} Z_{N_1}(1) Z_{N_2}(2) \tag{56}$$

where $Z_{N_i}(i)$ is the partition function of the pure substance of species i.

In this case

$$Z_{N_i}(i) = \lambda^{-3N_i}(eV/N_i)^{N_i} \tag{57}$$

We may deduce the chemical potential μ_i from Equation 56 and obtain

$$\mu_i = \mu_i^\circ + kT \ln x_i \tag{58}$$

where μ_i° is the chemical potential of species i in the pure state. Hence, the absolute activity is

$$\frac{\gamma_i}{\gamma_i^\circ} = x_i \tag{59}$$

We refer to a mixture for which Equation 58 or 59 is satisfied as an ideal mixture. Clearly this concept of an ideal mixture is an idealization that is not realized in practice. Real mixtures will show deviations from the ideal results. However, the properties of an ideal mixture are convenient reference states for thermodynamic properties. For example, it is conventional to use excess thermodynamic properties of mixing that are defined as the difference between the thermodynamic property of the mixture and those of an ideal mixture of the components at the same temperature and pressure.

The extension of the theory of pure fluids (developed in the preceding sections) to mixtures is straightforward. For a mixture of m species

$$\Phi(\mathbf{r}_1 \ldots \mathbf{r}_N) = \sum_{i<j} u(\alpha_i, \alpha_j; r_{ij}) + \sum_{i<j<k} u(\alpha_i, \alpha_j, \alpha_k; r_{ij}, r_{ik}, r_{kj}) + \ldots \tag{60}$$

where $\alpha_i = \lambda$ if molecule i is of species $\lambda (1 \leq \lambda \leq m)$ and $u_{\alpha\beta}(r) = u(\alpha, \beta; r)$ and $u_{\alpha\beta\gamma}(r, s, t) = u(\alpha, \beta, \gamma; r, s, t)$ are the pair and triplet interactions between molecules of species α and β and α, β, and γ, respectively. We note that $u_{\alpha\beta}(r) = u_{\beta\alpha}(r)$, etc.

As in the case for pure fluids, the perturbation theory is useful both as a calculational tool and as a guide to the development of empirical equations. If we use a mixture of hard spheres as our reference fluid

$$\frac{A - A_0}{NkT} = -2\pi\rho \sum_{ij} x_i x_j d_{ij}^2 g_0^{ij}(d_{ij})[d_{ij} - \delta_{ij}] + \sum_{n=1}^{\infty} (\beta\epsilon)^n A_n/NkT \tag{61}$$

where $x_i = N_i/N$, $N = \sum_i N_i$,

$$d_{ij} = \frac{d_{ii} + d_{jj}}{2} \tag{62}$$

d_{ii} are the diameters of the hard spheres,

$$\delta_{ij} = \int_0^{\sigma_{ij}} [1 - \exp\{-\beta u_{ij}(z)\}]dz \tag{63}$$

and A_0 and g_0^{ij} are the hard-sphere mixture free energy and RDF's. Following Barker and Henderson (*12*), we can choose $d_{ii} = \delta_{ii}$. Thus

$$\frac{NkT}{A - A_0} = -4\pi\rho x_1 x_2 d_{12}^2 g_0^{12}(d_{12})[d_{12} - \delta_{12}] + \sum_{n=1}^{\infty} (\beta\epsilon)^n A_n/NkT \tag{64}$$

The first term on the right-hand side of Equation 64 is very small in both the total and excess thermodynamic properties and can be neglected.

Table IV. Comparison with MC Calculations ($\xi_{12} = 1.0$) at $P = 0$ and $x = 1/2$ (G^E and H^E are in Joules/Mole and V^E in Cubic Centimeters/Mole)

System	*Property*	*Simulation*[a]	*Perturbation Theory*[b]
Ar + Kr	G^E	+46 ± 7	+39
(115.8 K)	H^E	−29 ± 17	−28
	V^E	−0.69 ± 0.06	−0.62
Ar + CH_4	G^E	−14 ± 6	−12
(91.0 K)	H^E	−60 ± 12	−30
	V^E	−0.22 ± 0.04	−0.12
CO + CH_4	G^E	+77 ± 7	+69
(91.0 K)	H^E	+15 ± 12	+24
	V^E	−0.76 ± 0.06	−0.63
Ar + N_2	G^E	+35 ± 5	+32
(83.8 K)	H^E	+16 ± 9	+35
	V^E	−0.25 ± 0.05	−0.23
Ar + CO	G^E	+26 ± 5	+25
(83.8 K)	H^E	+37 ± 15	+30
	V^E	−0.17 ± 0.05	−0.17
O_2 + N_2	G^E	+38 ± 5	+36
(83.8 K)	H^E	+39 ± 15	+42
	V^E	−0.28 ± 0.06	−0.25

[a] Calculated by McDonald (*26*).
[b] Calculated by Grundke et al. (*25*).

The first-order term

$$A_1/NkT \neq 2\pi\rho \sum_{ij} x_i x_j \int w_{ij}(r) g_0^{ij}(r) r^2 dr \qquad (65)$$

is calculated easily using the PY results (*23, 24*) for the $g_0^{ij}(r)$. If $w_{ij}(r)$ is given by Equation 31, an analytical expression can be obtained for A_1.

An expression for A_2 which is useful for computation has not been obtained yet. However, even truncated at first order, the perturbation theory provides a useful theory of mixtures. Grundke et al. (*25*) have used first-order perturbation theory with the 6:12 potential to calculate the excess thermodynamic properties of mixtures. As seen in Table IV, the agreement of their results with the computer simulations of McDonald (*26*) is excellent. The pure-fluid potential parameters used in these calculations are given in Table V. For these simulated mixtures,

$$\epsilon_{12} = \sqrt{\epsilon_{11}\epsilon_{22}} \qquad (66)$$

$$\sigma_{12} = \frac{1}{2}(\sigma_{11} + \sigma_{22}) \qquad (67)$$

In Table VI, the first-order perturbation theory results for the excess thermodynamic properties are compared with experimental results for several mixtures. For these calculations, Grundke et al. (*25*) used the 6:12 potential with the pure-fluid potential parameters listed in Table VII. Further they assumed Equation 67 and

$$\epsilon_{12} = \xi_{12}\sqrt{\epsilon_{11}\epsilon_{22}} \qquad (68)$$

and adjusted ξ_{12} so that either G^E or H^E at $x = 1/2$ was equal to the experimental value. The perturbation theory results agree well with experiment.

Table V. Potential Parameters Used in Table IV

Substance	ϵ/k*(K)*	σ*(Å)*
Argon	119.8	3.405
Krypton	167.0	3.633
Methane	152.0	3.74
Nitrogen	101.3	3.612
Oxygen	119.8	3.36
Carbon monoxide	104.2	3.62

Table VI. Comparison of Theory and Experiment at $P = 0$ and $x = 1/2$ (G^E and H^E are in Joules/Mole and V^E in Cubic Centimeters/Mole)

System	*Property*	*Experimental*	*Perturbation Theory*[a]	*VDW*	*VDW-1*	*SH*
Ar + Kr (116 K)	ξ_{12}	0.996	0.983	0.976	0.990	0.991
	G^E	84	84	84	84	84
	H^E		45	74	20	44
	V^E	−0.52	−0.47	−0.13	−0.92	−0.78
Kr + Xe (161 K)	ξ_{12}		0.978	0.971	0.984	0.986
	G^E	115	115	115	115	115
	H^E		69	105	30	85
	V^E	−0.70	−0.50	−0.17	−1.06	−0.75
$Ar + N_2$ (84 K)	ξ_{12}	0.999	1.001	1.000	1.004	1.004
	G^E	34	34	34	34	34
	H^E	51	35	37	36	50
	V^E	−0.18	−0.27	−0.12	−0.43	−0.33
Ar + CO (84 K)	ξ_{12}	0.986	0.986	0.988	0.989	0.990
	G^E	57	57	57	57	57
	H^E		79	67	86	104
	V^E	0.10	−0.07	−0.06	−0.12	0.01
$Ar + CH_4$ (91 K)	ξ_{12}	0.988	0.972	0.969	0.975	0.977
	G^E	74	74	74	74	74
	H^E	103	89	80	86	125
	V^E	0.17	0.03	−0.06	−0.04	0.18
$O_2 + Ar$ (84 K)	ξ_{12}		0.987	0.988	0.988	0.988
	G^E	37	37	37	37	37
	H^E	60	52	42	58	58
	V^E	0.14	0.06	0.02	0.09	0.09
$O_2 + N_2$ (78 K)	ξ_{12}		1.002	1.000	1.005	1.006
	G^E	42	42	42	42	42
	H^E	44	43	44	45	61
	V^E	−0.21	−0.26	−0.13	−0.42	−0.31
$N_2 + CO$ (84 K)	ξ_{12}	0.986	0.990	0.991	0.991	0.991
	G^E	23	23	23	23	23
	H^E		34	27	37	38
	V^E	0.13	0.07	0.03	0.10	0.11
$CO + CH_4$ (91 K)	ξ_{12}	0.998	0.983	0.978	0.992	0.992
	G^E	115	115	115	115	115
	H^E	105	96	109	89	92
	V^E	−0.32	−0.48	−0.13	−0.84	−0.82

[a] Calculated by Grundke et al. (*25*).

Table VII. Potential Parameters for Pure Substances

Substance	$\epsilon/(\epsilon\ Argon)$	$\sigma/(\sigma\ Argon)$
Argon	1	1
Krypton	1.387	1.070
Xenon	1.919	1.167
Nitrogen	0.836	1.063
Oxygen	1.022	0.995
Carbon monoxide	0.881	1.070
Methane	1.266	1.099
Carbon tetrafluoride	1.288	1.365

Empirical equations of state can be devised from Equation 65. The conventional approach is to assume that

$$\int_{\sigma_{ij}}^{\infty} w_{ij}(r)\, g_0^{ij}(r)\, r^2 dr \propto \epsilon_{ij}\sigma_{ij}^3 \tag{69}$$

Equation 69 is very nearly satisfied for most mixtures. Thus, neglecting higher-order terms

$$A = A_0 - N^2 a/V \tag{70}$$

where

$$a = \sum_{ij} x_j x_i a_{ij} \tag{71}$$

Differentiating gives Equation 47. Van der Waals suggested using Equation 39 for P_0 with

$$b = \sum_{ij} x_i x_j b_{ij} \tag{72}$$

Equations 71 and 72 can be rewritten in terms of the potential parameters ϵ and σ. The result is

$$\epsilon\sigma^3 = \sum_{ij} x_i x_j \epsilon_{ij} \sigma_{ij}^3 \tag{73}$$

$$\sigma^3 = \sum_{ij} x_i x_j \sigma_{ij}^3 \tag{74}$$

Results calculated using Equations 39, 70, 71, and 72 are listed in Table VI. Although not as good as for the perturbation theory, the agreement of the VDW results with experiment is quite good. Since the VDW theory does not give good, absolute properties, considerable cancellation of errors must have occurred.

A further possibility would be to use the VDW procedure but replace Equation 39 with some better hard-sphere isotherm such as Equation 15. Results of this procedure which are labelled VDW-1 (to indicate that a single-component fluid is used to obtain an expression for A_0 and P_0) are given in Table VI. The results are somewhat worse than the VDW results.

Equations 71 and 72 or 73 and 74 can be used with any equation of state and not just with the VDW or LHW equations. The physical content of these equations is that, within the VDW-1 approximation, the equation of state of the mixture is (apart from the entropy of mixing terms) identical to that of a single-component fluid specified by the parameters given by Equations 71 and 72 or 73 and 74.

A much better procedure is that of Snider and Herrington (*27*) who used the PY result for P_0 for a mixture of hard spheres. For a binary mixture the PY result is:

$$P_0/\rho kT = (\rho_1 + \rho_2)\,\frac{1 + \xi + \xi^2}{(1 - \xi)^3} - \frac{18}{\pi}\,\frac{\eta_1\eta_2}{(1 - \xi)^3}\,(d_{11} - d_{22})^2(d_{11} + d_{22} + d_{11}d_{22}X) \qquad (75)$$

where $\rho_i = N_i/V$, $\eta_i = \pi\rho_i/6$,

$$\xi = \frac{\pi}{6}\,\rho[x_1d_{11}^3 + x_2d_{22}^3] \qquad (76)$$

and

$$X = \frac{\pi}{6}\,\rho[x_1d_{11}^2 + x_2d_{22}^2] \qquad (77)$$

Using the expression of Mansoori et al. (*28*), which is a generalization of the Carnahan–Starling expression (*8*), for P_0 is appropriate provided that the ratio d_{22}/d_{11} is not exceedingly large (*29*).

Results calculated by means of the Snider–Herrington (SH) procedure are shown in Table VI. These results are much better than the VDW or VDW-1 results. They are comparable in accuracy with those of the perturbation theory.

The RK equation (Equation 45) has been applied to mixtures. However, such applications retain the outmoded VDW term (Equations 39 and 72 or 74). Clearly it is time to use Equation 75. Unfortunately, the deadline imposed by this conference has prevented calculations comparable with those in Table VI by such a procedure. However, it is anticipated that they would be pleasing.

Summary

I have attempted to survey some of the recent progress in the theory of liquids, emphasizing those concepts and results which I feel are most interesting to chemical engineers.

The major advance during the past decade is the realization that the structure and thermodynamics of a dense fluid are perturbations about those of a hard-core (usually hard-sphere) fluid with a suitably chosen dimension (usually a diameter). Thus, the first task in the development of any theory or any empirical equation of state is to ensure that the hard-core reference fluid is described adequately.

Unfortunately, almost no empirical equations of state contain an adequate description of the hard-sphere fluid. The VDW term, $(V - Nb)^{-1}$, seems to be a part of most, if not all, the popular empirical equations of state. Carnahan and Starling (*14*), Gubbins (*30*), Prausnitz (*31*), and no doubt, others have called for the replacement of this term in chemical-engineering practice. It is time that their advice was followed. The VDW term $(V - Nb)^{-1}$ is an inadequate description of a single-component, hard-sphere fluid. It is even less suitable to describe mixtures or fluids composed of nonspherical molecules. Fortunately, the replacement of $(V - Nb)^{-1}$ is easily accomplished since, in all the popular empirical equations of state, the contributions of the repulsive and attractive forces are separated clearly. Moreover, the form of the term representing the contribution of the attractive forces generally is written in a theoretically reasonable form. The replacement of $(V - Nb)^{-1}$ by a more reliable expression results not only in an intellectually more satisfying equation but usually in a more reliable one also.

To be sure it is possible to retain $(V - Nb)^{-1}$ and parameterize the term representing the attractive forces so as to obtain a reliable description of the equation of state of a fluid. However, one must work harder since the parameterization of this term must account not only for the effect of the attractive forces but also must compensate for the errors in $(V - Nb)^{-1}$. One example of such a compensation is the empirical equation of state of Martin (*32*):

$$P = \frac{NkT}{V - Nb} - \frac{N^2\mathrm{a}(1 + dT)}{(V + Nc)^2} \tag{78}$$

The part of the second term which is linear in T is, in reality, nothing more than a correction to the hard-sphere equation of state which is described inadequately by the first term.

Such cancellations of errors must be even more abundant in empirical equations of state for fluids composed of nonspherical molecules if $(V - Nb)^{-1}$ is retained. The fact that such equations of state can be and have been developed and used is a tribute to the cleverness and industry of their proponents. However, such equations of state inevitably must contain excessive numbers of parameters and can be no more than interpolation formulae which either must break down for some derivative (such as the heat capacity) of the property being fitted or which cannot be used with any confidence to calculate the properties of the fluid either outside the region in which the fit was made or in regions for which experimental data are not available.

If I may quote a phrase common in some political circles: "Why not the best?" The term $(V - Nb)^{-1}$ is not the best.

Glossary of Symbols

A = Helmholtz free energy
A_0 = Helmholtz free energy of a hard-sphere fluid
A_n = nth order term in the temperature expansion of the free energy
a_{ij} = parameter for molecules of species i and j in the VDW and similar equations of state of a mixture. Subscripts dropped for a pure fluid
b_{ij} = $2\pi d_{ij}^3/3$, parameter for molecules of species i and j in the VDW and similar equations of state of a mixture. Subscripts dropped for a pure fluid.
c = parameter in modifications of the van der Waals equation of state for a pure fluid (Equation 44) or a parameter in Equations 22 to 25
d_{ij} = diameter for molecules of species i and j in a hard-sphere mixture. Subscripts are dropped for a pure fluid
E_i = energy of ith state
$F_0(r_1, r_2)$ = hard-sphere distribution function in second-order perturbation theory
$G(s)$ = Laplace transform of RDF
$g(r)$ = radial distribution function (RDF)
$g_0^{ij}(r)$ = RDF for molecules of species i and j in hard-sphere mixture. Superscripts are dropped for pure fluid
$g(\mathbf{r}_1 \ldots \mathbf{r}_h)$ = h-body distribution function
$\mathcal{H}$ = Hamiltonian
h = Planck's constant
k = Boltzmann's constant

$L(s)$ = polynomial in the Laplace transform of the Percus–Yevick radial distribution function of a hard-sphere fluid.

m_i = molecular mass of a molecule of species *i*. Subscript dropped for pure fluid.

MRK = modified Redlich–Kwong equation

MSA = mean spherical approximation

$N = \sum_i N_i$, total number of molecules

N_i = number of molecules of species *i*

PY = Percus–Yevick

P = pressure

P_0 = pressure of a hard-sphere fluid

P_c = critical pressure

$P^* = P\sigma^3/\epsilon$, reduced pressure

p_i = *i*th generalized momenta

q_i = *i*th generalized coordinate

RK = Redlich–Kwong equation

RDF = radial distribution function

$\mathbf{r}_i$ = vector position of the *i*th molecule

r_{ij} = scalar distance between the *i*th and *j*th molecule

S = entropy

$S(s)$ = polynomial in the Laplace transform of the Percus–Yevick RDF of a hard-sphere fluid

SH = Snider–Herrington

s = Laplace transform variable

T = temperature

T_c = critical temperature

$T^* = kT/\epsilon$, reduced temperature

U = internal energy

$u_{ij}(r)$ = intermolecular pair potential for a pair of molecules of species *i* and *j*. Subscripts are dropped for pure fluid.

$u_0(r)$ = hard-sphere intermolecular pair potential

V = volume

V_c = critical volume

$v_c^* = V/N\sigma^3$, reduced volume

v_n = *n*th order coefficient in the temperature expansion of the mean spherical approximation free energy for a Yukawa fluid

$w_{ij}(r)$ = soft "tail" in the intermolecular pair potential. Subscripts dropped for pure fluid.

X = term, defined by Equation 77, in the Percus–Yevick expression for the pressure of a hard-sphere mixture

x = length of a spherocylinder divided by its diameter or reduced intermolecular distance, r/σ

x_i = N_i/N, concentration of molecules of species i

Z_N = partition function

z = range parameter of the Yaukawa potential

z_c = $P_c V_c / NkT_c$

α = parameter, defined by Equation 54, in the equation of state hard spherocylinders

α_i = parameter specifying species of molecule i

β = $1/kT$

δ_{ij} = effective hard-sphere diameter used in perturbation theory

ϵ_{ij} = depth of intermolecular pair potential between molecules of species i and j. The subscript is dropped for pure fluid

η = $\pi \rho d^3/6$

η_i = $\pi \rho_i/6$

λ_i = $h/\sqrt{2\pi m_i kT}$ or $\exp(\beta\mu_i)$. The subscript is dropped for pure fluid.

μ_i = chemical potential of species i. The subscript is dropped for pure fluid.

ξ = term, defined by Equation 76, in the Percus–Yevick expression for the pressure of a hard-sphere mixture

ξ_{ij} = parameter specifying deviation of ϵ_{ij} from $\sqrt{\epsilon_{ii}\epsilon_{jj}}$

ρ_i = N_i/V, density of molecules of species i. The subscript is dropped for pure fluid.

ρ_c = critical density

ρ^* = $N\sigma^3/V$, reduced density

σ_{ij} = distance between molecules of species i and j at which intermolecular pair potential vanishes

$\Phi(\mathbf{r}_1, \ldots \mathbf{r}_N)$ = potential energy of the system

$\varphi(x)$ = $u(r)/\epsilon$, reduced intermolecular pair potential

Acknowledgments

I am grateful to K. Gubbins, T. Leland, Jr., and J. Prausnitz for providing references and suggestions which I have found useful. Also I am grateful to F. Buff for a stimulating discussion and for pointing out Ref. *18*.

Literature Cited

1. Barker, J. A.; Henderson, D. *Rev. Mod. Phys.* **1976,** *48,* 587.
2. Wertheim, M. *Phys. Rev. Lett.* **1963,** *10,* 321.
3. Wertheim, M. *J. Math. Phys. (N.Y.)* **1964,** *5,* 643.
4. Thiele, E. *J. Chem. Phys.* **1963,** *39,* 474.
5. Smith, W. R.; Henderson, D. *Mol. Phys.* **1970,** *19,* 411.
6. Waisman, E. *Mol. Phys.* **1973,** *25,* 45.
7. Henderson, D., unpublished data.
8. Carnahan, N. F.; Starling, K. E. *J. Chem. Phys.* **1969,** *51,* 635.
9. Barker, J. A.; Henderson, D. *Mol. Phys.* **1971,** *21,* 187.
10. Barker, J. A.; Henderson, D. *Annu. Rev. Phys. Chem.* **1972,** *23,* 439.
11. Barker, J. A.; Henderson, D. *J. Chem. Phys.* **1967,** *47,* 2856.
12. Ibid., 4714.
13. Hirschfelder, J. O.; Curtiss, C. F.; Bird, R. B. "Molecular Theory of Gases and Liquids"; Wiley: New York, 1954.
14. Carnahan, N. F.; Starling, K. E. *AIChE J.* **1972,** *18,* 1184.
15. Hsu, C. C.; McKetta, J. J. *J. Chem. Eng. Data* **1964,** *9,* 45.
16. Redlich, O.; Kwong, J. N. S. *Chem. Rev.* **1949,** *44,* 233.
17. Longuet–Higgins, H. C.; Widom, B. *Mol. Phys.* **1964,** *8,* 549.
18. Jäger, G. *Wien. Ber.* **1896,** *105,* 15.
19. Guggenheim, E. A. *Mol. Phys.* **1965,** *9,* 43.
20. Ibid., 199.
21. Rigby, M. *J. Phys. Chem.* **1972,** *76,* 2014.
22. Gibbons, R. M. *Mol. Phys.* **1969,** *17,* 18.
23. Lebowitz, J. L. *Phys. Rev.* **1964,** *133,* A895.
24. Leonard, P. J.; Henderson, D.; Barker, J. A. *Mol. Phys.* **1971,** *21,* 107.
25. Grundke, E. W.; Henderson, D.; Barker, J. A.; Leonard, P. J. *Mol. Phys.* **1973,** *25,* 883.
26. McDonald, I. R. *Mol. Phys.* **1972,** *23,* 41.
27. Snider, N. S.; Herrington, T. M. *J. Chem. Phys.* **1967,** *47,* 2248.
28. Mansoori, G. A.; Carnahan, N. F.; Starling, K. E.; Leland, T. W., Jr. *J. Chem. Phys.* **1971,** *54,* 1523.
29. Waisman, E.; Henderson, D.; Lebowitz, J. L. *Mol. Phys.* **1976,** *32,* 1373.
30. Gubbins, K. E. *AIChE J.* **1973,** *19,* 688.
31. Vera, J. H.; Prausnitz, J. *Chem. Eng. J.* **1971,** *3,* 1.
32. Martin, J. J. *Ind. Eng. Chem.* **1967,** *59,* 34.
33. Barker, J. A.; Henderson, D. *Acc. Chem. Res.* **1971,** *4,* 305.

RECEIVED August 8, 1978.

2

Equation of State for Fluids: The Topology of the Vapor–Liquid-Phase Transition and the Critical Point

ARNOLD E. LIU[1]

Mathematisches Institut, Georg-August-Universität, D-3400 Göttingen, Federal Republic of Germany

A very wide-range differential equation of state for fluids is proposed. It describes not only the behavior of fluids on their stable vapor and liquid branches, but also in the coexistence envelope. The equation of state is compatible with the ideal gas law at low densities and with scaled equations of state as $T \rightarrow 1^-$*, where—passing through the power law for the critical isotherm—it evolves to the classical virial equation of state for supercritical temperatures. The topology of the vapor–liquid-phase transition is examined. High precision representations of real fluid behavior can be obtained with a small number of adjustable parameters.*

This chapter presents a new formulation of the equation of state for fluids at subcritical and critical temperatures, $T \leq 1$. Unlike many other equations of state, this equation of state is defined as a differential equation, and is designed to describe not only the behavior of isotherms on their stable vapor and liquid branches, but also in the two-phase transition region. Interesting insights concerning the nature of metastable and absolutely instable phases are obtained also.

The equation of state introduced in this chapter is compatible with the ideal gas law at low pressures and densities, and with the scaling law in the critical region, where—passing through the power law for the critical isotherm—it evolves into the classical virial expansion for supercritical temperatures.

[1] Current address: Quantum Dynamics, Inc., 19458 Ventura Blvd., Tarzana, CA 91356.

0-8412-0500-0/79/33-182-031$05.25/1

Owing to the very simple and intuitively clear definition of the equation of state, the topology of the vapor–liquid-phase transition and critical point is examined easily using the methods of dynamic system and bifurcation theory.

The equation of state can be fitted to real fluids, and yields very precise representations of ρ, P, T, and $dP/d\rho$ data.

The equation of state should, of course, be applicable to the analogous magnetic-phase transition.

The Differential Equation of State

Let ρ, P, and T denote the reduced density, the reduced pressure, and the reduced temperature of a fluid, respectively. We shall describe the slope $dP/d\rho$ of the subcritical isotherm $T < 1$ at the point (ρ, P) as

$$\frac{dP}{d\rho} = \frac{[P_\sigma(T) - P]f(\rho; T)}{[\rho_G(T) - \rho][\rho_D(T) - \rho][\rho_L(T) - \rho]} \tag{1}$$

where $P_\sigma(T)$ = reduced equilibrium vapor pressure; $\rho_G(T)$ = reduced saturated vapor density; $\rho_L(T)$ = reduced saturated liquid density; $\rho_D(T) = \frac{1}{2}[\rho_G(T) + \rho_L(T)]$; and $f(\rho; T)$ is an analytic function of ρ for each fixed temperature T, convex in the neighborhood of the coexistence envelope, with

$$f(\rho_G; T) = \tfrac{1}{2}[\rho_L(T) - \rho_G(T)]^2 = f(\rho_L; T) \tag{2}$$

Condition 2, a very strong constraint, ensures the proper behavior of the slope $dP/d\rho$ on the phase boundary. Note that when $P = P_\sigma$, then $dP/d\rho = 0$ for all regular points of Equation 1, i.e. Equation 1 yields the correct isothermal behavior in the two-phase transition region of a fluid (*see* Figures 1 and 2).

For the purpose of topological analysis of Differential Equation 1, we considered the equivalent dynamic system

$$\frac{d\rho}{d\lambda} = [\rho_G(T) - \rho][\rho_D(T) - \rho][\rho_L(T) - \rho] \tag{3a}$$

$$\frac{dP}{d\lambda} = [P_\sigma(T) - P]f(\rho; T) \tag{3b}$$

(Originally, a more general form of System 3, with $f = f(\rho, P; T)$ was considered in topological analyses. However, fitting Equation 1 to real fluids and certain mathematical–physical considerations indicate that the function f is independent of the pressure P, which, for the sake of brevity, is excluded from the present discussions).

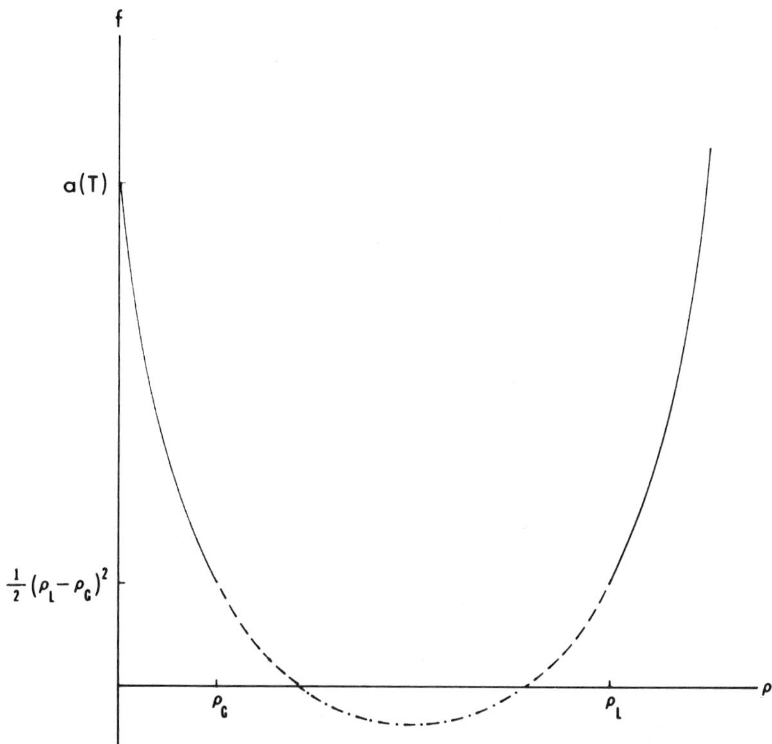

Figure 1. Behavior of f(ρ; T), T < 1

For each subcritical temperature $T < 1$, dynamic System 3 has a simple equilibrium point, i.e. $d\rho/d\lambda = 0 = dP/d\lambda$, at $(\rho_G(T), P_\sigma(T))$, $(\rho_D(T), P_\sigma(T))$, and $(\rho_L(T), P_\sigma(T))$. Using well-established methods of dynamic system theory—for simple equilibrium points it suffices to examine the eigenvalues of System 3—one then determines that both the saturated vapor point $(\rho_G(T), P_\sigma(T))$ and the saturated liquid point $(\rho_L(T), P_\sigma(T))$ are (orbitally stable with respect to λ) dicritical nodes, i.e. that each solution path $\{\rho(\lambda), P(\lambda)\}$ approaches the equilibrium point from a definite direction and, conversely, each direction corresponds to exactly one path (*see* Figure 2).

Note that if dynamic System 3 is perturbed temporarily so that equilibrium points do not occur, then the stable vapor and liquid branches can be continued into their respective metastable regions. (Such conditions can be induced, e.g., by the application of anomalous pressure near the phase boundary.)

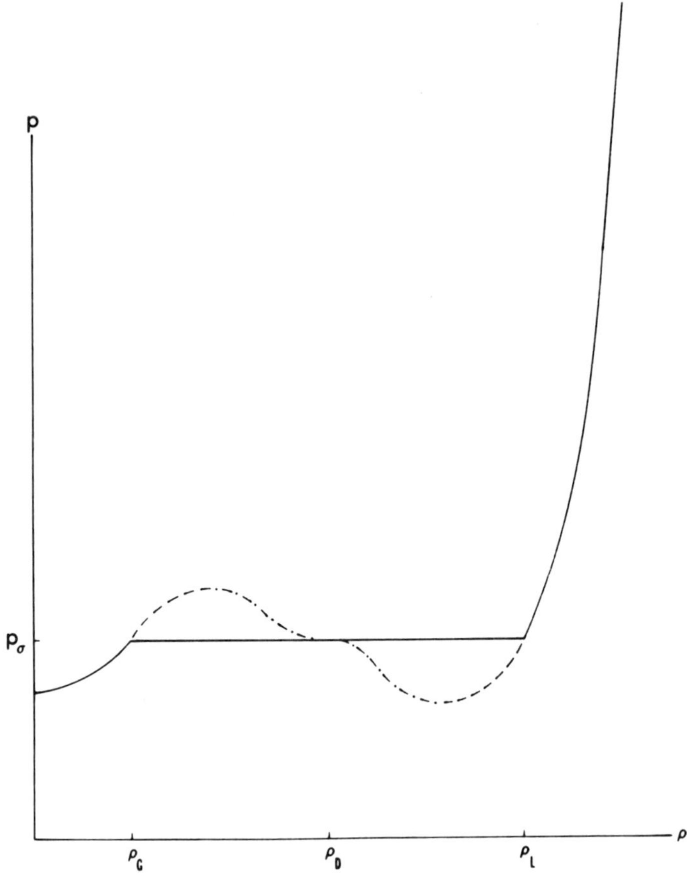

Figure 2. Isothermal behavior, T < *1*

Certain mathematical–physical considerations and the subsequent fitting of f(ρ; T) allow us to conclude that the coexistence envelope diameter point ($\rho_D(T)$, $\rho_\sigma(T)$) is an (orbitally unstable) improper node, i.e. that all solution paths leaving ($\rho_D(T)$, $\rho_\sigma(T)$) will do so only in directions 0 and π, $\pi/2$ and $3\pi/2$ (*see* Figure 2). As $T \rightarrow 1^-$, the three equilibrium points ($\rho_G(T)$, $\rho_\sigma(T)$), ($\rho_D(T)$, $P_\sigma(T)$), and ($\rho_L(T)$, $P_\sigma(T)$) converge to the critical point (1, 1). This multiple equilibrium point is an orbitally stable, but structurally (topologically) unstable, multiple node. The parameter T thus can be considered as a bifurcation parameter, and $T = 1$ as a bifurcation value of dynamic System 3.

Examination of a Specific Form of the Differential Equation of State

In this section we examine some elementary properties of Equation of State 1 for the simplest form of the convex function $f(\rho; T)$, namely

$$f(\rho; T) = a(T) + b(T)\rho + c(T)\rho^2 \tag{4}$$

where $a(T)$ and $c(T)$ are positive functions of T, while $b(T)$ is negative. The general behavior of Equation 4—and its relation to subcritical isothermal behavior—are depicted in Figures 1 and 2, wherein the solid lines correspond to the stable vapor and liquid branches of the isotherm T, the dashed lines correspond to the respective metastable phases, and the dashed–dotted lines correspond to absolutely unstable phases.

Note that there are two solutions ρ_1 and ρ_2 of the equation $f(\rho; T) = 0$ such that $\rho_G < \rho_1 < \rho_2 < \rho_L$; these solutions determine the limits of the metastable vapor and liquid phases, respectively. At $T = 1$ the equation $f(\rho; 1) = 0$ has a double solution $\rho_1 = 1 = \rho_2$, of course.

At low densities and pressures, $(\rho, P) \rightarrow (0, 0)$, Equations 1 and 4 yield

$$\frac{dP}{d\rho} = \frac{P_\sigma(T)a(T)}{\rho_G(T)\rho_D(T)\rho_L(T)} \tag{5}$$

Thus, if the ideal gas law $P = RT\rho$ is to be fulfilled at low densities and pressures, we must have

$$a(T) = RT\,\frac{\rho_G(T)\rho_D(T)\rho_L(T)}{P_\sigma(T)} \tag{6}$$

As $T \rightarrow 1^-$, the fraction on the right-hand side of Equation 6 becomes equal to unity, and $a(T) \rightarrow RT$. Thus, as $T \rightarrow 1^-$, the differential Equation of State 1 evolves continuously to a virial-type expansion

$$\frac{dP}{d\rho} = RT + b(T)\rho + c(T)\rho^2 \tag{7}$$

for supercritical temperatures $T > 1$, i.e. only the function $f(\rho; T)$ determines the slope $dP/d\rho$ of the isotherm at supercritical temperatures.

The vapor–liquid-phase transition can be interpreted mathematically as the intersection(s) of virial-type expansions with the singular curves given by $(\rho_G(T), P_\sigma(T))$, $(\rho_D(T), P_\sigma(T))$, and $(\rho_L(T), P_\sigma(T))$.

The preceding arguments can, of course, be extended to any convex higher-order polynomial $f(\rho; T) = a(T) + b(T)\rho + c(T)\rho^2 + \ldots$ At supercritical temperatures, one then obtains $b(T)/2$ as the second virial coefficient, $c(T)/3$ as the third virial coefficient, etc.; it is well known that such analytic expressions correctly describe the behavior of isotherms at supercritical temperatures.

The relatively simple Form 4 for $f(\rho; T)$ allows Equation 1 to be integrated readily. The differential equation

$$\frac{dP}{d\rho} = \frac{[P_\sigma - P][a + b\rho + c\rho^2]}{[\rho_G - \rho][\rho_D - \rho][\rho_L - \rho]} \tag{8}$$

can be expressed as

$$\frac{dP}{P_\sigma - P} = \left\{ \frac{A}{\rho_G - \rho} + \frac{B}{\rho_D - \rho} + \frac{C}{\rho_L - \rho} \right\} d\rho, \tag{9}$$

where

$$\begin{aligned} A &= 2[a + b\rho_G + c\rho_G^2]/[\rho_L - \rho_G]^2 = 1 \\ B &= -4[a + b\rho_D + c\rho_D^2]/[\rho_L - \rho_G]^2 = \delta(T) - 2 \\ C &= 2[a + b\rho_L + c\rho_L^2]/[\rho_L - \rho_G]^2 = 1 \end{aligned}$$

and are integrated to yield

$$\ln (P_\sigma - P) = A \ln (\rho_G - \rho) + B \ln (\rho_D - \rho) + C \ln (\rho_L - \rho) + \ln D, \tag{10}$$

where $\ln D$ is the integration constant, or

$$|P_\sigma - P| = |D(T) [\rho_G - \rho][\rho_D - \rho]^{\delta(T)-2}[\rho_L - \rho]| \tag{11}$$

The constant $D(T)$, is, of course, different on the vapor and liquid branches, $D_G(T) \neq D_L(T)$, but $D_G(T) \rightarrow D_L(T)$ as $T \rightarrow 1^-$. Thus, at the critical temperature $T = 1$, by setting $D = D_G(1) = D_L(1)$ and $\delta = \delta(1)$, we obtain the well-known power law

$$|1 - P| = L\,|1 - \rho|^\delta \tag{12}$$

for the critical isotherm.

Combining Equations 1 and 11, we find that the slope $dP/d\rho$ of a subcritical isotherm can be expressed as

$$\frac{dP}{d\rho} = D(T)\,[\rho_D(T) - \rho]^{\delta(T)-3} f(\rho; T) \tag{13}$$

with $f(\rho; T)$ as defined in Equation 4. On the coexistence boundary $f(\rho_\sigma; T) = \frac{1}{2}[\rho_L(T) - \rho_G(T)]^2$ by Equation 2, and thus

$$\frac{dP}{d\rho} = D(T)\{[\rho_D(T) - \rho_\sigma(T)]^{\delta(T)-3}\}\{\tfrac{1}{2}[\rho_L(T) - \rho_G(T)]^2\}$$
$$= \tfrac{1}{2}D(T)[\rho_L(T) - \rho_G(T)]^{\delta(T)-1}$$
$$\sim [1 - T]^{\beta(\delta(T)-1)} \tag{14}$$

Hence, as $T \rightarrow 1^-$, $\delta(T) \rightarrow \delta$ and we obtain as a limit Widom's scaling equality

$$\gamma' = \beta(\delta - 1) \tag{15}$$

Furthermore, since ΔC_v behaves essentially the same as C_v on the phase boundary as $T \rightarrow 1^-$, the analogous calculation of

$$\Delta C_v = \frac{T}{\rho^2}\left(\frac{\partial P}{\partial \rho}\right)_\sigma \left(\frac{\partial \rho_\sigma}{\partial T}\right) \tag{16}$$

yields the Griffiths scaling equality

$$\alpha' = 2 - \beta(\delta + 1), \tag{17}$$

or, equivalently—by substitution of Equality 15—

$$\alpha' = 2 - 2\beta - \gamma' \tag{18}$$

Fitting the Differential Equation of State to Real Fluids

Fitting Equation 1 to a real fluid requires not only knowledge of the (ρ, P, T) behavior of the fluid, but also of the isothermal slope $dP/d\rho$ and of the coexistence envelope. Therefore it was decided expedient to use data which had been fitted already, so that detailed $dP/d\rho$ data was already available. Obviously, however, the fit of Equation 1 will reflect any error contained in such prefitted data.

For each data point (ρ, P) on the subcritical isotherm $T \leq 1$ the quotient

$$Q = \frac{P_\sigma - P}{[\rho_G - \rho][\rho_D - \rho][\rho_L - \rho]} \tag{19}$$

can be calculated, and then divided into the corresponding slope $dP/d\rho$ to obtain the value

$$f = \frac{dP/d\rho}{Q} \tag{20}$$

The various f's thus obtained then can be fitted to the function $\hat{f}(\rho; T)$. It was decided to attempt to fit the (ρ, f) data thus obtained to the simple function Form 4

$$\hat{f}(\rho; T) = a(T) + b(T)\rho + c(T)\rho^2$$

discussed in the preceding section, since fluids are known to obey Power Law 12 for the critical isotherm, and one reasonably expect Form *4* to be valid in the neighborhood of the coexistence envelope (at the very least).

A preliminary fit was made for oxygen, using the data calculated and tabulated by Weber (*1*), who used a variety of different equations to represent the measured data in different regions. (This subsequently introduced some scatter in the fit of Equation 1.)

The isobaric data (namely the 0.10125, 1, 5, 10, 20, and 30 MN/m^2 isobars) tabulated in Table XIV of the above referenced work was rearranged into 35 subcritical isotherms (ranging from 70 K = 0.453 T_c to 154 K = 0.996 T_c in 2 K intervals), each consisting of six (ρ, P, $dP/d\rho$) points. Using, additionally, the coexistence data compiled in Table XIII, six (ρ, f) points were calculated for each isotherm in the manner described above; another two (ρ, f) points were calculated using Equation 2. The eight (ρ, f) points thus obtained were fitted then to Equation 4 using an unweighted least-squares fitting routine, and conformed quite well on both the vapor and liquid branches. (This would appear quite significant, since most of the fitting data was on liquid branches.) The difference between the fitted $\hat{f}$ and the calculated f remained consistently below the 1% level, with only several exceptions—apparently caused by the transition between Weber's representations. Inasmuch as the stated accuracy of Weber's $dP/d\rho$ is "an assumed 1% error," Equation 1 together with Equation 4 yield accurate representations of the tabulated data. The function $a(T)$ varied quite smoothly with the temperature, while there was some scatter in $b(T)$ and $c(T)$ at lower temperatures, the scatter in $b(T)$ occurring inversely to the scatter in $c(T)$, it was noted that smoothing $b(T)$ tended to smooth $c(T)$.

A second fit of Equation 1—including a detailed error analysis of the Integrated Equation of State 11—was performed using the methane data tabulated by Goodwin (*2*). Methane coexistence data was calculated using Equations 3, 4a, and 4b of the above referenced work. (It should be noted, however, that Goodwin uses a critical density of 10.0 mol/L, which is somewhat lower than the values determined by Ricci and Scafè (*3*) and Jansoone, Gielen, de Boelpaep, and Verbeke (*4*).) Calculated isothermal data was taken directly from Goodwin's Table XI, and consisted of 13 subcritical isotherms ranging from 95 K to 190 K

($0.498T_c$ to $0.997T_c$), with 7 to 46 (ρ, P, $dP/d\rho$) points each. The same fitting procedure was used as that for oxygen, and once again good fits to Equation 4 were achieved; typical plot fits are shown in Figure 3. The fitted isothermal slopes were duplicated within 1% of those tabulated by Goodwin. Since all of Goodwin's isothermal data was calculated using a single equation of state, the functions $a(T)$, $b(T)$, and $c(T)$ varied smoothly with the temperature. (It was noticed, however, that the fitted $\hat{f}$ for the $T = 0.997T_c$ isotherm was slightly shifted from about the $\rho = \rho^D$ axis, which may reflect the slightly low ρ_D calculated by means of Goodwin's equations.)

Using $a(T)$, $b(T)$, $c(T)$, and the coexistence data, the values of $A(T)$, $B(T)$, and $C(T)$ for Equations 10 and 11 were calculated. $A(T)$ and $C(T)$ were indeed equal to 1 (within small fractions of 1%) over most of the temperature range. However, at $T \simeq 0.95T_c$, $A(T)$ began decreasing, while $C(T)$ began increasing; this is apparently caused by inaccuracies in Goodwin's calculated data in the critical region. Since, for symmetry reasons, it is precisely in the critical region that Equation 4 should be most applicable, we simply let $A(T) = 1 = C(T)$ for the entire temperature range. (The closeness of $A(T)$ and $C(T)$ to the value 1 can thus be construed as a measure of the accuracy of the fitting data used.)

$B(T) = \delta(T) - 2$ varied quite smoothly with the temperature, increasing to a value of $B(T) = \delta(T) - 2 = 2.109$ at $T = 0.997T_c$. At the critical temperature the extrapolated value of the fitted $B(T)$ was $B(1) = 2.251$, corresponding to a critical exponent of $\delta = 4.251$. The behavior of $B(t)$, where $t = 1 - T$, is shown in Figure 4a. While the result $\delta = 4.251$ is slightly lower than the $\delta = 4.450$ determined by Jansoone et al. for methane, it falls within the generally accepted range of values for δ; it is possible, however, that more precise fitting data in the critical region—where Goodwin's error is as high as 1%—might yield a slightly higher value for δ.

Using $A(T) = 1 = C(T)$ and fitted values of $B(T) = \delta(T) - 2$, we compared the results of Equation 11 with the calculated (ρ, P) data tabulated in Goodwin's Table XI, and then with the experimentally measured data (which Goodwin compiled from the works of numerous researchers) in Table IV. For each subcritical temperature we first calculated the constants $D_G(T)$ and $D_L(T)$ using Equation 11 and representative (ρ, P) points on the vapor and/or liquid branch. The behavior of $D_G(t)$ and $D_L(t)$, where $t = 1 - T$, is shown in Figures 4b and 4c.

Using Equation 11 and the thus calculated values of $D_G(T)$ and $D_L(T)$, we calculated the value of P for each ρ tabulated in Table II. Such $\rho \rightarrow P$ calculations duplicated Goodwin's calculated pressures with

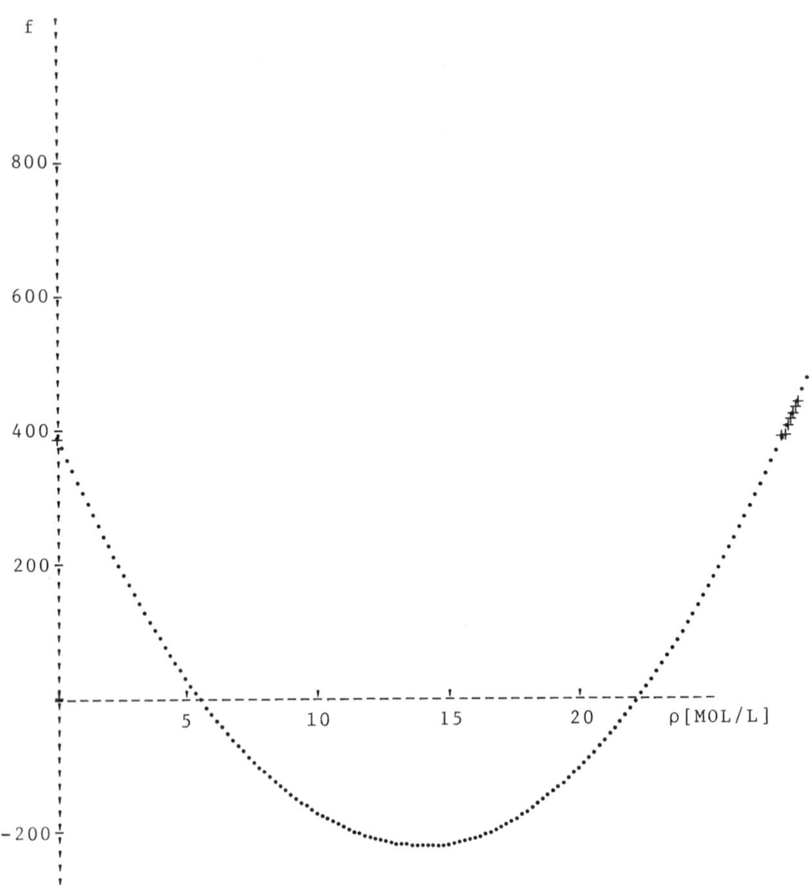

Figure 3a. CH_4 plot of ρ vs. f(ρ; T); T = 0.499 T_c

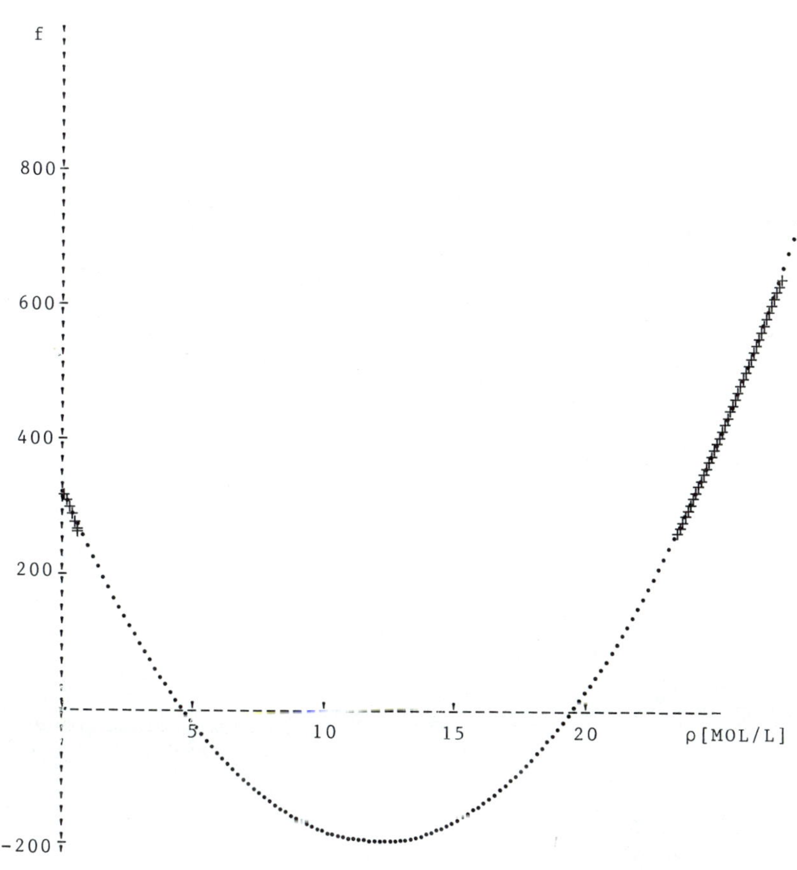

Figure 3b. CH_4 *plot of* ρ *vs.* $f(\rho; T)$; $T = 0.735\ T_c$

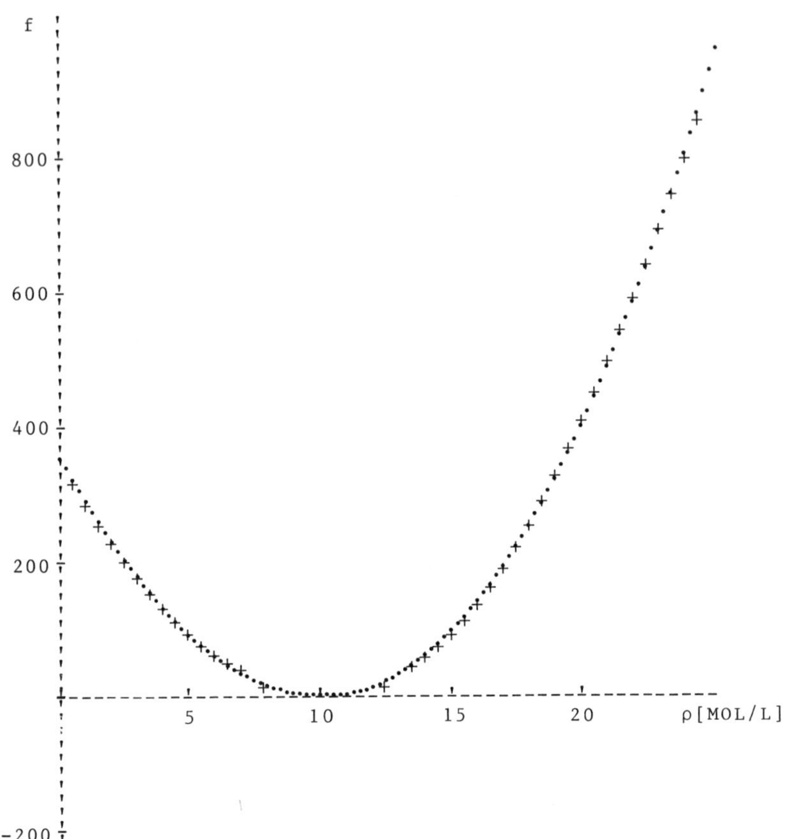

Figure 3c. CH_4 plot of ρ vs. f(ρ; T); T = 0.997 T_c

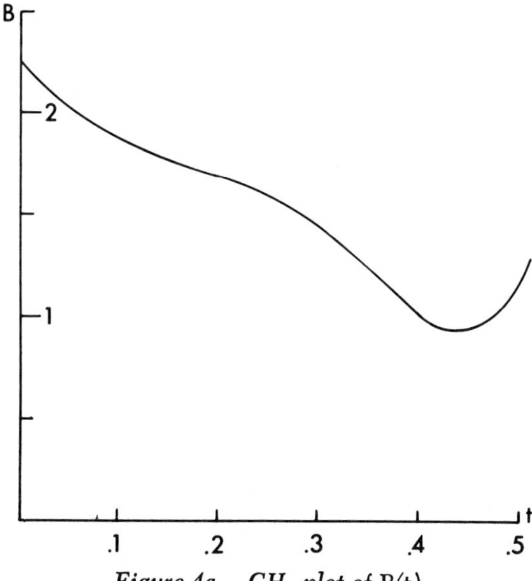

Figure 4a. CH_4 *plot of* B(t)

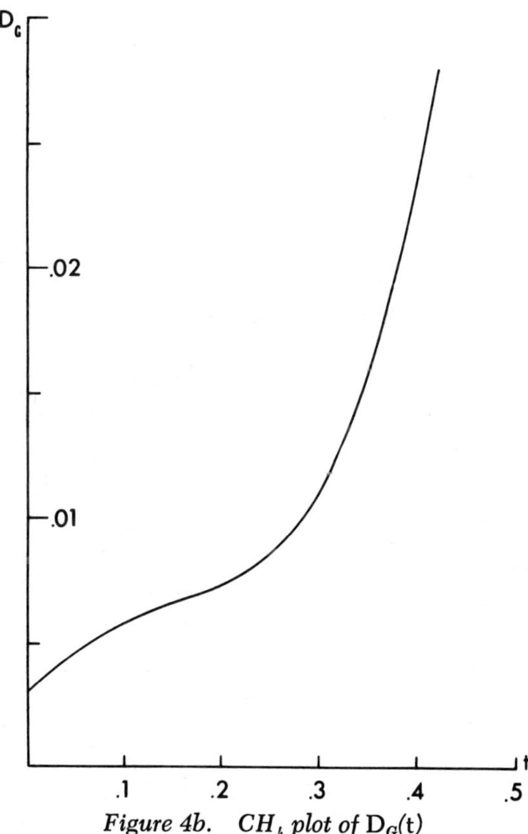

Figure 4b. CH_4 *plot of* D_G(t)

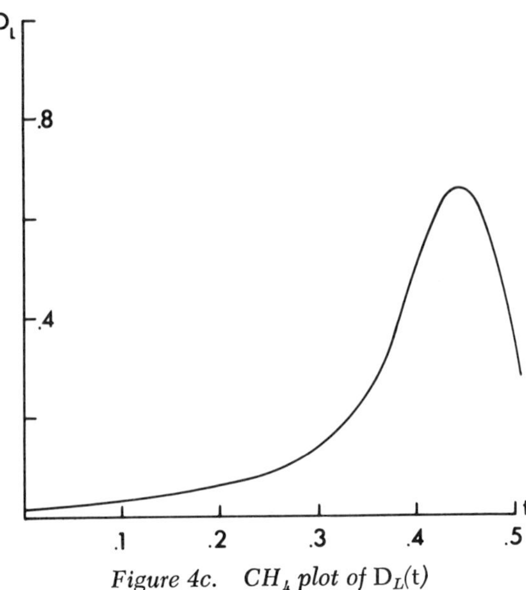

Figure 4c. CH_4 plot of $D_L(t)$

errors on the order of 0.1% or less on both the vapor and liquid branches, with greater error occurring only at very low densities for near critical temperatures.

For each value of P we then used a modified half-interval search to solve Equation 11 for ρ. For 276 experimentally measured subcritical points, the rms error in ρ was only 0.0567%, with systematic error occurring only at very low densities for near-critical temperatures. (Such points are omitted from the preceding rms error calculation. However, constraining $a(T)$ to the values given by Equation 6 and the inclusion of a small residual expression in Equation 4 extend the validity of Equation 1 to extremely low densities. The fact that such accurate results were achieved using only fitted data reflects very favorably on the general accuracy of Goodwin's data tabulation.)

Conclusions and Comments

The Differential Equation of State 1 provides not only a good qualitative description of isothermal behavior at subcritical temperatures $T \leq 1$, but also yields accurate quantitative representations of experimentally measured data. It describes not only the stable vapor and liquid branches, but also the two-phase transition region, additionally yielding information on the nature of metastable and absolutely unstable phases. A complete and simple description of the vapor–liquid-phase transition and the critical point also is provided by the differential equation of state.

Intrinsically compatible with the ideal gas law at low densities and with the scaling law at $T \rightarrow 1^-$, the differential equation of state evolves to the virial expansion for supercritical temperatures $T > 1$.

The simple Quadratic Representation 4 of $f(\rho; T)$ is definitely applicable in a large neighborhood of the coexistence boundary, and evolves to a form equivalent to a virial expansion with second and third virial coefficients at supercritical temperatures.

The use of the chemical potential μ instead of the pressure P might, for symmetry reasons, yield slightly better results. (For the same reasons, the differential equation of state should yield highly accurate results for the analogous magnetic-phase transition). However, the relatively good accessibility of (ρ, P, T) data and their ease of application make the present equation of state very attractive.

The Integrated Equation of State 11 yields very good representations of (ρ, P, T) data and provides a clear view of the evolution of the subcritical equation of state to the power law for the critical isotherm.

Thus, the Differential Equation of State 1 provides not only a clear and unified picture of the evolution of the subcritical equation of state to the power law for the critical isotherm, and thence to the virial equation of state for supercritical temperatures, but also some very fundamental insights into the vapor–liquid-phase transition process and its associated singularities.

Glossary of Symbols

P = reduced pressure
ρ = reduced density
T = reduced temperature
$P\sigma(T)$ = equilibrium vapor pressure at temperature T
$\rho_G(T)$ = saturated vapor density at temperature T
$\rho_L(T)$ = saturated liquid density at temperature T
$\rho_D(T) = \frac{1}{2}\,[\rho_G(T) + \rho_L(T)]$
$\rho\sigma(T)$ = either/both of the saturation densities $\rho_G(T), \rho_L(T)$

Acknowledgments

A. Levelt Sengers, M. E. Fisher, and R. B. Griffiths provided many useful comments on the requirements that equations of state must fulfill, as well as on the testing and fitting of the present equations of state. It would have been very difficult to test the equation of state without the extensive data compilations of L. A. Weber and R. D. Goodwin.

Literature Cited

1. Weber, L. A. "P-V-T, Thermodynamic and Related Properties of Oxygen from the Triple Point to 300 K at Pressures to 33 MN/m²," *J. Res. Natl. Bur. Stand., Sect. A* **1970,** *74*(1), 93–129.
2. Goodwin, R. D. "The Thermophysical Properties of Methane from 90 to 500 K at Pressures to 700 Bar," *Natl. Bur. Stand. (U.S.), Tech. Note* **1974,** *653.*
3. Ricci, F. P.; Scafè, E. "Orthobaric Density of CH_4 in the Critical Region," *Phys. Lett. A* **1969,** *29*(11), 650–651.
4. Jansoone, V.; Gielen, H.; De Boelpaep, J.; Verbeke, O. B.; "The Pressure–Temperature–Volume Relationship of Methane Near the Critical Point," *Physica* **1970,** *46,* 213–221.

RECEIVED September 21, 1978.

3

Cubic Equations of State: An Interpretive Review

MICHAEL M. ABBOTT

Department of Chemical and Environmental Engineering,
Rensselaer Polytechnic Institute, Troy, NY 12181

All cubic equations of state proposed to date are special cases of a general five-parameter expression. The apparent flexibility of such an expression is partly illusory, however, because of the inherent limitations imposed by its density dependence. Historically, the greatest successes with cubic equations had been achieved with variants of the Redlich–Kwong equation of state. Recent work, e.g., that of Peng and Robinson, has demonstrated the suitability of other cubic forms, and has inspired renewed efforts to identify the best cubic expression. One must approach such efforts with realistic expectations, and with an understanding of what cubic equations can and cannot do.

The Cubic Equation of State

Of the many proposed forms of the analytic equation of state, the polynomials in volume are of particular practical importance; e.g., this class includes truncations of the virial equation in density, the preferred working forms of the virial equation of state. Efficient root-finding techniques are available for the solution of polynomial equations, and moreover the number of roots is always known.

The simplest useful polynomial equation of state is one that is cubic in molar volume, for such an expression is capable of yielding the ideal gas equation in the limit as $V \rightarrow \infty$, and of representing both liquid- and vapor-like volumes for sufficiently low temperatures. If we require that the equation be explicit in pressure, then algebraic arguments lead us to a five-parameter expression of the form (*1*)

$$P = \frac{RT(V^2 + \alpha V + \beta)}{V^3 + \lambda V^2 + \mu V + \nu}$$

0-8412-0500-0/79/33-182-047$06.00/1

Parameters α, β, λ, μ, and ν can in principle all be functions of temperature T and, for mixtures, of composition.

Liquids are relatively incompressible. Therefore, if our equation is to represent liquid-like volumes, it must generate a steep portion of isotherm for appropriately small values of V. This behavior readily is incorporated by introduction of a zero into the denominator of the equation at $V = b$, where b is a small, positive number:

$$V^3 + \lambda V^2 + \mu V + \nu \equiv (V - b)(V^2 + \delta V + \epsilon)$$

Finally, we may eliminate parameters α and β in favor of two new quantities, Θ and η, via the definitions

$$\Theta \equiv RT(\delta - \alpha)$$

$$\eta \equiv (\beta - \epsilon)/(\delta - \alpha)$$

Combining the above equations then yields

$$P = \frac{RT}{V - b} - \frac{\Theta(V - \eta)}{(V - b)(V^2 + \delta V + \epsilon)} \quad (1)$$

Parameters b, δ, ϵ, Θ, and η depend generally on T and composition; their numerical values of course depend upon the identities of the chemical species in the system.

Equation 1 can be considered a generalization of the van der Waals (VDW) equation, to which it reduces as the simplest nontrivial special case. Scores of specializations of Equation 1 have been proposed since van der Waals' time; a few of them are listed in Table I and categorized according to the values (or types of functions) assumed for parameters Θ, η, δ, and ϵ. Significantly, all of the modern equations have a temperature-dependent Θ. Also, most cubic equations incorporate the constraint $\eta = b$, and in addition have zero values for certain parameters and/or specified relationships among some of the parameters b, η, δ, and ϵ.

Table I. Classification of Some Cubic Equations of State

Equation	Θ	η	δ	ϵ
van der Waals (1873)	a	b	0	0
Berthelot (1900)	a/T	b	0	0
Clausius (1880)	a/T	b	$2c$	c^2
Redlich–Kwong (1949)	$a/T^{1/2}$	b	b	0
Wilson[a] (1964)	$\Theta_W(T)$	b	b	0
Peng–Robinson (1976)	$\Theta_{PR}(T)$	b	$2b$	$-b^2$
Lee–Erbar–Edmister (1973)	$\Theta_{LEE}(T)$	$\eta(T)$	b	0

[a] Similarly, Barner et al. (1966) and Soave (1972).

The Virial Form of the Cubic Equation

The predictive capabilities of an equation of state at low densities may be tested conveniently by comparing experimental values of the virial coefficients against those implied by the (empirical) equation of state. Alternatively, realistic, low-density behavior may be built into an empirical equation of state by direct incorporation of expressions for one or more of the virial coefficients. Thus it is useful to express Equation 1 in the virial form

$$Z = 1 + B/V + C/V^2 + D/V^3 + \ldots$$

Here B is the second virial coefficient, C is the third, etc. For a mixture containing specified substances, the virial coefficients depend on T and composition only; moreover, the composition dependence of mixture virial coefficients is known.

The Z - explicit form of Equation 1 is

$$Z = \frac{V}{V-b} - \frac{(\Theta/RT)\,V\,(V-\eta)}{(V-b)\,(V^2+\delta V+\epsilon)} \tag{2}$$

Expanding the right side of Equation 2 in inverse powers of V and comparing the result with the virial equation, we find that

$$B = b - \frac{\Theta}{RT} \tag{3a}$$

$$C = b^2 - \frac{\Theta b}{RT} + \frac{\Theta}{RT}\,(\delta + \eta) \tag{3b}$$

$$D = b^3 - \frac{\Theta b^2}{RT} + \frac{\Theta b}{RT}\,(\delta + \eta) - \frac{\Theta}{RT}\,(\delta^2 - \epsilon + \eta\delta) \tag{3c}$$

etc.

Since the virial coefficients depend on T and composition only, the equation-of-state parameters can depend at most on T and composition, as already noted. The second virial coefficient B is the only one for which a decent data base and reliable estimation procedures are available; according to Equation 3a, values for B (as implied by our equation of state) are determined completely by specification of parameters b and Θ.

Unconstrained, Dimensionless Forms of the Cubic Equation

Application of Equation 1 or 2 requires the availability of numerical values for the equation-of-state parameters. Restricting ourselves for the moment to systems containing a single component, and anticipating the eventual use of corresponding-states concepts, we introduce a set of dimensionless parameters via the definitions

$$b \equiv (RT_c/P_c)\hat{b} \tag{4a}$$

$$\delta \equiv (RT_c/P_c)\hat{\delta} \tag{4b}$$

$$\eta \equiv (RT_c/P_c)\hat{\eta} \tag{4c}$$

$$\epsilon \equiv (R^2T_c^2/P_c^2)\hat{\epsilon} \tag{4d}$$

$$\Theta \equiv (R^2T_c^2/P_c)\hat{\Theta} \tag{4e}$$

Here T_c and P_c are the critical temperature and pressure. The dimensionless parameters [designated by a circumflex (^)] depend at most on reduced temperature T_r and on the identity of the substance considered.

If the two-parameter theorem of corresponding states applied, these parameters would be independent of chemical species.

Equations 1 and 2 can be put into reduced form by using Equation 4 and the additional definitions

$$P_r \equiv P/P_c \tag{5a}$$

$$T_r \equiv T/T_c \tag{5b}$$

$$V_r \equiv V/V_c \tag{5c}$$

$$\zeta_c \equiv P_cV_c/RT_c \tag{5d}$$

Parameter ζ_c is an apparent critical compressibility factor. We give it a special symbol to preclude its general identification with the experimental critical compressibility factor Z_c, to which, in practical applications of cubic equations of state, it is often assumed not equal. In such cases, the reduced volume V_r is defined always with respect to a (possibly hypothetical) critical volume V_c defined in terms of P_c, T_c, and ζ_c via Equation 5.

Combining Equations 1 and 2 with Equations 4 and 5 then yields the equivalent expressions

$$P_r = \frac{T_r}{\zeta_cV_r - \hat{b}} - \frac{\hat{\Theta}(\zeta_cV_r - \hat{\eta})}{(\zeta_cV_r - \hat{b})(\zeta_c^2V_r^2 + \hat{\delta}\zeta_cV_r + \hat{\epsilon})} \tag{6}$$

and

$$Z = \frac{\zeta_cV_r}{\zeta_cV_r - \hat{b}} - \frac{\hat{\Theta}\zeta_cV_r(\zeta_cV_r - \hat{\eta})}{T_r(\zeta_cV_r - \hat{b})(\zeta_c^2V_r^2 + \hat{\delta}\zeta_cV_r + \hat{\epsilon})} \tag{7}$$

The expressions for the virial coefficients may be written also in dimensionless form. Defining dimensionless virial coefficients by

$$\hat{B} \equiv BP_cRT_c \tag{8a}$$

$$\hat{C} \equiv CP_c^2/R^2T_c^2 \tag{8b}$$

$$\hat{D} \equiv DP_c^3/R^3T_c^3 \tag{8c}$$

etc., we can write Equation 3 as

$$\hat{B} = \hat{b} - \frac{\hat{\Theta}}{T_r} \tag{9a}$$

$$\hat{C} = \hat{b}^2 - \frac{\hat{\Theta}\hat{b}}{T_r} + \frac{\hat{\Theta}}{T_r}(\hat{\delta} + \hat{\eta}) \tag{9b}$$

$$\hat{D} = \hat{b}^3 - \frac{\hat{\Theta}\hat{b}^2}{T_r} + \frac{\hat{\Theta}\hat{b}}{T_r}(\hat{\delta} + \hat{\eta}) - \frac{\hat{\Theta}}{T_r}(\hat{\delta}^2 - \hat{\epsilon} + \hat{\eta}\hat{\delta}) \tag{9c}$$

Incorporation of the Classical Critical Constraints

Numerical values for the equation-of-state parameters may be established in many ways, and, since no equation of state is perfect, different values are obtained depending upon the methods used. One class of methods, which we can call "brute-force" methods, involves the use of nonlinear regression techniques to determine by analysis of experimental data the best values of the parameters for representation of a particular property (or properties) over a limited range of temperature and pressure. The equations which result can be quite precise for their intended purpose. Another much older approach is to impose a few selected mathematical or numerical constraints on the equation of state and to determine numerical values for the parameters by solving the resulting system of equations. These techniques, which we can call "algebraic" methods, lead to equations of state generally less precise than those resulting from the brute-force approach, but also often less likely to generate absurd results outside the intended range of application of the equation.

Martin (*2*) discusses and applies some of the features of real-fluid behavior which lend themselves to convenient expression as algebraic constraints for equation-of-state studies. The two constraints probably most often used are the classical conditions

$$(\partial P/\partial V)_{T,\mathrm{cr}} = 0 \tag{10}$$

$$(\partial^2 P/\partial V^2)_{T,\mathrm{cr}} = 0 \tag{11}$$

which assert that the critical isotherm has a point of horizontal inflection at the critical state (cr). A requirement equivalent to Equations 10 and 11 is that Equation 6, when expanded and specialized to the critical state ($T_r = P_r = 1$), has three equal roots; that is, it must be of the form

$$(V_r - 1)^3 \equiv V_r^3 - 3V_r^2 + 3V_r - 1 = 0 \qquad (12)$$

Thus we obtain, after some algebraic manipulations, the three equations

$$\hat{\delta}_c = \hat{b}_c + 1 - 3\zeta_c \qquad (13)$$

$$\hat{\epsilon}_c = \hat{\delta}_c(\hat{b}_c + 1) - \hat{\Theta}_c + 3\zeta_c^2 \qquad (14)$$

$$\hat{\eta}_c = \frac{\zeta_c^3 - (\hat{b}_c + 1)\hat{\epsilon}}{\Theta_c} \qquad (15)$$

Since for a pure material all parameters can in principle be functions of T_r, I have used the subscript c to designate a value for a parameter evaluated at $T_r = 1$.

If ζ_c is regarded as an empirical equation-of-state parameter, different in general from the experimental critical compressibility factor Z_c, then Equations 13, 14, and 15 constitute a system of three equations in six unknowns. Thus, three additional constraints must be imposed just to make the system determinate for the single temperature $T_r = 1$. If ζ_c is identified with Z_c, then the number of unknowns is reduced by one, but we are still left with two degrees of freedom. As suggested by the form of Equations 13, 14, and 15, when one deals with the more general case ($\zeta_c \neq Z_c$), parameters $\hat{b}_c$, $\hat{\Theta}_c$, and ζ_c are best treated as independent variables; that is, equation-of-state building usually proceeds most smoothly if these parameters are the ones fixed by the additional constraints.

As noted earlier in this chapter, most of the cubic equations of state proposed to now incorporate the constraint $\eta = b$; for such cases, Equations 13, 14, and 15 can be arranged to yield the three expressions

$$\hat{\Theta}_c = (\hat{b}_c + 1 - \zeta_c)^3 \qquad (16)$$

$$\hat{\delta}_c = \hat{b}_c + 1 - 3\zeta_c \qquad (17)$$

$$\hat{\epsilon}_c = (\zeta_c - \hat{b}_c)^3 + \hat{b}_c(3\zeta_c - 2\hat{b}_c - 1) \qquad (18)$$

Here we have only two independent variables, e.g. $\hat{b}_c$ and ζ_c. If ζ_c is identified with Z_c, then only a single degree of freedom remains, so that if, e.g. $\hat{b}_c$ is specified, then numerical values immediately follow for all of the remaining parameters at $T_r = 1$. The generic cubic equation in which $\eta = b$ has been treated by Abbott (*3*) for cases in which b, δ, and ϵ are constants, but for which Θ is a function of T.

Studies of the Critical Isotherm

Many practitioners classify equations of state according to the number of parameters (or constants) they contain. Moreover, it commonly is supposed that a four-parameter equation is inherently better than a three-parameter equation, that a five-parameter expression is better than one containing four, etc. While it is true that the judicious introduction of an extra parameter or two often can improve dramatically the performance of an equation of state for particular applications, it is equally true that one often quickly reaches a point of diminishing (or even negative) returns, particularly with the simpler equations of state. A controlling factor is the density dependence of the equation; with an equation of specified functional form in density or volume, one can get only so far by adjusting the temperature-dependent parameters. Thus, the cubic equation of state has certain inherent limitations which exist solely because it is a cubic equation.

A few of the features and limitations of the cubic equation can be illustrated neatly by considering the critical isotherm. Since T is fixed, we need not concern ourselves with the temperature dependence of the parameters; moreover, since $T = T_c$, we can use the critical constraints treated in the preceding section. The example which follows is inspired by a study reported by J. J. Martin (*2*), and I shall use the same data used by Martin: values of P_r and V_r for argon at $T_r = 1$, read to within about $\pm$ 1% from graphs prepared by Costolnick and Thodos (*4*).

We adopt the critical derivative constraints given by Equations 10 and 11; thus, the values of the equation-of-state parameters are not all independent, but are related by Equations 13, 14, and 15. Further, to ensure good behavior of the equation at low densities, we require that the equation generate an acceptable value for the second virial coefficient $\hat{B}_c$. Since by Equation 9a

$$\hat{B}_c = \hat{b}_c - \hat{\Theta}_c, \tag{19}$$

and since $\hat{B}_c$ is very nearly $-1/3$ for argon, this requirement adds another mathematical constraint to our set, viz.,

$$\hat{\Theta}_c = \hat{b}_c + 1/3 \tag{20}$$

Equations 13, 14, 15, and 20 thus reduce the degrees of freedom to either one (if ζ_c is identified with Z_c), or two (if ζ_c is treated as an additional parameter). For the former case, specification of $\hat{b}_c$ fixes the values of all other parameters, and thus determines the entire isotherm. This is similar for the latter case, for specification of $\hat{b}_c$ and ζ_c.

Table II contains the results of six case studies, in which the effects of varying the magnitude of $\hat{b}_c$ for two values of ζ_c are illustrated. In cases Ia, Ib, and Ic, ζ_c is fixed at $\zeta_c = Z_c(\exp) = 0.291$; In cases IIa, IIb, and IIc, $\zeta_c = 1/3$, a value inspired by the popular Redlich–Kwong (RK) equation and by some other modern cubic equations of state. In the calculations, V is an experimental value, and V_r is determined from the equation

$$V_r = \frac{VP_c}{\zeta_c RT_c}$$

Thus, for $\zeta_c = Z_c(\exp) = 0.291$, we have

$$V_r = V_r(\exp)$$

but, for $\zeta_c = 1/3$, we have

$$V_r = \frac{0.291}{(1/3)}\ V_r(\exp) = 0.873\ V_r(\exp)$$

Cases Ia, Ib, and Ic illustrate the difficulties which can arise if one incorporates experimental values of Z_c into a cubic equation of state. Excellent representation of the critical isotherm is obtained up to a reduced density of about 0.4, because of the realistic value for B_c built into the equation. Between $\rho_r = 0.4$ and 1.0, the predicted pressures are a little low, but not unreasonably so. For supercritical densities, however, the quality of the representation deteriorates badly, increasingly so the smaller one makes parameter $\hat{b}_c$. A large value of $\hat{b}_c$ is not the answer to the problem; as indicated by the results of Case Ic, a value of $\hat{b}_c = 0.150$ improves the predictions up to $\rho_r = 1.8$, but yields negative pressures for the two highest densities. The reason for this behavior is that for this case the limiting reduced volume (the value of V_r for which $P_r \to \infty$) is $V_r = \hat{b}_c/\zeta_c = 0.150/0.291 = 0.5155$, a number larger than the last two volume entries in the table; for $V_r < 0.5155$, one lands on a physically meaningless branch of the isotherm.

Cases IIa, IIb, and IIc show the effects of treating ζ_c as an adjustable parameter, larger in these cases than $Z_c(\exp)$ by about 15%. Numerical

Table II. Results of Parametric Studies of the Critical Isotherm of Argon

$V_r(exp)$	$P_r(exp)$	$P_r(calc)$ for Case[a, b, c] Ia	Ib	Ic	IIa	IIb	IIc
10	0.305	0.305	0.305	0.305	0.306	0.306	0.306
5	0.538	0.538	0.537	0.537	0.542	0.542	0.542
2.5	0.830	0.827	0.826	0.824	0.841	0.841	0.840
1.667	0.970	0.954	0.954	0.952	0.969	0.969	0.969
1.25	0.999	0.995	0.995	0.994	0.999	0.999	0.999
1.0	1.000	1.000	1.000	1.000	1.003	1.003	1.003
0.8333	1.00	1.004	1.005	1.006	1.054	1.057	1.061
0.7145	1.10	1.033	1.040	1.052	1.230	1.247	1.280
0.625	1.38	1.110	1.144	1.236	1.621	1.693	1.855
0.5556	2.15	1.263	1.404	2.156	2.329	2.568	3.225
0.50	4.00	1.543	2.096	(−3.541)	3.493	4.186	6.880
0.4545	7.80	2.051	5.237	(−0.572)	5.297	7.189	23.55

[a] For all cases, $\hat{B}_c = \hat{b}_c - \hat{\theta}_c = -1/3$.

[b] For case Ia, $\hat{b}_c = 0.100$; Ib, $\hat{b}_c = 0.125$; Ic, $\hat{b}_c = 0.150$ } $\zeta_c = Z_c(\text{exp}) = 0.291$, $V_r = V_r(\text{exp})$

[c] For case IIa, $\hat{b}_c = 0.075$; IIb, $\hat{b}_c = 0.100$; IIc, $\hat{b}_c = 0.125$ } $\zeta_c = 1/3$, $V_r = 0.873\ V_r(\text{exp})$

effects of the change in ζ_c can be seen by comparing entries for Case IIb with Case Ia ($\hat{b}_c = 0.100$) or Case IIc with Case Ib ($\hat{b}_c = 0.125$). The net effect on the representation of the isotherm is to produce predicted pressures greater than the experimental values for reduced densities up to about 0.4, but also to introduce a horizontal shift in the curve at supercritical densities, thus yielding for appropriate values of $\hat{b}_c$ (in this case, for $\hat{b}_c$ somewhat less than 0.100) a "compromise fit" of the steep portion of the isotherm. This, in fact, is exactly what one observes with the RK equation, and it is one of the reasons for the success of that expression and its numerous progeny. The price one pays, of course, is a less-than-perfect picture of the critical region (in the present case the critical volume is 15% too large), but an analytic equation of state is incapable of precise description of critical phenomena anyhow. More serious is the damage done with respect to representation of saturated liquid volumes, for unless $\hat{b}$ is considered a bizarre function of T_r, the effect of the large V_c is felt in the liquid phase down to the lowest reduced temperatures.

Schematic illustrations of some of the effects just described are shown in Figure 1.

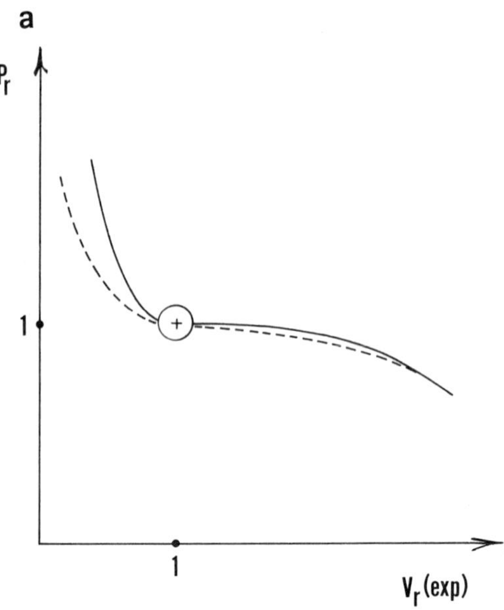

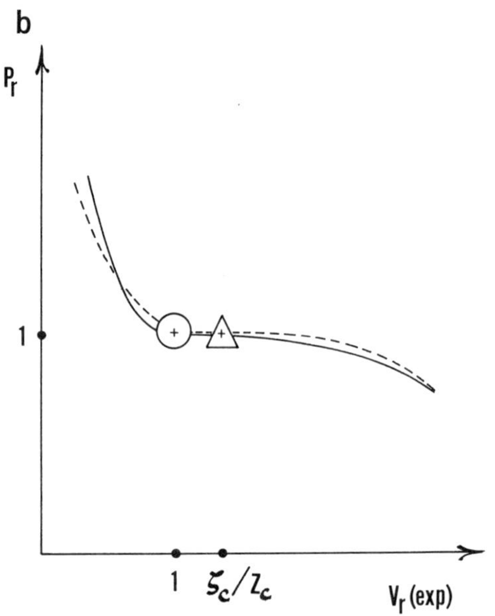

Figure 1. Schematic of the critical isotherm for Ar, illustrating Cases Ia and IIb from Table II: (——), experiment; (– – –), calculations with the cubic equation. In Figure 1a, the critical points (c.p.'s) coincide at ○. In Figure 1b, the computed c.p. (△) is displaced horizontally from the experimental c.p. (○).

A Priori Estimates of the Behavior of $\hat{b}$ *and* $\hat{\Theta}$

We may note two features of all successful cubic equations: first the presence of the ubiquitous VDW repulsive term $RT/(V - b)$, and second, the presence of a temperature-dependent parameter $\Theta(T)$. It is this second feature which distinguishes the VDW equation (for which $\Theta = \text{constant} = a$) from virtually all of its offspring, and it is the development of suitable expressions for $\Theta(T)$ which has resulted in most of the significant modern advances in cubic equation-of-statery.

The parameters b and Θ (or their dimensionless counterparts $\hat{b}$ and $\hat{\Theta}$) are thus of major importance; parameter δ places a close third, followed in order by ϵ and η ($\epsilon = 0$ and $\eta = b$ are the most common assignments for these quantities). Significantly, b and Θ are also susceptible to simple empirical interpretation, and hence to rough estimation. This is illustrated schematically in Figure 2, a plot of Z vs. P_r (for fixed T_r).

In the limit as $P_r \to 0$, we have the well-known relationship

$$\lim_{P_r \to 0} \left(\frac{\partial Z}{\partial P_r}\right)_{T_r} = \frac{\hat{B}}{T_r} \tag{21}$$

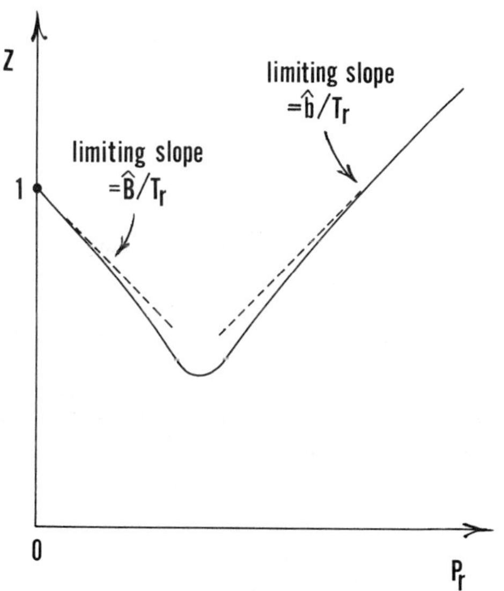

Figure 2. Schematic of an isotherm on a Z vs. P_r *diagram, illustrating the complementary relationship between* $\hat{B}$ *and* $\hat{b}$

Similarly, in the limit as $P_r \to \infty$, we find, from Equations 6 and 7,

$$\lim_{P_r \to \infty} \left(\frac{\partial Z}{\partial P_r}\right)_{T_r} = \lim_{P_r \to \infty} \left(\frac{Z}{P_r}\right) = \frac{\hat{b}}{T_r} \tag{22}$$

Equations 21 and 22 suggest a complementary relationship between $\hat{B}$ and $\hat{b}$ in terms of the limiting slopes of an isotherm on a Z vs. P_r diagram. The connection between these slopes and $\hat{\Theta}$ is established by Equation 9a, which relates $\hat{\Theta}$ to $\hat{B}$ and $\hat{b}$.

Analysis of high-pressure *PVT* data or of generalized charts by Equation 22 reveals that $\hat{b}$ is about 1/10, usually a little less. Unfortunately, it is very difficult to detect from such data trends of $\hat{b}$ with either temperature or molecular complexity. If liquid-phase isotherms were infinitely steep, then one would expect $\hat{b}$ to increase with T_r and to decrease with molecular complexity (as reflected, e.g., by decreasing values of Z_c, or by increasing values of the acentric factor ω). However, real liquids are compressible, and, moreover, the apparent dependence of $\hat{b}$ on T_r and, say, ω is known to be influenced strongly by the method and by the data base used for determination of numerical values of the equation-of-state parameters. The best one can say is that for practically any application $\hat{b}$ is of the order 1/10, and that its temperature dependence is usually less important than that of $\hat{\Theta}$.

The qualitative behavior of $\hat{\Theta}$ is rationalized most easily from Equations 9a and 19:

$$\hat{B} = \hat{b} - \frac{\hat{\Theta}}{T_r}$$

$$\hat{\Theta}_c = \hat{b}_c - \hat{B}_c$$

For simple fluids (Ar, Kr, and Xe) $\hat{B}_c$ is very nearly $-1/3$, and, as stated above, $\hat{b}_c$ is about 1/10. Thus, by Equation 19, $\hat{\Theta}_c$ for such substances should be about 0.43, or a little less. For more complex fluids, $\hat{B}_c$ is negative but larger in absolute value; however, $\hat{b}_c$ is probably slightly smaller than for the simple fluids. Ignoring the possible compensating effect of $\hat{b}_c$, we conclude that $\hat{\Theta}_c$ for most fluids should be in the range of about 0.40–0.60. Values outside of this range will result usually in grossly inaccurate values of $\hat{B}_c$, with correspondingly inaccurate representation of isotherms in the low-density region.

Equation 9a allows us to say something about the probable dependence of $\hat{\Theta}$ on T_r and on ω. Since $\hat{B}$ varies with T_r approximately as $T_r^{-3/2}$ to T_r^{-2}, and since $\hat{b}$ is probably no more than a weak function of T_r, then we conclude from Equation 9a that $\hat{\Theta}$ should decrease with T_r; any other temperature dependence must give an imprecise picture of *B*, and of the derived properties which depend upon the temperature derivatives of P or of Z. At subcritical temperature levels, where the $\hat{\Theta}$ term is the greatest contributor to $\hat{B}$, $\hat{B}$ is known to increase in absolute value with increasing ω. Hence we conclude from Equation 9a that $\hat{\Theta}$ should increase also with ω, at least for low reduced temperatures.

The observations just made with respect to $\hat{b}$ and $\hat{\Theta}$ are necessarily qualitative, but they can serve as useful guidelines in equation-of-state building, and as criteria for comparison and assessment of cubic equations of state. Most of the cubic equations proposed to date, whatever the actual basis for their development, conform more or less to these guidelines.

Some Contemporary Specializations of the Cubic Equation

Any person who seriously has attempted to follow the vast literature on the RK equation must be impressed (if not occasionally appalled) at the effort expended in attempts to make that expression do all of the things that a good equation of state must do. A review article by Horvath (*5*) by now badly outdated, lists 112 citations, many of them dealing with specific variants of the RK equation. Because of the availability of this large body of documented experience, the RK equation therefore serves nicely as a vehicle for a discussion of the techniques applied by specialists in developing high-performance cubic equations of state.

The original RK equation is (*6*)

$$P = \frac{RT}{V - b} - \frac{a}{T^{1/2}V(V + b)} \tag{23}$$

As shown in Table I, Equation 23 is recovered from the general cubic equation via the assignments

$$\Theta = a/T^{1/2}$$

$$\eta = \delta = b$$

$$\epsilon = 0$$

Here parameters a and b are constants. If Equation 23 is subjected to the classical critical constraints, then Equations 16, 17, and 18 apply. However, $\delta = b$, and we find from Equation 17 that $\zeta_c = 1/3$; also, since $\epsilon = 0$, we find from Equation 18 that $\hat{b} = (2^{1/3} - 1)/3 = 0.0866$. Finally, Equation 16 yields $\hat{\Theta}_c = (2^{1/3} + 1)^3/27 = 0.4275$. The magnitudes of $\hat{b}$ and $\hat{\Theta}_c$, as well as the dependence of $\hat{\Theta}$ on T_r, are reasonable.

The remarkable success of the RK equation arises mainly from two factors, neither of them immediately obvious. The first relates to the second virial coefficient. Figure 3 shows a plot of the RK $\hat{B}$, superposed

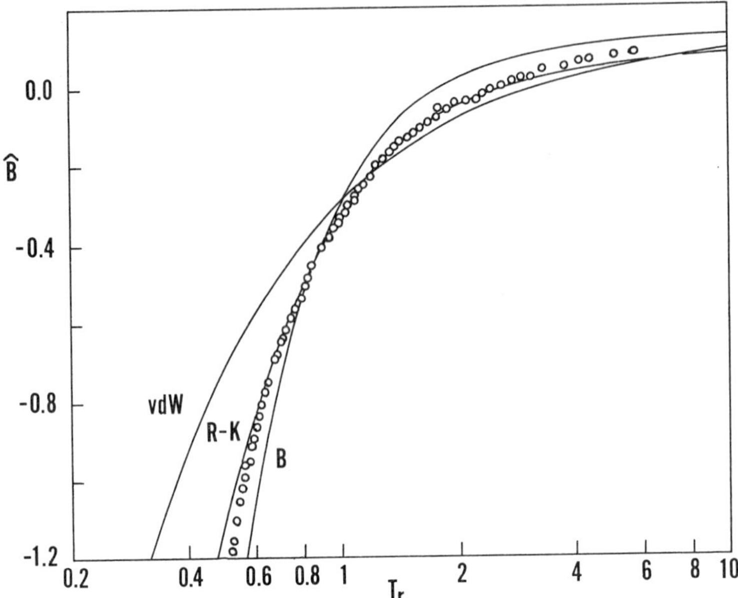

Figure 3. $\hat{B}$ *vs.* T_r *for Ar, Kr, and Xe: (○), data. Lines are computed from the Berthelot (B) equation, the Redlich–Kwong (RK) equation, and the van der Waals (VDW) equation.*

on $\hat{B}$ data for Ar, Kr, and Xe. The agreement is excellent, and guarantees good performance for simple fluids at low densities. A second attribute of the equation is its ability to provide a reasonable (if imprecise) description of the high-density region for simple fluids at supercritical pressures. This is illustrated in Table III, in which are reproduced RK calculations given by Martin (2) for the critical isotherm of argon. The effect in operation here was described earlier in this chapter. Although the true critical isotherm is decidedly not cubic in volume, the result (in this case) of the large value of ζ_c is to position the calculated curve so that it provides a "compromise fit" of the true isotherm.

It is difficult today to appreciate the magnitude of the advance that the RK equation represented at the time it was developed. To put this into perspective, I have included in Figure 3 and Table III curves and numerical values derived from the equations of van der Waals and Berthelot, two of the then-most-popular cubic equations of state. The results speak for themselves.

It soon became clear that the RK equation, despite its merits, was not really satisfactory for many kinds of calculations. A major problem with the original equation is that, in its generalized form, it contains only

Table III. The Critical Isotherm of Argon, as Calculated from the Redlich–Kwong, Berthelot, and van der Waals Equations

		P_r*(calc)*	
V_r*(exp)*	P_r*(exp)*	*Redlich–Kwong*[a]	*van der Waals and Berthelot*[b]
10	0.305	0.305	0.309
5	0.538	0.540	0.552
2.5	0.830	0.839	0.863
1.667	0.970	0.968	0.984
1.25	0.999	0.999	1.000
1.0	1.000	1.003	1.042
0.8333	1.00	1.060	1.337
0.7145	1.10	1.267	2.301
0.625	1.38	1.748	4.829
0.5556	2.15	2.681	11.12
0.50	4.00	4.336	28.85
0.4545	7.80	7.157	113.6

[a] For the RK equation, $V_r = 0.873 V_r(\text{exp})$.
[b] For the VDW and Berthelot equations, $V_r = 0.776 V_r(\text{exp})$.

two corresponding-states parameters: T_c and P_c. Thus, although the equation performs relatively well for the simple fluids Ar, Kr, and Xe (for which $\omega = 0$), it does not perform well for complex fluids with nonzero acentric factor. An early attempt at removing this difficulty was made by G. M. Wilson (*7, 8*), who proposed a new expression for $\hat{\Theta}$:

$$\hat{\Theta}_W = 0.4275\left[1 + (1.57 + 1.62\omega)\left(\frac{1}{T_r} - 1\right)\right]T_r \qquad (24)$$

Equation 24 regenerates the classical value of $\hat{\Theta}_c = 0.4275$ at $T_r = 1$ and, since $\hat{b}$ is unchanged, the Wilson modification of the RK equation yields a pure-component critical point for $\zeta_c = 1/3$. The numerical values of the constants in Equation 24 were established by forcing the equation of state to give reasonable values of the terminal slope of the vapor-pressure curve, a quantity which correlates with ω. Although Equation 24 results in a $\hat{B}$ which varies as T_r^{-1}, numerical agreement with experimental values (at least for the simple fluids) is satisfactory in the range T_r = 0.7 to 1.5; outside this range, predicted values of $\hat{B}$ are algebraically greater than those yielded by the original equation. An undesirable feature of Equation 24 is that it produces negative values for $\hat{\Theta}$ at high reduced temperatures. However, Wilson found that his modification gave satisfactory estimates of residual enthalpies and of binary vapor–liquid equilibrium for several systems of practical importance.

Barner et al. (*9*), in an effort to provide a simple equation of state suitable for computation of vapor enthalpies, proposed another expression for $\hat{\Theta}$:

$$\hat{\Theta}_B = 0.4275 \left(\frac{1 + 4.73\omega^{3/2} T_r^{-3/2}}{1 + 4.73\omega^{3/2}} \right) T_r^{-1/2} \qquad (25)$$

Equation 25 was developed from an empirical representation of the second virial coefficient correlation of Pitzer and Curl (*1*); parameter $\hat{b}$ was left unchanged at its classical value of 0.0866. Because of the substantial improvement in the prediction of $\hat{B}$ and its temperature derivatives for nonsimple fluids, the Barner modification of the RK equation gave improved estimates of enthalpy deviations for nonpolar vapors and for vapor-phase mixtures of hydrocarbons. However, the new equation was unsuitable for fugacity calculations.

In a significant departure from conventional practice, Chueh and Prausnitz (*11, 12*) proposed that the critical constraints on the RK equation be relaxed, and that parameters $\hat{b}$ and $\hat{\Theta}_c$ be treated as empirical constants, determined separately for the liquid phase and for the vapor phase of a given substance. The conventional RK expression for $\Theta(T)$ was retained; the application was to vapor–liquid equilibrium calculations, in which the vapor-phase version of the equation was used for computation of vapor-phase fugacity coefficients, but in which the liquid-phase version was used only for Poynting corrections. Thus, they proposed that

$$\hat{b} = \begin{cases} \hat{b}^l \text{ (liquid)} \\ \hat{b}^v \text{ (vapor)} \end{cases}$$

$$\hat{\Theta}_c = \begin{cases} \hat{\Theta}_c^{\,l} \text{ (liquid)} \\ \hat{\Theta}_c^{\,v} \text{ (vapor)} \end{cases}$$

$$\hat{\Theta} = \hat{\Theta}_c / T_r^{1/2}$$

Best values for $\hat{b}$ and $\hat{\Theta}_c$ were determined by regression of pure-component volumetric data for saturated liquids and vapors. The procedure was applied to 19 fluids, and the results are an interesting indicator of the limitations of the two-parameter, generalized version of the RK equation in the saturation region. Chueh and Prausnitz found $\hat{b}$ and $\hat{\Theta}_c$ to be different from their classical values (but of the same orders of magnitude), different for different substances, and different for the liquid and vapor phases of a given substance. Moreover, trends emerged: $\hat{b}^l$ tended to decrease with increasing ω, while $\hat{b}^v$ showed the opposite effect; similarly, $\hat{\Theta}_c^{\,l}$ tended to decrease with ω, with $\hat{\Theta}_c^{\,v}$ exhibiting the opposite trend.

The new ideas exposed in these studies (and in a few others like them) inspired great activity in the modification of the RK equation: parameters a and b were treated as functions of temperature; various constraints on the equation were relaxed; new ones were added; and many extensions to mixture behavior were proposed. This period (1967–1977) may be well remembered as the Redlich–Kwong Decade. In the midst of all this, Soave (*13*) published a RK variant which appeared to be a great improvement over other versions. Not surprisingly, the idea behind the new variation was a twist on an older one—the one exploited by Wilson in 1964.

Soave reasoned, as had Wilson, that if an equation of state is to be used for VLE calculations, then one should require that it also represent pure-component VLE data; that is, that vapor-pressure information for pure substances should somehow be built into the equation. Wilson had done this indirectly, through the limiting slope of the vapor-pressure curve; Soave did it more directly, by forcing the equation to reproduce experimental vapor pressures of pure nonpolar substances at $T_r = 0.7$. Soave retained the classical value of $\hat{b}$, but he represented his values of $\hat{\Theta}$ by a better-behaved function:

$$\hat{\Theta}_S = 0.4275\,[1 + (0.480 + 1.574\omega - 0.776\omega^2)(1 - T_r^{1/2})]^2 \quad (26)$$

Values of $\hat{B}$ for simple fluids generated by the Soave equation are practically identical with those computed from Wilson's equation up to $T_r = 1$. At supercritical temperatures, however, they are in much better agreement with experiment than are Wilson's $\hat{B}$'s. For $\omega = 0$, Equation 26 produces a zero value of $\hat{\Theta}$ at $T_r = 9.51$; above this temperature, Θ increases with T_r. This anomalous behavior at very high T_r leads to errors in the representation of VLE for hydrogen-containing systems; revised expressions for $\hat{\Theta}_S$ have been developed for hydrogen. [*See* Adler et al. (*14*) for a brief discussion of work done at Pennsylvania State University in evaluating and modifying the Soave equation for the prediction of VLE.]

Of the RK variants which have received extensive testing, Soave's equation appears to be one of the very best proposed to date. As long as one deals with systems containing hydrocarbons and certain light gases (e.g., N_2, CO_2, and H_2S), it provides acceptable predictions of gas densities, gas-phase enthalpy deviations, and VLE (*15*). It does not yield good estimates of liquid densities, however, and it is this shortcoming which prompted the recent development of the last of the modern cubic equations to be treated in this chapter.

Soave's equation predicts liquid-phase molar volumes which are too large; the deviations are greatest for substances with large ω (i.e., with small Z_c). Peng and Robinson (*16*) reasoned that this results from the large value of $\zeta_c = 1/3$ imposed upon the RK equation by the critical constraints, and they sought a simple form of the cubic equation, comparable in accuracy with Soave's equation, but with a smaller value of ζ_c. The expression they developed can be written as

$$P = \frac{RT}{V - b} - \frac{\Theta(T)}{V^2 + 2bV - b^2}$$

The Peng–Robinson equation is recoverable from Equation 1 by the assignments

$$\begin{aligned} \Theta &= \Theta(T) \\ \eta &= b \\ \delta &= 2b \\ \epsilon &= -b^2 \end{aligned} \tag{27}$$

Parameter b is treated as a constant for a given material, but Θ is a function of T, to be determined. Application of Equations 16, 17, and 18 yields the values $\zeta_c = 0.30740$, $\hat{b} = 0.07780$, and $\hat{\Theta}_c = 0.45725$.

An expression for $\hat{\Theta}$ was developed for Equation 27 by a procedure similar to that followed by Soave, and a correlation of identical functional form was obtained:

$$\hat{\Theta}_{PR} = 0.45725\,[1 + (0.37464 + 1.54226\omega - 0.26992\omega^2)\,(1 - T_r^{1/2})]^2 \tag{28}$$

Peng and Robinson report the results of comparisons of their equation with Soave's. They find that the two equations give similar values for gas densities and gas-phase enthalpy deviations, but that the Peng–Robinson equation yields improved correlation of pure-component vapor pressures and better estimates of liquid densities.

Extensions to Mixtures

Although equation-of-state building normally proceeds by considering the properties of pure fluids, actual application is usually to the estimation of properties of mixtures, or of species in solution. A complete catalog of approaches to mixtures adopted by past and present investigators would be inappropriate here; good discussions are given by Reid et al. (*15*). What follows is a summary evaluation and partial justification of the methods which appear to be in current favor, and which seem most likely to be used in future applications of cubic equations of state to mixture calculations.

The most direct extension of an equation of state to mixtures is through using mixing rules for the parameters. These are recipes (usually empirical) which relate mixture parameters to (a) composition, (b) pure-component parameters, and (possibly) (c) additional parameters intended to characterize interactions among unlike molecular species. Only a few types of mixing rules have gained wide acceptance for use with cubic equations; they all may be considered special cases of the general quadratic mixing rule

$$\pi_{\mathrm{m}} = \sum\sum x_i x_j \pi_{ij} \tag{29}$$

Here, π_{m} is the value of parameter π for a mixture, and the quantity π_{ij} characterizes interactions between species i and j. If $j = i$, then $\pi_{ij} = \pi_{ii} (\equiv \pi_i)$, the parameter for pure i. Once a scheme is devised for assignment of numbers to pure-component parameters, the value of π_i is unambiguous.

The difficulty in the application of Equation 29 is in the assignment of values to π_{ij} when $i \neq j$. While such a parameter is said to represent the effects of unlike-pair interactions, it is often difficult to give this idea a clear physical or mathematical statement. Thus one usually resorts to combination rules, which relate π_{ij} to the pure-component parameters and (possibly) to an empirical interaction parameter, a number (by optimistic definition) usually either of order zero or of order unity.

If we adopt an arithmetic-average combination rule

$$\pi_{ij} = \tfrac{1}{2} (\pi_i + \pi_j) \tag{30}$$

then Equation 29 reduces to a linear mixing rule:

$$\pi_{\mathrm{m}} = \sum x_i \pi_i \tag{31}$$

If instead we adopt a geometric-average combination rule

$$\pi_{ij} = (\pi_i \pi_j)^{1/2} \tag{32}$$

then Equation 29 reduces to the mixing rule:

$$\pi_{\mathrm{m}} = (\sum x_i \pi_i^{1/2})^2 \tag{33}$$

Equations 31 and 33 both have the desirable feature of producing mixture parameters from pure-component parameters alone. They also have the unfortunate disadvantage (when, e.g., applied to parameter Θ), of often yielding less-than-adequate results, except when used for systems containing similar chemical species.

What now seems to be conventional practice in this area can be summarized briefly. Virtually all investigators assume a linear mixing rule for parameter b

$$b_{\mathrm{m}} = \sum x_i b_i \tag{34}$$

but retain the general quadratic mixing rule for Θ:

$$\Theta_{\mathrm{m}} = \sum\sum x_i x_j \Theta_{ij} \tag{35}$$

This approach in fact has a reasonable basis, a basis suggested by Equation 3a. If the cubic equation is to be useful at low densities, then the composition dependence of its parameters should be compatible with that of the second virial coefficient, which is known to be quadratic in the mole fractions. Thus, if b_{m} is assumed linear in composition, then Θ_{m} should be quadratic in composition.

Having adopted Equation 35 for Θ_{m}, one still must provide means for estimating Θ_{ij}. Neither of the simple combination rules Equation 30 or 32 suffice; however, both of them serve as reasonable points of departure. Modern practice seems to favor using Equation 32, modified by including an interaction parameter k_{ij}:

$$\Theta_{ij} = (1 - k_{ij})\,(\Theta_{ii}\Theta_{jj})^{1/2} \tag{36}$$

The calculation of Θ_{m} thus reduces to estimating values of k_{ij} for each pair of species. By convention, $k_{ij} = 0$ if $i = j$; if $i \neq j$, then k_{ij} is (usually) a small number, which may be either positive or negative. One possible difficulty with Equation 36 must be noted. Because of the square-root operation in the combination rule, values for the pure-component Θ's must be so constrained as to always yield positive values for the Θ_{ii} for all anticipated ranges of T_{r} and ω.

The result of a calculation can be quite sensitive to the values for the k_{ij}. Although these quantities can be correlated at times against combinations of properties for pure species i and j (e.g., critical-volume ratios), they are best treated as purely empirical parameters, values of which are (ideally) "backed out" of good experimental mixture data for the type of property which is to be represented. Thus, if accurate calculation of low-to-moderate-pressure volumetric properties is required, then the k_{ij} could be estimated from available data on mixture second virial coefficients for the constituent binaries. Alternatively, if application to multicomponent VLE calculations is envisioned, then the k_{ij} would be best estimated from available VLE data on the constituent binaries. (It should be noted that, by assumption, no more than binary data are needed to completely characterize a multicomponent system via the methods outlined here.)

Mixing rules for the remaining equation-of-state parameters (η, δ, and ϵ) may be rationalized in several ways. If these parameters are zero, or if they are specified functions of b (*see* some of the special cases listed in Table I), then there is no difficulty. If they are regarded as completely independent parameters, then Equations 3b and 3c offer some guidelines. The third virial coefficient C is cubic in composition; if b is linear and Θ is quadratic in composition, then Equation 3b suggests that η and δ be linear functions of composition. Similar examination of Equation 3c suggests that parameter ϵ be at worst quadratic in composition. In fact, common sense is probably the best guide here, for if one assumes a general quadratic dependence of ϵ on composition, then he is faced with the prospect of defining (and evaluating!) a whole new class of interaction parameters, e.g., "l_{ij}," where

$$\epsilon_{ij} = (1 - l_{ij})(\epsilon_{ii}\epsilon_{jj})^{1/2}$$

The accuracy inherent in a cubic equation of state can hardly justify using two sets of interaction parameters. Thus the only reasonable procedure with ϵ is to use Equation 29 in conjunction with either of the simple combination rules, Equation 30 or 32, yielding the alternative mixing rules

$$\epsilon_m = \sum x_i \epsilon_i$$

or

$$\epsilon_m = (\sum x_i \epsilon_i^{1/2})^2$$

If ϵ is observed to assume negative values (and it may well, for it is a "fine-tuning" parameter, subject to no neat physical interpretation or special constraint), then the last equation is clearly inappropriate.

We therefore arrive at the following provisional set of mixing rules for η, δ, and ϵ, which are useful in the absence of any assumed relationships among b, η, δ, and ϵ:

$$\eta_m = \sum x_i \eta_i \quad (37)$$

$$\delta_m = \sum x_i \delta_i \quad (38)$$

$$\epsilon_m = \sum x_i \epsilon_i \quad (39)$$

Equations 34 through 39 constitute a rational set of mixing rules, based in part upon the accumulated experience of many investigators with special cases of the cubic equation. Although the rules are reasonable, and in some cases (notably for b) have even been accorded theoretical significance, they must be regarded in fact as no more than useful empiricisms, for the cubic equation itself is no more than a useful (albeit often surprisingly powerful) approximator of real-fluid behavior.

Conclusion

Virtually all modern efforts at devising better versions of the cubic equation of state have been either directly or indirectly inspired by the example of the RK equation. Hence, many experts routinely label any new cubic equation as another variation on the RK equation. Some are, and some are not, but the attitude is expressed often enough that one well might ask if there are possibilities for significant advances within the cubic equation framework. The answer is probably a qualified "yes"—qualified because one must approach such efforts with an appreciation of what a cubic equation can and cannot do.

There undoubtedly exists a best (and as yet undiscovered) cubic equation of state, but best only in a coarse statistical sense. Such an equation, constructed so as to avoid manifestly unreasonable predictions of all possible thermodynamic properties of interest, probably would not produce really adequate estimates of any single property (except perhaps over very limited ranges of the variables of state). The search for this equation, if successful, would yield an expression of limited usefulness as a cubic equation. [It could, of course, provide the basis for more precise, noncubic expressions, obtained e.g. by coupling the cubic equation with deviation functions (*see*, e.g., Gray et al. (*17*) and Redlich (*18*)).]

There is, however, room for possible advancement in the area of special-purpose cubic equations, equations developed for use in a particular kind of calculation over a well-defined range of variables. If one relaxes the critical constraints, e.g., the cubic equation gains considerable flexibility. Thus Chaudron et al. (*19*) and Simonet and Behar (*20*) have developed unconstrained modifications of the RK equation, in which both $\hat{\Theta}$ and $\hat{b}$ are treated as functions of T_r and ω; good results are claimed for representation of volumetric properties over wide ranges of conditions.

All equations of state have inherent advantages and liabilities. In many industrial applications, where repetitive calculations are the rule and where one typically deals with systems containing large numbers of chemical species, a practical balance must be struck between accuracy on the one hand, and simplicity and generalizability on the other. The cubic equations, despite their shortcomings, meet this requirement, and will continue to enjoy their current popularity until appropriate substitutes are found.

Glossary of Symbols

B, C, D = virial coefficients
$b, \Theta, \eta, \delta, \epsilon$ = parameters in Equation 1
ζ_c = apparent critical compressibility factor
k_{ij} = an interaction parameter (*see* Equation 36)
P = pressure
π = general symbol for an equation-of-state parameter
R = gas constant
ρ = molar density
T = absolute temperature
V = molar volume
x_i = mole fraction of species i
Z = compressibility factor $\equiv PV/RT$
ω = acentric factor

Subscripts, Superscripts, etc.

c, cr = critical state
(calc) = denotes a calculated value
(exp) = denotes an experimental value
i, j = identifies species i, j
l = denotes the liquid state
m = denotes a mixture property
r = reduced value of P, V, T, or ρ
v = denotes the vapor state
$\hat{}$ = dimensionless (reduced) quantity

Acknowledgment

I am pleased to acknowledge helpful discussions with H. C. Van Ness and the generous support of the Department of Chemical and Environmental Engineering at Rensselaer Polytechnic Institute.

Literature Cited

1. Robinson, R. L., personal communication to the author, Sept. 1977.
2. Martin, J. J. "Equations of State," *Ind. Eng. Chem.* **1967,** 59, 34.
3. Abbott, M. M. "Cubic Equations of State," *AIChE J.* **1973,** *19,* 596.
4. Costolnick, J. J.; Thodos, G. "A Reduced Equation of State for Gaseous and Liquid Substances," *AIChE J.* **1963,** 9, 269.

5. Horvath, A. L. "Redlich-Kwong Equation of State: Review for Chemical Engineering Calculations," *Chem. Eng. Sci.* **1974,** *29,* 1334.
6. Redlich, O.; Kwong, J. N. S. "On the Thermodynamics of Solutions. V. An Equation of State. Fugacities of Gaseous Solutions," *Chem. Rev.* **1949,** *44,* 233.
7. Wilson, G. M. "Vapor–Liquid Equilibria, Correlation by Means of a Modified Redlich–Kwong Equation of State," *Adv. Cryog. Eng.* **1964,** *9,* 168.
8. Wilson, G. M. "Calculation of Enthalpy Data from a Modified Redlich-Kwong Equation of State," *Adv. Cryog. Eng.* **1966,** *11,* 392.
9. Barner, H. E.; Pigford, R. L.; Schreiner, W. C. "A Modified Redlich–Kwong Equation of State," paper presented at 31st Midyear API Meeting, Houston, 1966.
10. Pitzer, K. S.; Curl, R. F., Jr. "The Volumetric and Thermodynamic Properties of Fluids. III. Empirical Equations for the Second Virial Coefficient," *J. Am. Chem. Soc.* **1957,** *79,* 2369.
11. Chueh, P. L.; Prausnitz, J. M. "Vapor–Liquid Equilibria at High Pressures: Calculation of Partial Molar Volumes in Nonpolar Liquid Mixtures," *AIChE J.* **1967,** *13,* 1099.
12. Chueh, P. L.; Prausnitz, J. M. "Vapor–Liquid Equilibria at High Pressures. Vapor-Phase Fugacity Coefficients in Nonpolar and Quantum Gas Mixtures," *Ind. Eng. Chem., Fundam.* **1967,** *6,* 492.
13. Soave, G. "Equilibrium Constants from a Modified Redlich–Kwong Equation of State," *Chem. Eng. Sci.* **1972,** *27,* 1197.
14. Adler, S. B.; Spencer, C. F.; Ozkardesh, H.; Kuo, C.-M. "Industrial Uses of Equations of State: A State-of-the-Art Review," *ACS Symp. Ser.* **1977,** *60,* 150.
15. Reid, R. C.; Prausnitz, J. M.; Sherwood, T. K. "The Properties of Gases and Liquids," 3rd ed.; McGraw-Hill: New York, 1977.
16. Peng, D.-Y.; Robinson, D. B. "A New Two-Constant Equation of State," *Ind. Eng. Chem., Fundam.* **1976,** *15,* 59.
17. Gray, R. D., Jr.; Rent, N. H.; Zudkevitch, D. "A Modified Redlich–Kwong Equation of State," *AIChE J.* **1970,** *16,* 991.
18. Redlich, O. "On the Three-Parameter Representation of the Equation of State," *Ind. Eng. Chem., Fundam.* **1975,** *14,* 257.
19. Chaudron, J.; Asselineau, L.; Renon, H. "A New Modification of the Redlich–Kwong Equation of State Based on the Analysis of a Large Set of Pure Component Data," *Chem. Eng. Sci.* **1973,** *28,* 839.
20. Simonet, R.; Behar, E. "A Modified Redlich–Kwong Equation of State for Accurately Representing Pure Components Data," *Chem. Eng. Sci.* **1976,** *31,* 37.

RECEIVED October 5, 1978.

4

Effective Molecular Diameters for Fluid Mixtures

JAMES I. C. CHANG, FRANK S. S. HWU, and THOMAS W. LELAND

Department of Chemical Engineering, Rice University,
P.O. Box 1892, Houston, TX 77001

A general method of predicting the effective molecular diameters and the thermodynamic properties for fluid mixtures based on the hard-sphere expansion conformal solution theory is developed. The method of Verlet and Weis produces effective hard-sphere diameters for use with this method for those fluids whose intermolecular potentials are known. For fluids with unknown potentials, a new method has been developed for obtaining the effective diameters from isochoric behavior of pure fluids. These methods have been extended to polar fluids by adding a new polar excess function, to account for polar contributions in a mixture. A new set of pseudo parameters has been developed for this purpose. The calculation of thermodynamic properties for several fluid mixtures including CH_4*–*CO_2 *has been carried out successfully.*

Applications of Equations of State for Hard Spheres

The development of an accurate equation of state for mixtures of hard spheres (*1*) has led to significant improvements in the computation of thermodynamic properties of mixtures, both through the development of new equations of state (*2*) and from improved conformal solution techniques (*3*). Properties predicted by these methods, however, are extremely sensitive to the numerical values assigned to the molecular

0-8412-0500-0/79/33-182-071$06.25/1

diameters and this chapter reports a new procedure for their determination in actual rather than theoretical fluids. The method is applicable to mixtures which contain both polar and nonpolar components.

It is important to realize that the diameters needed for thermodynamic calculations do not necessarily represent a true minimum attainable separation distance between molecules. The objective is rather to determine optimal or "effective" diameters which give best results when used with a particular method of dealing with the contributions of molecular attraction. In this chapter the effective diameters sought are to be used specifically with the hard-sphere expansion (HSE) conformal solution theory of Mansoori and Leland (*3*). This theory generates the proper pseudo parameters for a pure reference fluid to be used in predicting the excess of any dimensionless property of a mixture over the calculated value of this property for a hard-sphere mixture. The value of this excess is obtained from a known value of this type of excess for a pure reference fluid evaluated at temperature and density conditions made dimensionless with the pseudo parameters. For example, if X_M represents any dimensionless property for a mixture of n nonpolar constituents at mole fractions $x_1, x_2, \ldots x_{n-1}$ at temperature T and density ρ, then:

$$X_M(T, \rho, x_1, x_2, \ldots x_{n-1}) = X_{HSM}(\rho d^3{}_{11}, \rho d^3{}_{22}, \ldots x_1, x_2, \ldots x_{n-1}) + X_{REF}(kT/\bar{\epsilon}, \rho \bar{d}^3) - X_{HS}(\rho \bar{d}^3). \quad (1)$$

In Equation 1 X_{HSM} is the value of X for a hard-sphere mixture of diameters d_{11}, d_{22} . . . etc., and X_{HS} is the value of X for a pure hard-sphere fluid with diameter $\bar{d}$. The value of X_{HS} is calculated from the Carnahan–Starling (CS) equation (*4*). X_{REF} represents the value of X as obtained from a reduced equation of state for the pure reference fluid evaluated at T and ρ made dimensionless by the pseudo parameters $\bar{\epsilon}$ and $\bar{d}^3$.

Several important principles concerning the determination of effective diameters for real fluids can be established by examining methods of obtaining them for theoretical fluids with exactly known potentials.

Effective Diameters for Perturbation Theories

A theoretical basis for the computation of effective hard-sphere diameters is now well developed for the perturbation theory. Its numerical evaluation requires exact knowledge of the intermolecular potential so that the method is not immediately applicable to real molecules for which this potential is usually unknown. Despite this difficulty, the principles involved are important in designing a procedure for real molecules with the HSE theory.

The pseudo parameters of the HSE theory are derived from an equation of state expanded in powers of $1/kT$ about a hard-sphere fluid, as is developed by the perturbation theory. Consequently, it is reasonable to expect that procedures for defining optimal diameters for the perturbation theory should work well with the HSE procedure. The first portion of this chapter shows that this is indeed correct. The Verlet–Weis (VW) (5) modification of the Weeks, Chandler, and Anderson (WCA) (6) procedure was used here to determine diameters in a mixture of Lennard–Jones (LJ) (12–6) fluids. These diameters then were used in the HSE procedure to predict the mixture properties.

The VW diameters are defined as:

$$d = d_B \left[1 + \left(\frac{\sigma_1}{2\sigma_o} \right) \delta_{VW} \right] \tag{2}$$

The d_B term is the Barker–Henderson (BH) (7) diameter:

$$d_B = \int_0^\infty [1 - e^{-\beta u_o(r)}] dr \tag{3}$$

where u_o is the repulsive portion of the intermolecular potential

$$\delta_{VW} = \int_0^\infty \left[\frac{r}{d_B} - 1 \right]^2 \frac{d}{dr} [e^{-\beta u_o(r)}] dr \tag{4}$$

and

$$\frac{\sigma_1}{2\sigma_o} = \frac{1 - (1\tfrac{7}{4})\, b + 1.362\, b^2 - 0.8751\, b^3}{(1 - b)^2} \tag{5}$$

where:

$$b = \frac{\pi}{6} (\rho d^3) \left[1 - \frac{\pi}{96} (\rho d^3) \right] \tag{6}$$

$$\beta - \frac{1}{kT}$$

The solution of these equations for d requires an iteration, usually using the BH diameter as a first trial.

The BH diameter is designed to include temperature-dependent, soft-repulsion contributions in the hard-sphere equation of state. The WCA procedure also does this and further modifies the diameter to be the proper value when using a hard-sphere distribution function in perturbation terms due to attraction. The VW procedure corrects the WCA result for some of the limitations of the Percus–Vevick (PY) approximation and expresses the result in a simple algebraic form. A table of d_B and δ_{VW} values is given so no integration is required.

Use of Verlet–Weis Diameters in the HSE Procedure

Table I, in the column headed HSE–VW, shows the results of using Equations 2 through 6 to define the diameters with the HSE method to calculate the properties of an equimolar mixture of LJ fluids. The reference is a pure LJ fluid. Other columns show comparison with the machine-calculated results of Singer and Singer (*8*) in column MC. The van der Waals (VDW) one-fluid theory (*9*) and the VDW two-fluid theory (*10*) are in columns VDW-1 and VDW-2. The GHBL column gives the Grundke, Henderson, Barker, Leonard (GHBL) (*11*) pertubation theory results with each diameter determined by Equation 3.

The column headed HSE uses an approximation made originally by Mansoori and Leland (3) that the diameter used in the hard sphere equations of state is $c_0\sigma$, the LJ σ parameter for each molecule multiplied by a universal constant for conformal fluids. This approximation then requires that d_{ij} be replaced by σ_{ij} in the equations defining the HSE pseudo parameters, Equations 10 and 11. The results in the HSE column use $c_0 = 0.98$, the value for LJ fluids obtained empirically by Mansoori and Leland. This procedure is correct only for a Kihara-type potential and it is not consistent with the LJ fluids in Table I. Furthermore, this causes only the high temperature limit of the repulsion effects to be included in the hard-sphere calculation. Soft repulsions are predicted by the reference fluid.

In comparing the last two columns, one can see that the density-dependent VW diameters, while perhaps not the optimal values for the HSE theory, are and better than the constant $c_0\sigma$ value and a temperature

Table I. Calculated Properties

$\epsilon_{22}/\epsilon_{12}$	*Property*	*MC*	*VdW-1*
	$\sigma_{22}/\sigma_{12} = 1.06$		
0.810	G^E/NkT	0.305	0.334
	H^E/NkT	0.34	0.35
	V^E	−1.29	−1.39
0.900	G^E/NkT	0.105	0.114
	H^E/NkT	0.15	0.15
	V^E	−0.41	−0.54
1.000	G^E/NkT	−0.005	−0.011
	H^E/NkT	0.03	
	V^E	0.02	−0.18
1.111	G^E/NkT	−0.030	−0.044
	H^E/NkT	−0.05	−0.089
	V^E	−0.10	−0.27
1.235	G^E/NkT	0.040	0.014
	H^E/NkT	−0.10	−0.12
	V^E/NkT	−0.68	−0.81

and density dependence is definitely needed. These columns also show the importance of defining a diameter which accounts for all the repulsion effects in the fluid.

Table I shows that the HSE theory is much better than either of the VDW theories, which justifies the direct computations of the hard-sphere contributions as opposed to their prediction by the reference fluid as is the case with the VDW theories. Furthermore, derivation of the HSE pseudo parameters considers $(1/kT)^2$ terms in the expansion about hard-sphere propertics, whereas in the VDW theories the expansion is truncated after the first order $1/kT$ term and all of the repulsion properties of the fluid are predicted by the reference fluid. The importance of including the second order $(1/kT)^2$ terms is indicated by the excellent results of the HSE–VW method for the enthalpy, in that this property is much less dependent on the choice of the diameter and the hard-sphere contribution.

The GHBL perturbation procedure is remarkably accurate and the HSE–VW method is only slightly better in its overall agreement with the machine-calculated results. This comparison is not completely valid in that the conformal solution theory uses pure component data while in the perturbation theory each term is calculated from molecular parameters.

The HSE Procedure for Mixtures of Polar and Nonpolar Molecules

One of the advantages of the HSE procedure is that contribution from intermolecular attraction can be predicted more accurately if it consists primarily of long-range interactions, while very short-range inter-

of an Equimolar LJ Mixture

VdW-2	*GHBL*	*HSE*	*VW–HSE*
	$\sigma_{22}/\sigma_{12} = 1.06$		
0.236	0.271	0.280	0.316
0.24	0.27	0.28	0.34
−1.13	−1.39	−1.27	−1.21
0.073	0.095	0.110	0.119
0.09	0.12	0.15	0.16
−0.47	−0.48	−0.47	−0.39
−0.007	−0.006	0.01	0.01
	−0.014	0.02	0.043
−0.18	−0.08	−0.01	−0.02
−0.006	−0.033	−0.04	−0.018
−0.03	−0.050	−0.07	−0.04
−0.21	−0.12	−0.26	−0.06
0.077	0.014	0.019	0.039
	−0.073	−0.07	−0.067
−0.57	−0.61	−0.67	−0.55

action effects are removed and evaluated in a separate computation. This is the case for mixtures containing polar molecules where the longer-range interactions are suitable for expansion in spherical harmonics and their multipole coefficients, the dipole and quadrapole moments, are experimentally accessible. This is not the case for the shorter-range asymmetric interactions involving the overlap coefficient and nonisotropic polarizability. The contributions of these to the potential are proportional to inverse separation distances to powers of from 12 to 24. These are comparable with the $1/r$ dependence of repulsion potentials and the results are simplified and improved if an effective diameter can be found to include these short-range asymmetric effects in the hard-sphere calculation.

For mixtures containing polar molecules the most effective route to thermodynamic properties is to determine first the molal residual Helmholtz free-energy function A, where:

$$A = \left(\frac{\overline{A} - \overline{A}^*}{\mathrm{N}kT}\right)_{T,\rho} \tag{7}$$

This occurs because the contribution from the asymmetric portion of the potentials is more conveniently calculated for this property. Other functions can be obtained from it by differentiation. For example, the compressibility factor is

$$z = 1 + \frac{\rho}{RT}\left(\frac{\partial A}{\partial \rho}\right)_T \tag{8}$$

The original derivation of the HSE procedure (*3*) can easily be extended to include polar components by considering that the excess over hard-sphere properties can be divided into two portions for the mixture and for a pure reference fluid. The first includes the contribution of the symmetric excess from all nonpolar interactions plus the leading symmetric contributions in an expansion in spherical harmonics for each interaction involving a polar molecule. The second portion of the excess is the contribution of the asymmetric part of all interaction potentials.

The expansion of potentials in spherical harmonics and the resulting contribution to thermodynamic properties were initially presented by Pople (*12*) and more recently developed in further studies (*13, 14*). The equation produced for the excess over hard-sphere properties shows that the first-order perturbation coefficient $(1/kT)$ involves only the symmetric portion of all potentials. The asymmetric contributions first appear in the coefficient of the second-order $(1/kT)^2$ term where they are weighted by the distribution functions for molecules which have only symmetric potentials. Consequently, the distribution functions for the mixture still can be represented by the mean density approximation:

$$g_{ij}^{M}\ (r, T, \rho, x_1, x_2 \ldots x_{n-1}) \simeq g_{REF}\left(\frac{r}{\sigma_{ij}}, \frac{\epsilon_{ij}}{kT}, \rho \bar{d}^3\right) \tag{9}$$

This assumption means, as in the original derivation, that coefficients of like powers of $(1/kT)$ which involve pair interactions can be made equal in the mixture and in the reference by the proper choice of the pseudo parameters. Equating these coefficients of $(1/kT)$ and $(1/kT)^2$ in the symmetric potential portion of each excess gives the same pseudo parameters as the original theory, Equations 10 and 11. The asymmetric excess has no coefficients of $(1/kT)$ and equating the coefficients of $(1/kT)^2$ that arise from pair interactions results in Equations 12, 13, and 14. Some coefficients of $(1/kT)^2$ terms in both the symmetric and in the asymmetric excesses arise from three-body interactions. In the symmetric excess these can be equated in mixture and reference introducing three-body potential parameters, making a superposition approximation for the triplet distribution function, and equating them in mixture and reference. Errors are not serious if one simply assumes that the three-body effects in mixture and reference are roughly comparable without a new parameter to force them to be equal. In the asymmetric excess, however, there are three-body coefficients in the mixture which have no counterpart in the pure asymmetric excess. These cannot be accounted for in the theory.

The pseudo parameters are as follows:

$$\bar{d}^3 = \frac{\left[\sum_i \sum_j x_i\, x_j\, \epsilon_{ij}\, d_{ij}^3\right]^2}{\left[\sum_i \sum_j x_i\, x_j\, d_{ij}^3\, \epsilon_{ij}^2\right]} \tag{10}$$

$$\bar{\epsilon} = \left[\frac{\sum_i \sum_j x_i\, x_j\, \epsilon_{ij}^2\, d_{ij}^3}{\sum_i \sum_j x_i\, x_j\, d_{ij}^3\, \epsilon_{ij}}\right] \tag{11}$$

To obtain pseudo critical values VCP and TCP replace:

d_{ij}^3 with $[\phi_{ir}\, V_{ci} + \phi_{jr}\, V_{cj}]\ \frac{1}{2}$

ϵ_{ij} with $\xi_{ij}\ \sqrt{T_{ci}\, \theta_{ir}\, T_{cj}\, \theta_{jr}}$

where θ_{ir} and ϕ_{ir} are shape factors which allow fluid *i* to be predicted by reference fluid *r* when the two fluids are nonconformal.

$$\overline{Q}^4 = \overline{d}^7 \sum_i \sum_j x_i x_j \left(\frac{Q_i^2 Q_j^2}{d_{ij}^7} \right) \tag{12}$$

$$(\overline{\mu Q})^2 = \overline{d}^8 \sum_i \sum_j x_i x_j \left(\frac{\mu_i^2 Q_i^2}{d_{ij}^8} \right) \tag{13}$$

$$\overline{\mu}^4 = \overline{d}^3 \sum_i \sum_j x_i x_j \left(\frac{\mu_i^2 \mu_j^2}{d_{ij}^3} \right) \tag{14}$$

From the pseudo critical parameters the excess residual Helmholtz function for this mixture is:

$$A^{\mathrm{EX}} = \left[A \left(\rho\overline{d}^3, \frac{\overline{\epsilon}}{kT} \right) - A^{\mathrm{HS}}(\rho\overline{d}^3) \right] - \pi \left[\frac{14}{5} \frac{\overline{Q}^4}{\overline{d}^7} \mathrm{I}_{10}(\rho\overline{d}^3) + \frac{(\overline{\mu Q})}{\overline{d}^8} \mathrm{I}_8(\rho\overline{d}^3) + \frac{2}{3} \frac{\overline{\mu}^4}{\overline{d}^3} \mathrm{I}_6(\rho\overline{d}^3) \right] \frac{\rho\overline{d}^3}{(kT)^2} \tag{15}$$

The first bracketed term is the symmetric excess which is predicted exactly as the excess in Equation 1. The second term is the asymmetric excess which is calculated directly. The I_6, I_8, and I_{10} functions are integrals involving hard-sphere distribution functions of the type

$$\mathrm{I}_n = \int_1^\infty g^{\mathrm{HS}}(y, \rho\overline{d}^3)\, y^{2-n} dy \tag{16}$$

Padé approximants for them in convenient form are given by Stell et al. (*14*). The $\overline{\mu}$, $\overline{\mu Q}$, and $\overline{Q}$ terms are the pseudo dipole, dipole–quadrapole, and quadrapole moments, respectively. They are combined as in a pure component to give the asymmetric excess of the mixture. The final result for the mixture is

$$\frac{\overline{A} - \overline{A}^*}{NkT} = A^{\mathrm{HSM}}(\rho, x_1, x_2 \ldots, d_1, d_2 \ldots) + A^{\mathrm{EX}} \tag{17}$$

where the A^{HSM} term is the residual Helmholtz free energy for the hard-sphere mixture calculated from the MCSL equation. The analytical form of this and the calculated results are presented by Mansoori et al. (*1*). The A^{EX} term is given by Equation 15.

Optimal Diameters for the HSE Theory from PVT Data

We will now discuss the problem of determining effective or optimal diameters for use with the HSE theory for real fluids when both the form of the intermolecular potential and its parameters are unknown but accurate equations of state which represent the *PVT* behavior over an extensive range are available for the pure components.

For nonpolar fluids and symmetric reference fluids for polar substances we will assume that the unknown potential function for each may be modeled with a symmetrical potential consisting of a hard-sphere repulsion potential for spheres of diameter d plus an excess which depends on (r/d) and a single energy parameter, ϵ, in the form $\epsilon\, f(r/d)$. If the fluids are nonspherical, ϵ is an average which may depend on temperature and to some extent on density. If the unknown true potential involves a soft repulsion, d may depend on both temperature and density.

It is convenient to relate this ϵ and d to critical constants through new types of shape factors so that for any fluid i:

$$\epsilon_{ii} = a_k\, \theta_{ik}\, (T_c)_i \tag{18}$$

and

$$d_{ii}^3 = b_k\, \varphi_{ik}\, (V_c)_i \tag{19}$$

where a_k and b_k are proportionality constants characteristic of the reference fluid k. The dimensionless shape factors θ_{ik} and φ_{ik} express the dependence on temperature and density and establish conformality with the reference fluid.

Although different numerically and evaluated in a different manner, these shape factors obey the same rules and operate in the same fashion as those developed earlier for the VDW one-fluid theory (*15, 28, 29*). In this manner the problem of finding the optimum diameter for use with a given reference fluid is solved by finding the proper shape factors for use with the reference fluid chosen.

All fluids which can be modeled by an unknown potential in terms of ϵ and d according to these assumptions are conformal and their dimensionless equations of state can be represented in terms of the reference fluid, k, in the form:

$$z = Z^{HS}(\rho d_i^3) - \psi_k\left(\rho d_i^3, \frac{\epsilon_i}{kT}\right) \tag{20}$$

The form of the ψ function is unspecified but is the same for all of the conformal fluids. The function for the reference then can be used to furnish properties of any other fluid as indicated in Equation 20. The value of ψ is always > 0. The Z^{HS} (ρd_i^3) term is the CS equation for hard spheres of diameter d_i.

Using Equations 18 and 19, an equivalent form of Equation 20 may be expressed in terms of known critical constants.

$$z_i = Z^{HS}(\rho \, b_k \, \varphi_{ik} \, V_{ci}) - \psi_k\left(\rho \, \varphi_{ik} \, V_{ci} \, \frac{T}{\theta_{ik} T_{ci}}\right) \tag{21}$$

Mixture properties are predicted from Equation 21 by using the hard-sphere equation for mixtures and using pseudo criticals to evaluate ψ_k from known values of it for the reference.

Substitution of Equations 18 and 19 into Equations 10 and 11 converts the pseudo parameters to pseudo criticals. The result is:

$$\text{VCP} = \frac{\left[\sum_i \sum_j x_i \, x_j \, \frac{1}{2} \, (\varphi_i \, V_{ci} + \varphi_j \, V_{cj}) \, \xi_{ij} \sqrt{\theta_i \, T_{ci} \, \theta_j \, T_{cj}}\right]^2}{\left[\sum_i \sum_j x_i \, x_j \, \frac{1}{2} \, (\varphi_i \, V_{ci} + \varphi_j \, V_{cj}) \, \xi_{ij}^2 \, (\theta_j \, T_{ci} \, \theta_i \, T_{cj})\right]} \tag{22}$$

$$\text{TCP} = \left[\frac{\sum_i \sum_j x_i \, x_j \, \frac{1}{2} \, (\varphi_i \, V_{ci} + \varphi_j \, V_{cj}) \, \xi_{ij}^2 \, (\theta_j \, T_{ci} \, \theta_i \, T_{cj})}{\sum_i \sum_j x_i \, x_j \, \frac{1}{2} \, (\varphi_i \, V_{ci} + \varphi_j \, V_{cj}) \, \xi_{ij} \sqrt{\theta_i \, T_{ci} \, \theta_j \, T_{cj}}}\right]. \tag{23}$$

The x terms are mole fractions and ξ_{ij} is the unlike pair interaction coefficient obtained by Mollerup and Rowlinson (*16, 17, 18*) for the VDW one-fluid procedure. The particular form of the unlike pair interactions was developed by Hsu (*19*) who tested several alternative formulations.

Criteria for the Optimal Diameters

From the derivation of the HSE method and the behavior of the VW diameters for a fluid with a known potential, it is possible to enumerate criteria which the diameters should fulfill for a fluid with an unknown potential. The diameters should be chosen for each fluid so that:

(1) the ψ function for each fluid at a constant density is represented as closely as possible by a quadratic form:

$$\psi \simeq -\left(\frac{a_1(\rho)}{T} + \frac{a_2(\rho)}{T^2}\right) \tag{24}$$

where $a_1(\rho)$ and $a_2(\rho)$ are functions of density only;

(2) the $a_1(\rho)$ and $a_2(\rho)$ coefficients in Equation 24 involve the effects of intermolecular attraction only;

(3) the Z^{HS} term in Equation 20 accounts for all repulsion effects plus the effects caused by attraction which have higher order temperature dependence than $(1/T)^2$ and cannot be accounted for in Equation 24 ; and

(4) at high densities where three-body interactions begin to make contributions to the ψ term of the order $(1/T)^2$, as much as possible of the contribution of these effects is removed from the $a_2(\rho)/T^2$ term and included in the Z^{HS} term.

These criteria are ideal conditions which cannot be realized completely by altering the value of *d*. In order to find the diameters which approach these criteria as closely as possible, it is necessary to have an accurate representation of the constant density isochoric behavior of each pure component over as wide a temperature range as possible.

Effective Diameters from a Pure Component Equation of State

In this study the BWR–Starling (BWR–S) (*20, 21*) equation of state was used, primarily because it has been fitted to a wide range of pure components. For some components there are other equations which represent isochoric behavior much better. However, the effect of the HSE procedure for mixtures when pure-component equations of state are known is to generate a more rigorous composition dependence in the reference-fluid equation of state. The BWR–S equation for mixtures has an empirical composition dependence which works well in some cases, particularly for hydrocarbon mixtures, but not so well for others, especially when nonhydrocarbons are involved. When the BWR–S equation is used for all pure component properties, it is interesting to compare the theoretically based composition dependence induced by the HSE theory with its empirical form in the BWR–S mixture equation.

For a pure component the BWR–S equation, or any empirical equation of state which has been fitted over the widest range of temperature and density conditions, can be represented conceptually as follows.

$$z = Z(\rho) + [Z^{+}(\rho, T) + Z^{-}(\rho, T)] \qquad (25)$$

where z is the compressibility factor. The term in brackets in Equation 25 includes all the temperature dependence and is the negative value of the ψ term in Equation 20. If represented at a constant density by an expansion in powers of $1/T$ it would need many terms of higher order than $(1/T)^2$ if the empirical equation of state was valid over the widest possible range of temperatures at this density. Consequently the ψ term in Equation 25 is defined as

$$\psi_\infty = -[Z^+(\rho, T) + Z^-(\rho, T)] \tag{26}$$

The ∞ subscript indicates that in an expanded form it would include all orders of $(1/T)$ in representing the value of ψ over the widest possible temperature range.

If the equation of state in Equation 25 were generated by an imaginary Kihara-type potential the $Z(\rho)$ term would represent the contribution of the hard core. Because the molecules with this type of potential also have a soft repulsion at separation distances slightly greater than the hard core diameter, the $Z(\rho)$ term does not include all repulsion effects. The soft repulsive potential contributions to the compressibility factor are positive and temperature dependent. Since the ψ term includes all of the temperature dependence of z, the soft repulsive contributions are included in it and are designated by $Z^+(\rho,T)$ in Equation 26. The major source of temperature dependence in the ψ term is caused by contributions from the attractive portion of the unknown intermolecular potential. These contributions, designated as $Z^-(\rho, T)$ in Equation 26 are large and negative, causing the value of ψ_∞ to be always positive.

It is important to point out that although any empirical equation of state fitted to a wide range of *PVT* properties can give the value of ψ_∞ and $Z(\rho)$, the term $Z^+(\rho, T)$ has no relation to and cannot be identified as a sum of the positive temperature-dependent terms which appear in the empirical equation. Likewise, $Z^-(\rho, T)$ has no relation to the combined negative temperature dependent terms.

If it were possible to separate ψ precisely into the true $Z^+(\rho, T)$ and $Z^-(\rho, T)$ contributions, the optimum diameter to meet the criteria discussed above should be given by the solution of:

$$Z^{\mathrm{HS}}(\rho d^3) = Z(\rho) + Z^+(\rho, T) + \alpha Z^-(\rho, T) \tag{27}$$

where α is the fraction of $Z^-(\rho, T)$ owing to the sum of all attractive terms of order $(1/T)^3$ and higher plus all three-body attraction terms of the order $(1/T)^2$ in an expansion of $Z^-(\rho, T)$ in powers of $(1/T)$. The diameter then is obtained by evaluating $Z^{\mathrm{HS}}(\rho d^3)$ from Equation 27 and equating the result to the CS equation so that

$$Z^{\mathrm{HS}}(\rho d^3) = \frac{1 + \eta + \eta^2 - \eta^3}{(1 - \eta)^3} \tag{28}$$

where:

$$\eta = \frac{\pi}{6} N \rho d^3 \tag{29}$$

and solving for d.

A method of approximating the right side of Equation 27 in order to solve for d in this way was developed in this study. From the equation of state only the $Z(\rho)$ term is obtainable. For example, from the BWR–S equation:

$$Z(\rho) = 1 + B_o\rho + b\rho^2 \tag{30}$$

where B_o and b_o are the constants of the temperature-independent terms of the BWR–S equation. Special procedures must be used to estimate the other terms in Equation 27. The value of these terms can be determined directly in two limiting cases. The first of these is a high-density or high-pressure limit and was studied by Bienkowski and Chao (*2*). As they have shown, in this limit:

$$\lim_{P \to \infty} [Z(\rho) + Z^+(\rho, T)] >> \alpha Z^-(\rho, T) \tag{31}$$

and in this limit $Z^-(\rho, T)$ in Equation 25 is also negligible in comparison with $Z^+(\rho, T)$. Consequently the solution for values of d from Equations 27 through 29 should approach the diameters obtained by Bienkowski and Chao. Unfortunately, conditions of interest are very far from this limit and it appears that the optimal diameters do not approach the limiting values monotonically.

The other limiting case is at a high-temperature limit where

$$\lim_{1/T \to 0} [Z^+(\rho, T) + \alpha Z^-(\rho, T)] = 0. \tag{32}$$

In this case the right side of Equation 27 becomes only $Z(\rho)$, which is furnished directly by the equation of state as in Equation 30. This is called the high-temperature limit and at some temperature conditions of interest, especially at low densities, the optimal diameters approach it closely. These diameters are always smaller than the high-density limit of Bienkowski and Chao.

These limits are very nearly upper and lower bounds for the optimal diameters although they do not closely approach the upper bound at any conceivable density of interest. A few cases at low density showed the optimal diameter very slightly below the high temperature limit. The discrepancy is easily within the experimental uncertainty, however.

Determination of Optimal Diameters from Isochores

We must now consider a more general method for use when these limiting conditions are not applicable. Determining $Z^{HS}(\rho d^3)$ with the optimal diameter at a given finite temperature and density is carried out by considering a limited temperature range along an isochore at the given density. This temperature range is selected to locate the given temperature as near to the center of the range as possible. Isochores extrapolate smoothly into the two-phase region and in the liquid phase at lower temperatures in the range these extrapolations may even produce negative compressibility factors without adverse effects on the solution for the diameters. Properties along the isochore can be obtained either from direct experimental data or from an equation of state which represents isochoric behavior well. If such an equation of state is used, the temperature range selected must be shifted to higher values if necessary to insure that $(\partial P/\partial \rho)_T$ as calculated by the equation is positive at each temperature value within the range.

The width of the range is selected ideally to determine at a given temperature and density, T and ρ, the first and second derivatives of the dimensionless property with respect to inverse temperatures and to predict the property at each temperature in the range with an accuracy within its experimental error by a quadratic function. For example, if the compressibility factor is being evaluated, the values of z at ρ at each point in the range about T are fit by least squares to:

$$z = a_o(\rho) + \frac{a_1(\rho)}{T} + \frac{a_2(\rho)}{T^2} \qquad (33)$$

In this work a range was selected consisting of eleven temperatures, 10°F apart, including the given temperature. If $(\partial P/\partial \rho)_T$ is positive at each temperature, the range consists of five temperatures above and five temperatures below the given value; otherwise, the range is shifted upward so that the lowest temperature in the range is nearer to the given temperature. If $(\partial P/\partial \rho)_T$ is negative at the given temperature the method is inoperable at the given conditions. Varying the width of the range did not affect the results as long as the conditions described for it were met. At every density studied, Equation 33 gave an excellent reproduction of z values along the 100°F range as defined here.

The value of ψ for the quadratic fit of the isochore in this limited range is defined as ψ_2 to indicate that it contains two inverse temperature terms. Consequently, from Equation 33:

$$\psi_2 = -\left(\frac{a_1(\rho)}{T} + \frac{a_2(\rho)}{T^2}\right) \qquad (34)$$

We can find the temperature dependence that is not accounted for by the quadratic fit by comparing ψ_2 with ψ_∞ and assuming that ψ_∞ describes the maximum possible range of temperatures. The difference is defined as

$$\delta = (\psi_\infty - \psi_2) \tag{35}$$

since the HSE pseudo critical values for the excess over the hard-sphere behavior were derived by considering only terms in $(1/T)$ and $(1/T)^2$ in its expansion. Furthermore, these terms involved only pairwise contributions from the attractive portion of the intermolecular potential. Consequently, at conditions where the coefficients a_1 and a_2 in Equation 34 contain predominatly attractive contributions of this type, the $\alpha Z^-(\rho, T)$ term in Equation 27 contains no triplet potential effects of order $(1/T)^2$. Consequently, it is entirely included among the interactions of order $(1/T)^3$ in the ψ_∞ expression. If an expansion of $Z^+(\rho, T)$ in powers of $1/T$ gives coefficients of $(1/T)$ and $(1/T)^2$ which are negligible in comparison with the attractive contributions to these terms, their presence will not appreciably affect the $\bar{\epsilon}$ pseudo attraction parameter predicted by the HSE theory. Consequently, the only soft-sphere contributions which need to be included in the hard-sphere term by adjusting the diameter are those in terms of order $(1/T)^3$ and higher in the ψ_∞ term.

In this case, the difference between ψ_∞ and ψ_2 represents all terms of higher order than $1/T^3$ which need to be combined into the hard-sphere result. This difference defines the δ parameter in Equation 35. If all terms of order $(1/T)^3$ were zero, the best $Z^{HS}(\rho d^3)$ value at these conditions would be the a_o leading term in Equation 33 for the quadratic fit. With corrections for the higher order terms, the best $Z^{HS}(\rho d^3)$ value is

$$Z^{HS}(\rho d^3) = (a_o - \delta) \tag{36}$$

The negative sign is the result of the contribution to z as given by $-\psi$ as defined in Equations 26 and 34. Equation 36 may be regarded as the best approximation to Equation 27 under these conditions.

The limits of validity of Equation 36 are indicated by the magnitude and sign of the $a_2(\rho)$ term in Equation 34. The assumptions leading to Equation 36 become invalid at high densities. At the lowest densities below $\rho V_c \simeq 0.6$, $\delta \simeq 0$, and $a_o \simeq Z(\rho)$, the high-temperature limit of the equation of state in Equation 30. The $a_2(\rho)$ in Equation 34 is small and negative. As densities increase above $\rho V_c = 0.6$ the absolute value of a_2 begins to increase while it is still negative. Presumably this means an attraction contribution is being represented. Positive contributions of the soft repulsion are apparently still negligible and Equation 36, which requires this, is still valid. This causes the value of δ in Equation 35 to be negative and Z^{HS} in Equation 36 increases.

As density increases further the absolute value of a_2 begins to decrease. Although it still remains negative at this point, the absolute value of δ is a maximum. The a_2 term, which is becoming less negative in this way, is considered to be altered by the onset of the positive contributions of soft repulsion which at these densities begins to affect the coefficient of $(1/T)^2$. This maximums in $|\delta|$ and $|a_2|$ occur at a reduced density of about 1.6

The reduced density of 1.6 is considered to be the upper limit of the validity of Equation 36. At densities higher than this $|\delta|$ and $|a_2|$ decrease rapidly and a_2 itself eventually becomes positive, interpreted as its domination by positive soft-repulsion effects. Diameters from Equation 36 give poor results in this region. There is no way that these soft effects can be separated from attraction effects and the optimal diameter cannot be calculated.

The diameters can be predicted once more at very high densities where a_2 has become very large and positive, indicating dominance of the second-order term by soft-repulsion effects. It is also very likely that three body contributions to this term are no longer negligible. The optimal diameters then are obtained by placing all of the second-order $(1/T)^2$ term in the hard-sphere equation since it is now repulsion dominated. The temperature range used for the quadratic fit is reduced from 100° to 50°F with the given temperature near the center of this shorter range. The objective is now to obtain an accurate representation of each z value in the range by a least-squares fit of the linear relation:

$$z = a_o' + \frac{a_1'}{T} \tag{37}$$

The a_1' term is always negative at readily accessible densities. Since none of the negative $1/T$ dependence should appear in the hard-sphere equation and Equation 37 represents the z values accurately in the shorter range, the best Z^{HS} result is:

$$Z^{HS}(\rho d^3) = a_o' \tag{38}$$

Equation 38 then is solved for the diameter. This linear fit method gives excellent results for the optimal diameter at reduced densities of about 2.4 and higher.

The reduced density region between 1.6 and 2.4 is thus an indeterminant region. As a first approximation, the best $Z^{HS}(\rho d^3)$ values in this region were assumed to be given by a spline-fit interpolation between points at $\rho_R > 2.4$ and those at $\rho_R < 1.6$. For liquids at low temperatures the indeterminant region is lengthened because the liquid no longer can be extrapolated to reduced densities near 1.6 because of the stability limit. Low density values at $\rho_R < 0.6$ are still obtainable by equating $Z^{HS}(\rho d^3)$ to the high-temperature limit of the equation of state.

The behavior of the quadratic and linear fit methods is shown in Figure 1. The interpretation of the a_2 coefficient behavior in terms of soft-repulsion effects in the quadratic fit $a_z(\rho)$ assumes that the data fitted

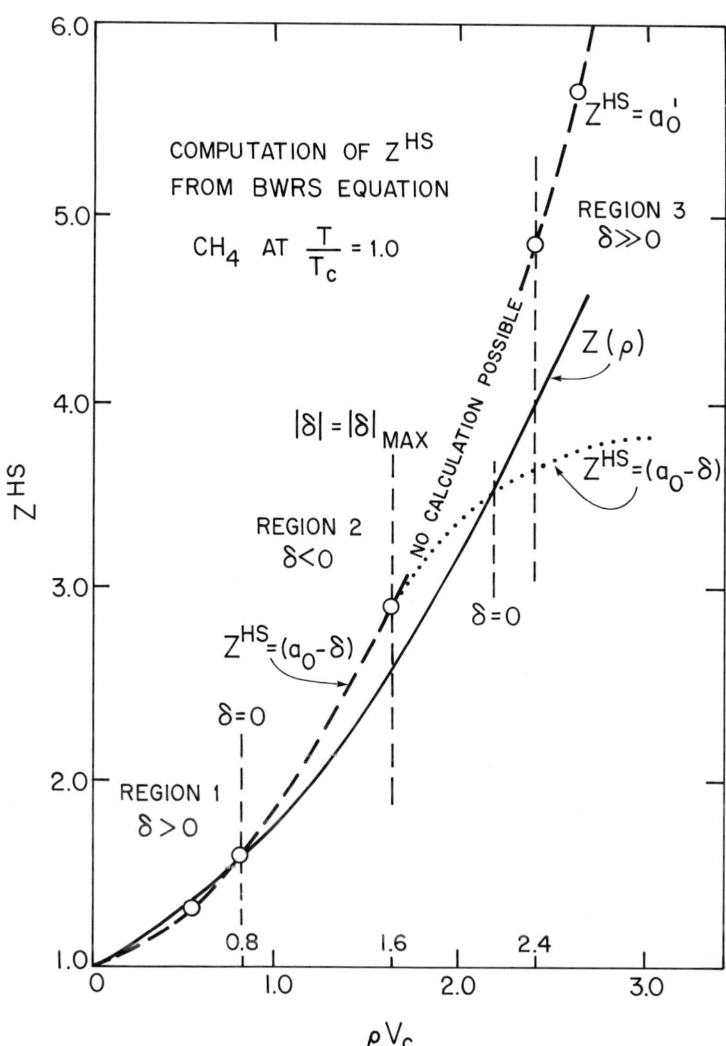

Figure 1. Computation of Z^{HS} from BWR–S equation (CH_4 at $T/T_c = 1.0$)

to the isochores gives a valid second derivative with respect to $1/T$. This may not be the case for values generated by the BWR–S equation. Regardless of this, the quadratic fit method below $\rho_R = 1.6$ and the linear fit method at $\rho_R > 2.4$ give excellent results for the optimal diameters. This was checked by increasing and decreasing $Z^{HS}(\rho d^3)$ about the predicted value and noting the effect on the predicted mixture properties. The weakest prediction for $Z^{HS}(\rho d^3)$ is in the spline-fit region. A typical result of this test on the predicted value is shown in Table II.

Table II. Effect of Z^{HS} on Calculated z in the Incalculable Region for Z^{HS}[a]

	Z^{HS}	% Error
	2.73	5.31
	3.39	1.77
$(a_0 - \delta) = 3.48 \rightarrow$		
	3.68[b]	0.29
	3.74[c]	0
	3.94	−0.99
	4.21	−2.23

[a] 50% CH_4; 50% C_3H_8. $T = 160°F$; $P = 5000$ psia.
[b] By spline fit between Z_{HS} at $\rho_R > 2.4$ and Z^{HS} at $\rho_R < 1.6$.
[c] Predicted by interpolation to 0% error.

Application of the Method

Although only compressibility factor calculations are used as an example in the explanation of the method, other properties can be predicted equally well. Because of the temperature and density dependence of the diameters and shape factors needed to relate them to critical constants it is best to determine separate values of them for each component. Three basic dimensionless properties should be determined. These are the ones best suited to the use of the HSE method with an equation of state in terms of temperature and density. These are the compressibility factor, z; the internal energy deviation $(\bar{U}^* - \bar{U})/RT$; and a dimensionless fugacity ratio, $\ln(f/\rho RT)$. All other desired properties can be obtained from them. The $\ln(f/\rho RT)$ and z are calculated similarly. The computation scheme is outlined as shown in Table III.

Table III. Shape Factors and Diameters for Nonconformal Fluids with Unknown Potentials Example for Compressibility Factor

1. *Calculate* d_i *and* d_r *from:*

$$Z_r^{HS}(\rho_r^\circ) = Z_i^{HS}(\rho_i^\circ)$$

Table III. Continued

obtained from pure-component equations of state by method reported in this chapter (*22*).

Conditions at which shape factors are evaluated (*28*):

$$\rho_r{}^\circ = \frac{\rho_m \mathrm{VCP}}{(V_c)_r} \quad \text{and} \quad \rho_i{}^\circ = \frac{\rho_m \mathrm{VCP}}{(V_c)_i \Phi_i}$$

ρ_m = mixture density

Solve for φ_i at trial VCP.

To force conformality betwen Component i and Reference r:

$$Z_r^{\mathrm{HS}}(\rho_r{}^\circ) = Z_i^{\mathrm{HS}}(\rho_i{}^\circ) = \frac{1 + \eta + \eta^2 - \eta^3}{(1-\eta)^3}$$

↑ Solve for η

Solve for d:

$$d_r = \left(\frac{\eta}{\frac{\pi}{6} \mathrm{N} \rho_r{}^\circ}\right)^{1/3} \quad \text{and} \quad d_i = \left(\frac{\eta}{\frac{\pi}{6} \mathrm{N} \rho_i{}^\circ}\right)^{1/3}$$

2. Calculate θ_{ir}:

to force conformality between Component i and Reference r:

$Z_i(\rho_i{}^\circ, T_i{}^\circ)$	=	$Z_r(\rho_r{}^\circ, T_r{}^\circ)$	+	Z^{EXA}
↑		↑		↑
from equation of state or known values for pure polar fluid		from equation of state for the pure nonpolar reference fluid		asymmetric excess calculated below

$$Z^{\mathrm{EXA}} = -\rho_i{}^\circ \frac{\partial}{\partial \rho_i{}^\circ} \left\{ \left[\frac{14\pi}{5} \frac{Q_i^4}{d_i^7} \mathrm{I}_{10}(\rho_i{}^\circ d_i^3) + \pi \frac{(\mu Q)_i^2}{d_i^8} \mathrm{I}_8(\rho_i{}^\circ d_i^3) + \frac{2\pi}{3} \frac{\mu_i^4}{d_i^3} \mathrm{I}_6(\rho_i{}^\circ d_i^3) \right] \frac{\rho_i{}^\circ d_{ii}^3}{(kT_i{}^0)^2} \right\}$$

$T_i{}^\circ$ = temperature at which θ_i must be evaluated in a mixture (*28*). Using previously calculated d_i, solve for $T_i{}^\circ$:

Calculate θ_{ir} at trial TCP value from $T_i{}^\circ$:

$$T_i{}^\circ = \frac{T_M}{\mathrm{TCP}} (T_c)_i \theta_{ir} \quad \text{and} \quad T_r{}^\circ = \frac{T_M}{\mathrm{TCP}} (T_c)_r$$

Now calculate new TCP and VCP and repeat until they are constant.

For $(\bar{U}^* - \bar{U})/RT$ no hard-sphere property calculations are made and the a_0 term of the quadratic fit along the compressibility factor isochores can be equated to $Z^{HS}(\rho d^3)$. This is then solved for the diameter used in the pseudo parameter computations.

Calculated results for methane and propane obtained by Hwu (*22*) with an ethane reference are presented in Table IV. The BWR–S equation is used for all pure components. The agreement is generally excellent.

Table IV. Compressibility Factors—Density[a]

P	*Exp*[b]	*Calc. HSE*	*Exp*[b]	*Calc. HSE*
(psia)	ρ *(lb–mol/ft³)*	ρ *(lb–mol/ft³)*	Z	Z
200	0.0322	0.0321	0.9351	0.9369
400	0.0692	0.0691	0.8692	0.8696
800	0.1648	0.1666	0.7313	0.7228
1000	0.2252	0.2333	0.6670	0.6448
3000	0.6760	0.6778	0.6676	0.6645
5000	0.7955	0.7955	0.9451	0.9446
7000	0.8636	0.8631	1.2189	1.2191
9000	0.9091	0.9101	1.4886	1.4866

[b] Sage and Lacey (*23*).

The poorer results at 1000 psia apparently are due to a weakness in the BWR–S reference equation at this point. It coincides with the minimum of the curve of z vs. P.

Table V shows the computations by Hwu for $\bar{H}^* - \bar{H}$ in a methane–propane mixture in comparison with the predictions of Mollerup (*16, 17, 18*) using the VDW one-fluid theory with shape factors. The improvement of the HSE method is very slight. The theoretical advantages of the HSE method for enthalpy calculations may be offset here by using a generally poorer reference equation of state than that used by Mollerup.

Table VI presents preliminary calculations by Chang (*24*) for a polar–nonpolar mixture. The highest pressures may be invalid because they were made before the method for evaluating the optimal diameters was developed. These computations use the high-temperature limit of the BWR–S equation for $Z^{HS}(\rho d^3)$ to obtain the diameter. It was hoped that comparison with the BWR–S equation would show a more distinct advantage of the theoretical composition dependence of the HSE method. In fact, the two methods give about the same results. Probably no conclusion about this can be drawn from the comparison because the constants determined by Hopke and Lin (*25*) for the BWR–S equation were obtained by fitting the equation to this binary. The results are given in the column headed BWRSE. The test of the improved composition

Table V. Enthalpy Calculations of $(\bar{H} - \bar{H}^*)$ [a]

T (°F)	P (psia)	*Expt.* [b] (Btu/lb)	*M–R* [c] (Btu/lb)	*HSE* (Btu/lb)
100	250	−20.4	−18.5	−19.3
	1250	−142.4	−142.4	−144.0
	1750	−143.6	−143.1	−144.4
50	750	−157.1	−154.9	−156.8
	1250	−156.1	−154.8	−156.3
	1750	−155.5	−154.3	−156.0
0	750	−169.3	−167.1	−168.3
	1250	−168.5	−166.2	−167.4
	1750	−167.1	−165.1	−166.5
−50	750	−180.7	−178.6	−178.4
	1250	−179.3	−177.2	−177.1
	1750	−178.0	−175.8	−176.0

[a] 23.4% CH_4, 76.6% C_3H_8.
[b] Yesavage–Powers (*24*).
[c] Mollerup–Rowlinson (*16, 17, 18*).

Table VI. Calculated Thermodynamic Properties for Methane–Carbon Dioxide Mixture at 100°F

P (psia)	Z *Expt.* (*23*)	Z *BWR–SE*	Z *HSE*	$(\bar{A} - \bar{A}^*)/RT$ *BWR–SE*	$(\bar{A} - \bar{A}^*)/RT$ *HSE*
		$x_1 = 0.2035$ [a]			
200	0.9512	0.9489	0.9500	−0.0507	−0.0503
600	0.8347	0.8400	0.8443	−0.1670	−0.1653
800	0.7830	0.7787	0.7836	−0.2357	−0.2331
1000	0.7160	0.7121	0.7165	−0.3150	−0.3116
2000	0.4438	0.4458	0.4518	−0.8184	−0.8080
3000	0.4958	0.4942	0.5033	−0.9831	−0.9613
4000	0.5921	0.5910	0.5992	−1.0336	−1.0153
5000	0.6947	0.6941	0.6999	−1.0561	−1.0389
7000	0.8982	0.9005	0.9087	−1.0722	−1.0556
9000	1.1012	1.1025	1.1058	−1.0724	−1.0557
		$x_1 = 0.4055$			
200	0.9606	0.9491	0.9595	−0.0469	−0.0408
600	0.8584	0.8748	0.8771	−0.1309	−0.1303
800	0.8360	0.8305	0.8336	−0.1807	−0.1791
1000	0.7921	0.7855	0.7873	−0.2308	−0.2336
2000	0.6100	0.6056	0.6067	−0.5268	−0.5302
3000	0.6055	0.6031	0.6067	−0.6928	−0.6870
5000	0.7755	0.7678	0.7748	−0.7838	−0.7747

[a] The x_1 = mole fraction methane.

dependence will have to await calculations when a third component is added, making a mixture not included in the fitting of the constants. It is encouraging that a theoretically based method for a mixture of a nonpolar symmetric potential fluid, methane, and a strong quadrapole fluid, CO_2, produces such good results. The nonpolar reference fluid was methane.

Conclusions

In summary the results show that it is indeed possible to extend the HSE method successfully to mixtures containing polar molecules. Methods have been developed to obtain effective diameters and shape factors which are optimal for use with the HSE theory. Although the determination of diameters for fluids with unknown potential functions with these methods is not possible at all densities, enough calculations can be made to allow a correlation by fitting the results to the VW equations for the optimal diameter with the perturbation theory. The success of the VW diameters for the HSE theory was confirmed.

The results obtained encourage future study and illustrate the power of conformal solution methods. It is reasonable to expect that the excellent accuracy obtained by Mollerup (*16, 17, 18*) with the VDW one-fluid theory for natural gas mixtures can be expected with the HSE theory for polar mixtures and other systems in which there are large dissimilarities between the components and the reference fluid.

Nomenclature

English Letters

a_1, a_2 = coefficients of $1/T$ and $(1/T)^2$ in an expansion of an equation of state

a_k = universal proportionality constant between ϵ and T_c for all fluids conformal with fluid k

A = dimensionless residual Helmholtz free energy function

$\bar{A}$ = molal Helmholtz free energy at a given T and P

$\bar{A}^*$ = molal Helmholtz free energy at T and P if the fluid obeyed the perfect gas law

b_k = universal proportionality constant between σ^3 and V_c for all fluids conformal with fluid k

b_o, B_o = constants in the BWR Starling equation of state

d = effective hard-sphere diameter

$\overline{d}^3$ = pseudo parameter used to form the reduced density in predicting the molecular attraction contribution to a mixture property

f = fugacity

g_{ij}^{M} = radial distribution function for an ij pair in a mixture with other constituents
g_{REF} = radial distribution function for a pair in a pure reference fluid
g^{HS} = radial distribution function for a pair of hard spheres in a hard-sphere fluid
$\overline{H}$ = molal enthalpy
k = Boltzmann's constant
N = Avogadro's number
P = pressure
Q = quadrapole moment
r = separation distance between a pair of molecular centers
R = gas constant
T = temperature
T_c = critical temperature
u = intermolecular pair potential
$\overline{U}$ = molal internal energy
V = volume
V_c = critical volume
x_i = mole fraction of constituent i in a mixture
X = any dimensionless thermodynamic property of a fluid
z = compressibility factor
Z = compressibility factor of a hard-sphere fluid
Z^{+} = contribution of soft repulsion to an equation of state expressed in terms of the compressibility factor
Z^{-} = contribution of intermolecular attraction to an equation of state expressed in terms of the compressibility factor

Greek Letters

α = fraction of the attractive contribution to expanded equations of state due to terms of order $(1/T)^3$ and higher and three-body interactions of order $(1/T)^2$ and higher
β = $1/kT$
δ = parameter used in obtaining the effective hard-sphere diameter from isochores
δ_{VW} = parameter in the Verleit–Weis equation for the effective hard-sphere diameter
ϵ = Lennard–Jones parameter for the algebraic minimum in the pair potential
$\overline{\epsilon}$ = pseudo parameter used to form the reduced temperature in predicting the molecular attraction contribution to mixture properties.
μ = dipole moment
π = 3.141516 - - - - -

η = dimensionless density in the Carnahan–Starling equation

ϕ_{ij} = shape factor coefficient of V_{ci} to make fluid i conformal with fluid j

ρ = density

ρ_r = reduced density, ρV_c

σ = Lennard–Jones parameter for the finite separation distance at which the intermolecular potential is zero

θ_{ij} = shape factor coefficient of T_{ci} to make fluid i conformal with a reference fluid j. (For a common reference the second subscript is sometimes omitted.)

ξ_{ij} = coefficient of the Berthelot combining rule for unlike ij pair interaction potentials

ψ = temperature dependent portion of a dimensionless equation of state

Acknowledgment

The theoretical studies concerning the test of the HSE method with mixtures of LJ molecules and the work involved in extending the method to polar molecules were supported by the National Science Foundation. The work on evaluating molecular diameters from PVT data and from pure-component equations of state was supported by the Gas Research Institute.

Literature Cited

1. Mansoori, G. A.; Carnahan, N. F.; Starling, K. E.; Leland, T. W. *J. Chem. Phys.* **1971,** *54,* 1523.
2. Bienkowski, P. R.; Deneholz, H. S.; Chao, K. C. *AIChE J.* **1973,** *19,* 167.
3. Mansoori, G. A.; Leland, T. W. *J. Chem. Soc., Faraday Trans. II* **1972,** *68,* 320.
4. Carnahan, N. F.; Starling, K. E. *J. Chem. Phys.* **1969,** *51,* 635.
5. Verlet, L.; Weis, J. J. *Phys. Rev. A* **1972,** *5,* 939.
6. Weeks, J. D.; Chandler, D.; Andersen, H. C. *J. Chem. Phys.* **1971,** *54,* 5237.
7. Barker, J. A.; Henderson, D. *J. Chem. Phys.* **1967,** *47,* 4714.
8. Singer, J. V. L.; Singer, K. *Mol. Phys.* **1972,** *24,* 357.
9. Leland, T. W.; Rowlinson, J. S.; Sather, G. A. *Trans. Faraday Soc.* **1968,** **64,** 1447.
10. Leland, T. W.; Rowlinson, J. S.; Watson, I. D. *Trans. Faraday Soc.* **1969,** *65,* 2034.
11. Grundke, E. W.; Henderson, D.; Barker, J. A.; Leonard, P. J. *Mol Phys.* **1973,** *25,* 883.
12. Pople, J. A. *Proc. R. Soc. London, Ser. A* **1954,** *221,* 508.
13. Mo, K. C.; Gubbins, K. E. *J. Chem. Phys.* **1975,** *63,* 1490.
14. Stell, G.; Rasaiah, J.; Narang, H. *Mol. Phys.* **1972,** *23,* 393.
15. Rowlinson, J. S.; Watson, I. D. *Chem. Eng. Sci.* **1969,** *24,* 1565.
16. Mollerup, J.; Rowlinson, J. S. *Chem. Eng. Sci.* **1974,** *29,* 1373.
17. Mollerup, J. *Adv. Cryog. Eng.* **1975,** *20,* 172.
18. Ibid., **1978,** *23,* 550.

19. Hsu, R. "Thermodynamic Properties of Mixtures by the Hard Sphere Expansion Theory," M.S. Thesis, Rice University, 1977.
20. Starling, K. E. *Hydrocarbon Process.* **1972,** *50.*
21. Ibid., **1973,** *51.*
22. Hwu, F. S. S. M.S. Thesis, Rice University, 1978.
23. Sage, B. H.; Lacey, W. H. "A.P.I. Project 37"; American Petroleum Institute: New York, 1950.
24. Chang, J. I. C. "Improvement of the Hard Sphere Expansion Conformal Solution Theory," M.S. Thesis, Rice University, 1978.
25. Lin, C. J.; Hopke, S. W. "Application of the BWRS Equation to Natural Gas Systems," presented at National A.I.Ch.E. Meeting, Tulsa, Oklahoma, March 1974.
26. Reamer, H. H.; Olds, R. H.; Sage, B. H.; Lacey, W. N. *Ind. Eng. Chem.* **1944,** *88.*
27. Bienkowski, P. R.; Chao, K. C. *J. Chem. Phys.* **1975,** *63,* 4217.
28. Leach, J. W.; Chappelear, P. S.; Leland, T. W. *Proc. Am. Pet. Inst., Sec. 3* **1966,** *46,* 223.
29. Leach, J. W.; Chappelear, P. S.; Leland, T. W. *AIChE J.* **1968,** *14,* 568.

RECEIVED September 21, 1978.

5

A Predictive Method for *PVT* and Phase Behavior of Liquids Containing Supercritical Components

PAUL M. MATHIAS and JOHN P. O'CONNELL

Department of Chemical Engineering, University of Florida, Gainesville, FL 32611

Concentration derivatives of fugacities are given in terms of statistical–mechanical direct-correlation function integrals which are insensitive to the detailed nature of the intermolecular forces in dense fluids. From this, a simple method is given for the properties of liquids containing supercritical components using only two pure-component parameters and a binary parameter for each pair. With good solvent density data, Henry's constants and easily estimated binary parameter values, we describe high-pressure vapor–liquid equilibrium in such systems as hydrogen–benzene and nitrogen–ammonia. Also, using a reference-solvent Henry's constant, values can be found for different pure and mixed solvents. These can be used for high-pressure vapor–liquid equilibrium, e.g., for hydrogen in coal oils knowing only hydrogen in quinoline and pure solvent densities.

Modern chemical processing demands much better estimates of physical properties than in previous eras because the costs of excessive design are now often too great. High-pressure systems are of particular concern, and they are becoming more prevalent as in coal gasification and liquefaction and in Fischer–Tropsch syntheses. A weak link in estimation techniques for these systems is the vapor–liquid equilibrium distribution of components at temperatures well above the critical temperature of one or more of the species present and below the critical temperature of the others.

0-8412-0500-0/79/33-182-097$05.00/1

Prausnitz (*1, 2*) has discussed this problem extensively, but the most successful techniques, which are based on either closed equations of state, such as discussed in this symposium, or on dilute liquid solution reference states such as in Prausnitz and Chueh (*3*), are limited to systems containing nonpolar species or dilute quantities of weakly polar substances. The purpose of this chapter is to describe a novel method for calculating the properties of liquids containing supercritical components which requires relatively few data and is of general applicability. Used with a vapor equation of state, the vapor–liquid equilibrium for these systems can be predicted to a high degree of accuracy even though the liquid may be 30 mol % or more of the supercritical species and the pressure more than 1000 bar.

General Expression

The basis of the method lies in the molecular theory which relates integrals of the statistical–mechanical direct correlation function to derivatives of the total pressure and the fugacity of each species with respect to the concentration of the species of the system (*4, 5, 6*). In equation form these are

$$\left.\frac{\partial \ln \hat{f}_i/x_i}{\partial \rho_j}\right]_{T,\rho\ k\neq j} = -\int c_{ij}(r;T,\underline{\rho})\,d\underline{r} \equiv -C_{ij}(T,\underline{\rho})/\rho \qquad (1)$$

and

$$\left.\frac{\partial P/RT}{\partial \rho_j}\right]_{T,\rho\ k\neq j} = \sum_{i=1}^{N} x_i(1 - C_{ij}) \qquad (2)$$

where $\rho_i \equiv x_i\rho$, x_i is the mole fraction, $\hat{f}_i$ is the fugacity, and c_{ij} is the direct correlation function. The essence of our method is to express the direct correlation function integrals, C_{ij}, in terms of reduced temperature and density, integrate Equation 2 from a suitable reference state T, P^r, ρ^r, $\underline{x}^r$ to the final state T, P^f, $\underline{x}^f$ to solve for the final density, ρ^f, and then integrate Equation 1 for all species in the system to obtain the final state fugacity, $\hat{f}_i^f$ from the reference state fugacity $(\hat{f}_i/x_i)^r$. Using a single dummy integrating variable, *t*, the equations become

$$\ln \hat{f}_i^f = \ln x_i^f + \ln (\hat{f}_i/x_i)^r + \ln (\rho^f/\rho^r) - \int_0^1 \left[\sum_{j=1}^{N} (\rho^f x_j^f - \rho^r x_j^r)\, C_{ij}(\tilde{\underline{\rho}}, \tilde{T})/\rho\right] dt \qquad (3)$$

and

$$\frac{P^f - P^r}{RT} = \sum_{i=1}^{N} (\rho^f x_i^f - \rho^r x_i^r) \left\{ 1 - \int_0^1 \left[\sum_{j=1}^{N} x_j C_{ij}(\underset{\sim}{\rho}) \right] dt \right\} \quad (4)$$

where, in the integrals

$$x_j = [\rho^r x_j^r + (\rho^f x_j^f - \rho^r x_j^r) t]/\rho \quad (5)$$

$$\rho = \rho^r + (\rho^f - \rho^r) t \quad (6)$$

To complete the calculation, the model for the C_{ij} (which is a function of T, ρ, and $\underset{\sim}{x}$ through $\underset{\sim}{\tilde{T}}$, $\tilde{T}_{ij}$, and $\tilde{\rho}$ defined below) needs to be specified, the choice of reference state made, and the input parameter and property values need to be given. Each of these shall be described below for the representative cases which we have examined.

Direct Correlation Function Integrals

We use a perturbation theory as the basis for a corresponding-states expression for the correlation function integrals, the major portion being that from rigid spheres and a minor contribution from an added second virial coefficient term. (This concept had been chosen previously by Bienkowski et al. (*7*) in another context while application of the principle of corresponding states for the correlation function integrals had been examined by Gubbins and O'Connell (*8*).)

$$C_{ij}(\underset{\sim}{\tilde{T}}, \tilde{T}_{ij}, \tilde{\rho}) = C_{ij}^{hs}(\underset{\sim}{\tilde{T}}, \tilde{\rho}) - 2\rho V_{ij}^* \left[\frac{B_{ij}}{V_{ij}^*} (\tilde{T}_{ij}, \underset{\sim}{\tilde{\rho}}) - \frac{B_{ij}^{hs}}{V_{ij}^*} (\underset{\sim}{\tilde{T}}, \tilde{\rho}) \right] \quad (7)$$

Specifically, we have used the Carnahan–Starling Equation (*9*) for rigid spheres. Equations 1 and 9–18 of Ref. 9 were used to obtain Equation 8 here.

$$\begin{aligned} -\frac{C_{ij}^{hs}}{\xi_0} = {} & (\sigma_i + \sigma_j)^3/(1 - \xi_3) + [3\sigma_i\sigma_j\xi_2[(\sigma_i + \sigma_j)^2 + \sigma_i\sigma_j] \\ & + 3\xi_1(\sigma_i\sigma_j)^2(\sigma_i + \sigma_j) + \xi_0(\sigma_i\sigma_j)^3]/(1 - \xi_3)^2 \\ & + 9(\sigma_i\sigma_j\xi_2)^3/(1 - \xi_3)^4 + \xi_2(\sigma_i\sigma_j)^2/(1 - \xi_3)^3 \\ & \times \{9\xi_2(\sigma_i + \sigma_j) + 6\xi_1\sigma_i\sigma_j + [6 + \xi_3(-15 + 9\xi_3)]/\xi_3 \\ & - (\sigma_i + \sigma_j)\xi_2[6 + \xi_3(-15 + 12\xi_3)]/\xi_3^2 \\ & + \xi_2^2\sigma_i\sigma_j[6 + \xi_3(-21 + \xi_3(26 - 14\xi_3))]/\xi_3^3(1 - \xi_3)\} \\ & + 6\xi_2(\sigma_i\sigma_j)^2 \ln(1 - \xi_3)\{\xi_3 - (\sigma_i + \sigma_j)\xi_2 + \xi_2^2\sigma_i\sigma_j/\xi_3\}/\xi_3^3 \end{aligned} \quad (8)$$

where

$$\xi_\alpha = \frac{\pi}{6}\,\rho \sum_{i=1}^{N} x_i \sigma_i^{\alpha} \tag{9}$$

Also,

$$B_{ij}/V_{ij}^* = 0.4996 - \frac{1.134}{\tilde{T}_{ij}} - \frac{0.4759}{\tilde{T}_{ij}^2} - \frac{0.0416}{\tilde{T}_{ij}^3} - \frac{0.00209}{\tilde{T}_{ij}^8} \qquad \tilde{T}_{ij} \leq 3.2 \tag{10a}$$

$$= 0.3301 - \frac{0.1376}{\tilde{T}_{ij}} - \frac{1.972}{\tilde{T}_{ij}^2} \qquad \tilde{T}_{ij} \geq 3.2 \tag{10b}$$

where

$$V_{ij}^* = [V_i^{*1/3} + V_j^{*1/3}]^3/8 \tag{11}$$

$$\tilde{T}_{ij} = T/[(T_i^* T_j)^{1/2}(1 - K_{ij})] \tag{12}$$

and

$$B_{ij}^{\mathrm{hs}} = 2/3\,\pi \left[\frac{\sigma_i + \sigma_j}{2}\right]^3 \tag{13}$$

with

$$\sigma_i^3 = \frac{3V_i^*}{2\pi}\,\{b_i + a_2 \exp[-a_4(\tilde{\rho} + a_1\tilde{T}_i)^2] - a_3 \exp[-a_5(\tilde{\rho} + a_1\tilde{T}_i - a_6)^2] + a_9 \exp[-a_{10}(\tilde{T}_i - a_{13})^2 - a_{10}a_{11}(\tilde{\rho} - a_{12})^2\} \tag{14}$$

$$b_i = a_7(\tilde{T}_i)^{a_8} \qquad \tilde{T}_i \geq 0.73 \tag{15a}$$

$$= a_{14} \exp[-a_{15}\tilde{T}_i] \qquad \tilde{T}_i \leq 0.73 \tag{15b}$$

$$\tilde{\rho} \equiv \rho \sum_{i=1}^{N} \sum_{j=1}^{N} x_i x_j V_{ij}^* \tag{16a}$$

$$\tilde{T}_i \equiv \tilde{T}_{ij} \qquad i = j \tag{16b}$$

Table I lists the values of the parameters a_1–a_{15}

Thus, to the degree that the corresponding-states theory for the direct correlation function integrals is accurate, the results are applicable to any system. In practice, we have found that when the species are at

Table I. Constants for Equation 14

$a_1 = 0.54008832$	$a_6 = 2.1595955$	$a_{11} = 0.48197123$
$a_2 = 1.2669802$	$a_7 = 0.64269552$	$a_{12} = 0.76696099$
$a_3 = 0.05132355$	$a_8 = 0.17565885$	$a_{13} = 0.76631363$
$a_4 = 2.9107424$	$a_9 = 0.18874824$	$a_{14} = 0.80965780$
$a_5 = 2.5167259$	$a_{10} = 17.952388$	$a_{15} = 0.24062803$

similar reduced densities and temperatures, the integrations of Equations 1 and 2 are too sensitive to slight errors in the correlation to yield accurate predictions of fugacities (a "small-difference-of-large-numbers" problem). However, for liquids with supercritical species (as well as certain others we are examining where the "natural" reference state is infinite dilution instead of pure component) the reduced states are separated sufficiently that good results are obtained.

Parameter Determination

The form of Equations 10b, 14, and 15a, b were obtained by fitting the isothermal compressibility data for the pure noble gases over all available ranges of reduced temperature and density to the relation

$$\frac{1}{\rho\kappa_T RT} = 1 - C^{\mathrm{hs}}(\rho\sigma^3) + 2\rho V^* \left[\frac{B}{V^*}\ (T/T^{\mathrm{c}}) - \frac{2\pi}{3}\ \frac{\sigma^3}{V^*}\right] \tag{17}$$

with $V^* = V_{\mathrm{c}}$ in Equations 14–17.

Then, the isothermal compressibility data on other pure substances, generally for $V/V^{\mathrm{c}} < 1/2$, were used to obtain their T^* and V^* values. Table II reports the results for the systems here. More extensive tables are reported elsewhere (*5, 6*). In general, the fitting is most sensitive to the value of V^*, so when few data are available, T^* can be estimated as the critical temperature and a one-parameter fit is used. A consequence of this is that the binary parameter, K_{12}, will compensate for any erroneous

Table II. Characteristic Parameters for Pure Fluids

Substance	T^* *(K)*	V^* *(cm³/g mol)*
Hydrogen	38.6	53.2
Carbon monoxide	132.9	93.1
Nitrogen	126.3	90.9
Benzene	571.9	256.9
Ammonia	360.2	65.1
Nitrobenzene	608.2	327.0
Tetralin	693.0	454.2
Quinoline	800.0	380.0

choice of T^*, and will minimize its effect on the results. Note that in all cases the density of the solvent component must be available. If the density is estimated when T^* and V^* are found, the same method also should be used in the phase-equilibrium calculations.

Reference State Properties

In Equation 3 it is necessary to have reference-state values for component density, $x_i\rho^r$, and fugacity $(\hat{f}_i/x_i)^r$. For pure-solvent components these are the pure saturated liquid density, ρ_i°, and the fugacity of the saturated vapor, $\hat{f}_i^{0\,\mathrm{sat}}$, (which equals that of the saturated liquid) respectively, both at the temperature of interest. For mixed solvents, the density also requires volume of mixing data for the solvents, and the solvent fugacities are found from the vapor–liquid equilibrium data on the solute-free mixture.

For the solute components, two binary cases commonly are encountered. One is where data on the system are available either at low or high pressures at the temperature of interest. Here, the value of $(\hat{f}_i/x_i)^r$ for the solute is Henry's constant, $H_{ij} \equiv \lim_{x_j \to 1} (y_i\hat{\Phi}_i P/x_i)$. At low pressures, the limit may be reached in the experiment. At high pressures, we have fitted the values of H_{ij} and K_{ij} to the vapor–liquid equilibrium data, obtaining parameters for the equation

$$\ln H_{ij} = A_{ij} + B_{ij}T + D_{ij}T^2 \tag{18}$$

The number of parameters depends on the number of temperatures for which the data are available. However, if data are not available for the particular solute–solvent $(i–j)$ pair we can make use of Henry's constant, H_{ik}, for the solute in another solvent, k. This value is used in a suitable transformation of Equation 3:

$$\ln (H_{ij}/H_{ik}) = \ln \frac{\rho_j^\circ}{\rho_k^\circ} - \int_0^1 [(\rho_j^\circ C_{ij} - \rho_k^\circ C_{ik})/\rho]dt \tag{19}$$

where

$$\rho = \rho_k^\circ + (\rho_j^\circ - \rho_k^\circ)t \tag{20}$$

Unfortunately, unlike the integrals of Equations 3 and 4, Equation 19 is sensitive to the values of K_{ij} and K_{ik} and we recommend that at least one experimental value of H_{ij} be used to determine K_{ij} with Equations 19 and 20. For a solute in a mixed solvent, we determine Henry's constant, H_{im}, from a selected reference solvent, k

$$\ln (\hat{f}_i/x_i)^{\mathrm{r}} = \ln H_{im} = \ln H_{ik} + \ln (\rho^{\mathrm{r}}/\rho_k{}^{\circ}) - \int_0^1 \left[\sum_{j=1}^{N_s} (\rho^{\mathrm{r}} x_j - \rho_j{}^{\circ} \delta_{jk}) C_{ij}/\rho \right] dt \tag{21}$$

where N_s is the number of solvents, δ_{ij} is the Kronecker delta ($\delta_{jk} = 1$ for $j = k$, $\delta_{jk} = 0$ for $j \neq k$), and

$$\rho = \rho_k{}^{\circ} + \left[\left(\sum_{j=1}^{N_s} \rho^{\mathrm{r}} x_j{}^{\mathrm{r}} \right) - \rho_k{}^{\circ} \right] t \tag{22}$$

$$x_j = [\rho_j{}^{\circ} (1 - t) \delta_{jk} + \rho^{\mathrm{r}} x_j{}^{\mathrm{r}} t]/\rho \tag{23}$$

It is recommended that component k be the dominant solvent component or one of similar chemical constitution. As before, the results are sensitive to the values of the K_{ij} except in the case of hydrogen.

Results

Figure 1 shows comparisons of the calculations and experiment (*10*) of the pressure-composition relations at various temperatures for hydrogen with benzene. Here we have chosen $K_{12} = 0.17$ and fitted the data to obtain parameters in Equation 18. Figure 2 shows a comparison of the densities for these solutions. Figure 3 shows $P - x$ data for nitrogen in ammonia calculated in the same way (*11*). Deviations from Henry's Law (dashed lines) are significant, but if only the Poynting Correction had been used (*1, 2*), the mole fractions would have been too low at a given pressure, particularly at the higher temperature.

Figure 4 shows results when the reference state uses data on another system. The $P - x$ behavior for hydrogen and tetralin (*12*) is obtained using estimated densities, pure component and binary parameters, and Henry's constant for hydrogen in quinoline. The last was found from correlating the high-pressure data of Chao (*12*).

Finally, Figure 5 shows values of the variation of Henry's constant for carbon monoxide in the mixed solvent nitrobenzene–benzene at 298 K. The binary parameters were chosen so that correct values were obtained for H_{12} when using H_{13} and for H_{13} when using H_{12} in the integrations of Equation 19.

The above results are only some of many so far investigated. Further exploration of the theory is being done and more complete comparisons will be forthcoming (*5, 6*).

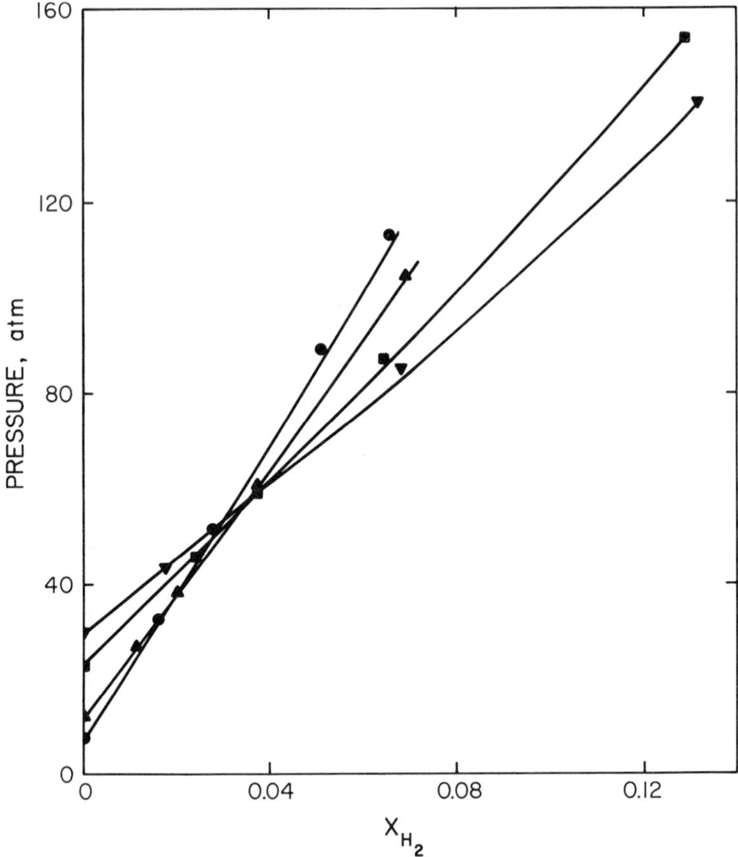

Figure 1. Solubility of hydrogen in benzene (10): (●), 433.15 K; (▲), 473.15 K; (■), 503.15 K; (▼), 523.15 K. (——), Calculated [$K_{12} = .17$].

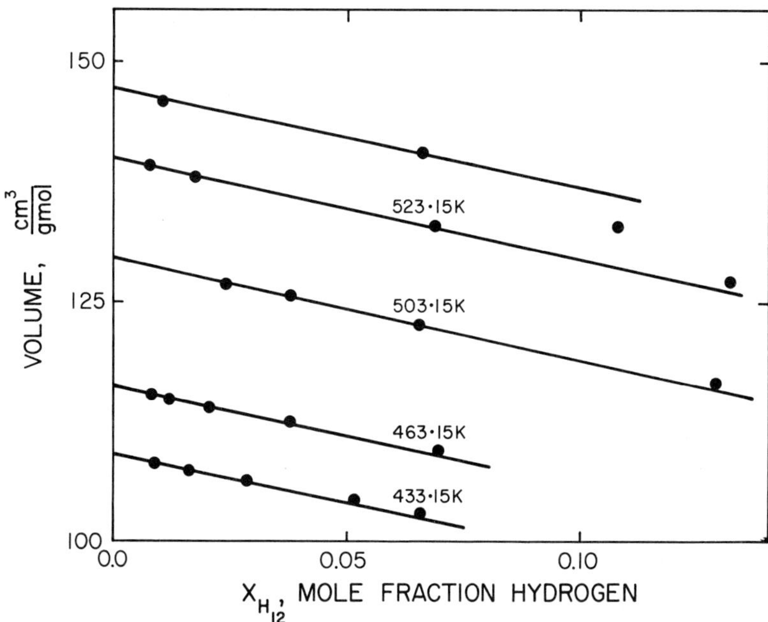

Figure 2. Molar volumes of hydrogen–benzene systems: (●), data of Connolly (10). (——), Calculated [$K_{12} = 0.17$].

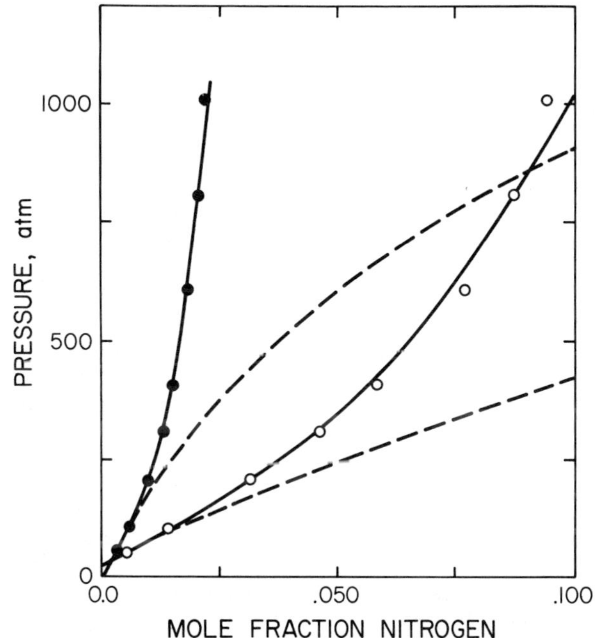

Figure 3. Solubility of nitrogen in ammonia: (●, ○), data of Wiebe et al. (11); (——), $K_{12} = 0.0$; and (– – –), ideal solution. (●), 273 K and (○), 333 K.

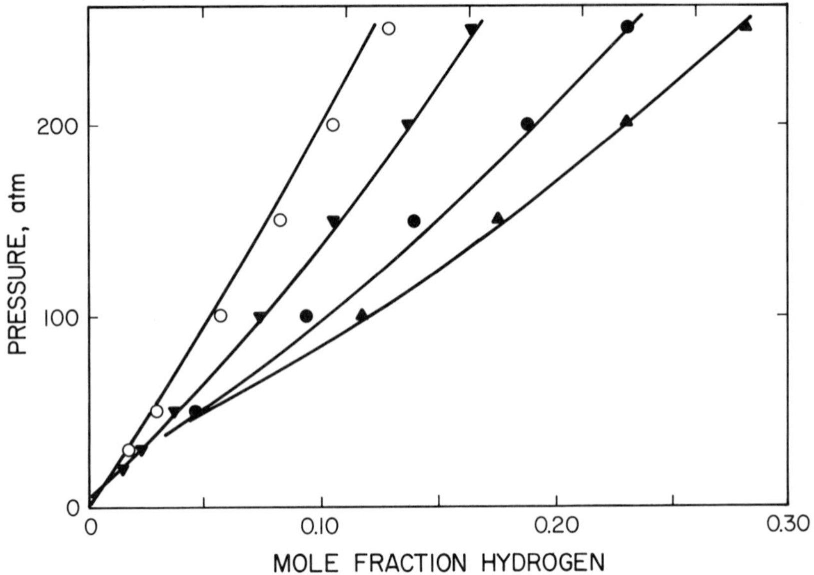

Figure 4. Solubility of hydrogen (1) in tetralin (3) using Henry's constant in quinoline, H_{12} *(12).* $K_{12} = K_{13} = 0$. *(○), 462.8* K; *(▼), 541.9* K; *(●), 621.8* K; *(▲), 662.3* K; $T_2^c = 794$ K; *and* $T_3^c = 716$ K.

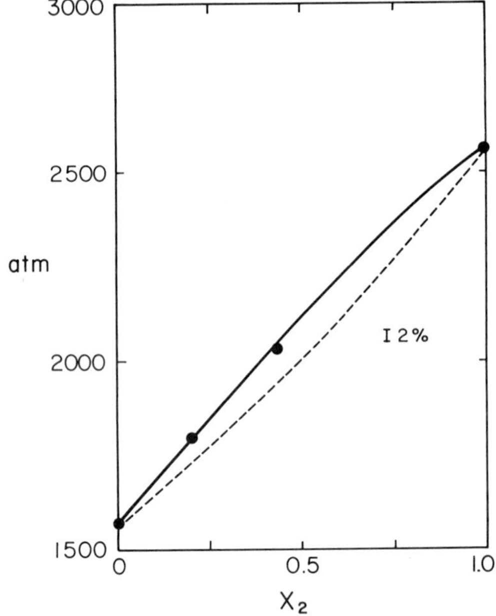

Figure 5. Henry's constant for carbon monoxide (1) in nitrobenzene (2) –benzene (3): (——), calculated; (– – –), ideal solution; and (●), data from international critical tables. T = 298 K.

Glossary of Symbols

A_{ij}, B_{ij}, D_{ij} = temperature parameters for Henry's constants
B_{ij} = second virial coefficient, m^3/mol, Equations 7, 10, and 13
c_{ij} = pair direct correlation function
C_{ij} = spacial integral of c_{ij} times density, Equation 1
f_i = component fugacity, KPa
H_{ij} = Henry's constant of solute i in solvent j, KPa
K_{12} = binary parameter, Equation 12
N = number of components
P = total pressure, KPa
r = separation of molecular centers, meters
R = universal gas constant, KJ/mol-K
t = dummy integrating variable, Equations 3–6 and 19–23
T = absolute temperature, K
T_i^* = characteristic temperature, K
$\tilde{T}_i = T/T_i^*$
$\tilde{T}_{ij}$ = Equation 12
V_i^* = characteristic volume, m^3/mol
V_{ij}^* = Equation 11
x_i = liquid-mole fraction
y_i = vapor-mole fraction
ρ_i = concentration of component i, mol/m^3
σ_i = molecular diameter of species i, Å
ξ_α = Equation 9
$\kappa_T = \partial \ln\rho/\partial P)_{T,x} \equiv$ isothermal compressibility, KPa^{-1}
$\hat{\Phi}_i = f_i/y_iP \equiv$ vapor fugacity coefficient
δ_{jk} = Kronecker delta = 1 for $j = k$; = 0 for $j \neq k$
ρ = molar density, mol/m^3

Superscripts

c = critical point
f = final state
hs = hard sphere
r = reference state
o = pure component
$\sim$ = reduced value

Subscripts

s = solvents only
___ = vector of values

Acknowledgment

We are grateful to the National Science Foundation for financial support, the NERDC of the State University System of Florida for use of their facilities, and to E. Buck and T. Krolikowski–Buck for helpful discussions.

Literature Cited

1. Prausnitz, J. M. *Adv. Chem. Eng.* **1968,** *7,* 139.
2. Prausnitz, J. M. "Molecular Thermodynamics of Fluid-Phase Equilibria"; Prentice-Hall: Englewood Cliffs, NJ, 1969.
3. Prausnitz, J. M.; Chueh, P. L. "Computer Calculations for High Pressure Vapor–Liquid Equilibria; Prentice-Hall: Englewood Cliffs, NJ, 1968.
4. O'Connell, J. P. *Mol. Phys.* **1971,** *20,* 27.
5. Mathias, P. M.; O'Connell, J. P., *Chem. Eng. Sci.,* in press.
6. Mathias, P. M.; O'Connell, J. P., *AIChE J.,* in press.
7. Bienkowski, P. R.; Denenholz, H. S.; Chao, K-C. *AIChE J.* **1973,** *19,* 167.
8. Gubbins, K. E.; O'Connell, J. P. *J. Chem. Phys.* **1974,** *60,* 3449.
9. Reed, T. M.; Gubbins, K. E. "Applied Statistical Mechanics"; McGraw-Hill: New York, 1973.
10. Connolly, J. F. *J. Chem. Phys.* **1962,** *36,* 2897.
11. Wiebe, R.; Gaddy, V. L. *J. Am. Chem. Soc.* **1937,** *59,* 1984.
12. Chao, K-C., EPRI-AF-466 (Research Project 367-11) Final Report, 1976.

Received October 5, 1978.

6

A New, Generalized Equation of State Valid Within the Critical Region

MOHAMMED S. NEHZAT[1], KENNETH R. HALL, and PHILIP T. EUBANK[2]

Chemical Engineering Department, Texas A&M University, College Station, TX 77843

An isochoric equation of state, applicable to pure components, is proposed based upon values of pressure and temperature taken at the vapor–liquid coexistence curve. Its validity, especially in the critical region, depends upon correlation of the two leading terms: the isochoric slope and the isochoric curvature. The proposed equation of state utilizes power law behavior for the difference between vapor and liquid isochoric slopes issuing from the same point on the coexistence cruve, and rectilinear behavior for the mean values. The curvature is a skewed sinusoidal curve as a function of density which approaches zero at zero density and twice the critical density and becomes zero slightly below the critical density. Values of properties for ethylene and water calculated from this equation of state compare favorably with data.

The critical region provides a severe test for the applicability of any equation of state, and often just as severe a test for the patience of the correlator. The reason is the almost pathological behavior of fluids as they approach their critical points. Many thermodynamic properties either become zero or else diverge to infinity at the critical point.

This study is another rather successful attempt to correlate the fluid properties in the critical region. We have chosen an isochoric equation of state with constant curvature to represent these properties and we

[1] Present address: Division of Energy, Arya–Mehr University of Technology, Isfahan, Iran.
[2] Author to whom correspondence should be addressed.

0-8412-0500-0/79/33-182-109$05.00/1

have imposed some of the ideas from the scaling hypothesis. Because this is a correlation, we have allowed data to influence the model. When the data did not provide conclusive guidance, we chose to make the correlation internally consistent.

We selected ethylene and water as test substances for the correlation. These two compounds are important commercial substances and they are also interesting from a scientific viewpoint. In addition, the critical region correlations for these compounds were in rather poor agreement with the data.

Previous Work

Andrews (*1*) first discovered the critical point of a fluid in 1869. Shortly thereafter in 1873, Van der Waals (*2*) presented his dissertation, "On the Continuity of the Gas and Liquid State." This and later work in the following twenty years provided the classical theory of the critical region for fluids. However, Verschaffelt in the early 1900's found the critical exponents β and δ to be about 0.35 and 4.26, respectively, compared with the classical values of 1/2 and 3. The surface tension exponent also was found to be near 1.25 instead of the classical value of 3/2. An excellent detailed historical review of this period has been given by Levelt Sengers (*3*).

In 1965, Widom (*4*) proposed a nonclassical model for the critical region, the scaling hypothesis. This model was remarkably successful and spawned a tremendous number of papers both applying and refining the model. Books by Stanley (*5*) and Ma (*6*) together with the comprehensive review by Levelt Sengers, Greer, and Sengers (*7*) provide the necessary background material. The critical region description by the scaling model was so successful that no serious challenges arose for 10 years.

Because of its mathematical complexity and lack of simple tractability in terms of measured thermodynamic properties (i.e., the heat of vaporization), scaled equations of state are popular with few experimental thermodynamicists and practicing engineers. While some question its correctness, most simply do not understand this field dominated by theoretical physicists and mathematicians. Despite the efforts of the Equation of State Section of the NBS in Washington to popularize the subject through useful applications, few engineers can use any of the results save such printed thermodynamic property tables as those for steam (*8*). Another complication is the necessity to blend the scaled equation of state with a second, analytical equation of state valid outside the critical region (*9*); within the critical region, the analytical terms are referred to as the "background."

Thermodynamically consistent, nonanalytical, empirical equations of state induced from experimental measurements can avoid the above difficulties. Since 1965, at least two laboratories actively were developing isochoric equations of state (Refs. *10, 11*). These workers had the benefit of the scaling work and included nonclassical behavior in the critical region for their equations. The equation presented in this chapter arose from utilizing the same basic strategy.

In 1976, Hall and Eubank (*12, 13*) published two papers which have direct bearing upon the present equation of state. In the first paper, they noted the rectilinear behavior for the mean of the vapor and liquid isochoric slopes issuing from the same point on the vapor pressure curve near the critical point and the power law behavior for the difference in these slopes. The second paper presented an empirical description of the critical region which generally agreed with the scaling model but differed in one significant way—the curvature of the vapor pressure curve.

Equation of State

The basic function of an isochoric equation of state is to describe isochores as they issue from the vapor pressure curve. Figure 1 illustrates

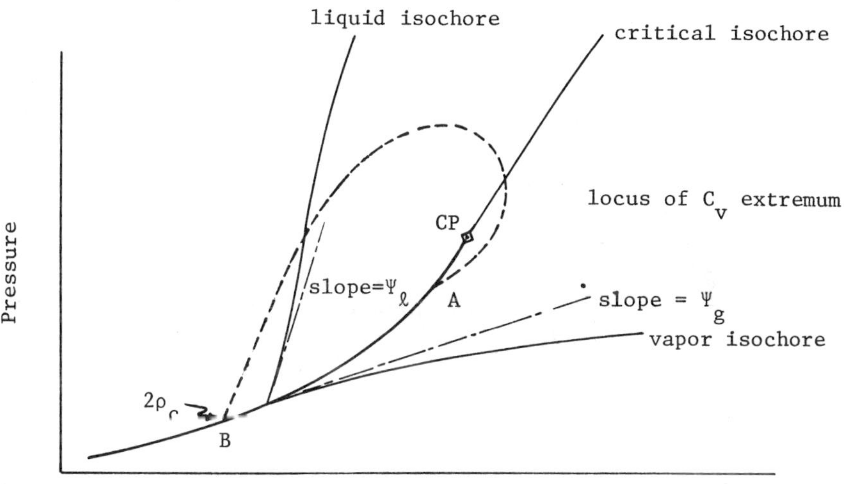

Figure 1. Coexistence curve with isochores and the isochoric slope, ψ_o. At points **A** *and* **B** *the curvature of the isochores is zero as well as along the locus indicated by the dotted line (the locus of the isochoric heat capacity extrema).*

the generally regular behavior of these curves on a P–T diagram. The basic formulation is a Taylor's series expansion about the vapor pressure:

$$P = P_\sigma + \left(\frac{\partial P}{\partial T}\right)_\rho\bigg|_\sigma (T - T_\sigma) + \frac{1}{2}\left(\frac{\partial^2 P}{\partial T^2}\right)_\rho\bigg|_\sigma (T - T_\sigma)^2 + \ldots \tag{1}$$

where P is pressure, T is temperature, ρ is density, and subscript σ denotes a saturation value (e.g., P_σ is the vapor pressure). In the near-critical region, which we arbitrarily shall define as within $|1 - T_R| \lesssim 0.01$ and $|1 - \rho_R| \lesssim 0.3$, Equation 1 represents the isochores adequately when truncated after the second-order term. Finally, we write the equation of state in reduced form for increased numerical tractability:

$$\frac{P}{P_c} = \frac{P_\sigma}{P_c} + \frac{T_c}{P_c}\left(\frac{\partial P}{\partial T}\right)_\rho\bigg|_\sigma \frac{(T - T_\sigma)}{T_c} + \frac{T_c^2}{2P_c}\left(\frac{\partial^2 P}{\partial T^2}\right)_\rho\bigg|_\sigma \frac{(T - T_\sigma)^2}{T_c^2} \tag{2}$$

or with the usual definitions of reduced variables:

$$P_R = P_{\sigma,R} + \left(\frac{\partial P_R}{\partial T_R}\right)_{\rho_R}\bigg|_\sigma (T - T_{\sigma,R}) + \frac{1}{2}\left(\frac{\partial^2 P_R}{\partial T_R^2}\right)_{\rho_R}\bigg|_\sigma (T - T_{\sigma,R})^2$$

$$= P_{\sigma,R} + \Psi_\sigma(T_R - T_{\sigma,R}) + \frac{1}{2}\Phi_\sigma(T_R - T_{\sigma,R})^2 \tag{3}$$

in which Ψ_σ and Φ_σ represent the reduced first and second derivatives, respectively.

Clearly, utilization of Equation 3 requires expressions for the saturation properties: P_σ, Ψ_σ, Φ_σ, and T_σ. Fortunately, in the critical region, these expressions are relatively simple. Taking the properties in order, the vapor pressure equation is the first variable to consider. We have chosen an expression proposed by Walton et al. (*14*) which is an integration of the truncated scaling equation:

$$\frac{d^2 P_{\sigma,R}}{d\tau^2} \approx c_1\tau^{-\Theta}(1 + c_2\tau^\eta) \tag{4}$$

in which $\tau = 1 - T_{\sigma,R}$ and θ and η are critical exponents. Integration of Equation 4 ultimately yields:

$$P_{\sigma,R} = 1 - \Psi_c\tau + k_7\tau^{2-\Theta} + k_8\tau^{2-\Theta+\eta} \tag{5}$$

where k_7 and k_8 are functions of c_1, c_2, θ, and η (we designate them k_7 and k_8 to maintain consistency with earlier publications), and Ψ_c is the value of Ψ_σ at the critical point. Equation 5 is an excellent representation for the vapor pressure in the critical region producing maximum deviations on the order of 10 ppm. Figure 2 illustrates this assertion for ethylene.

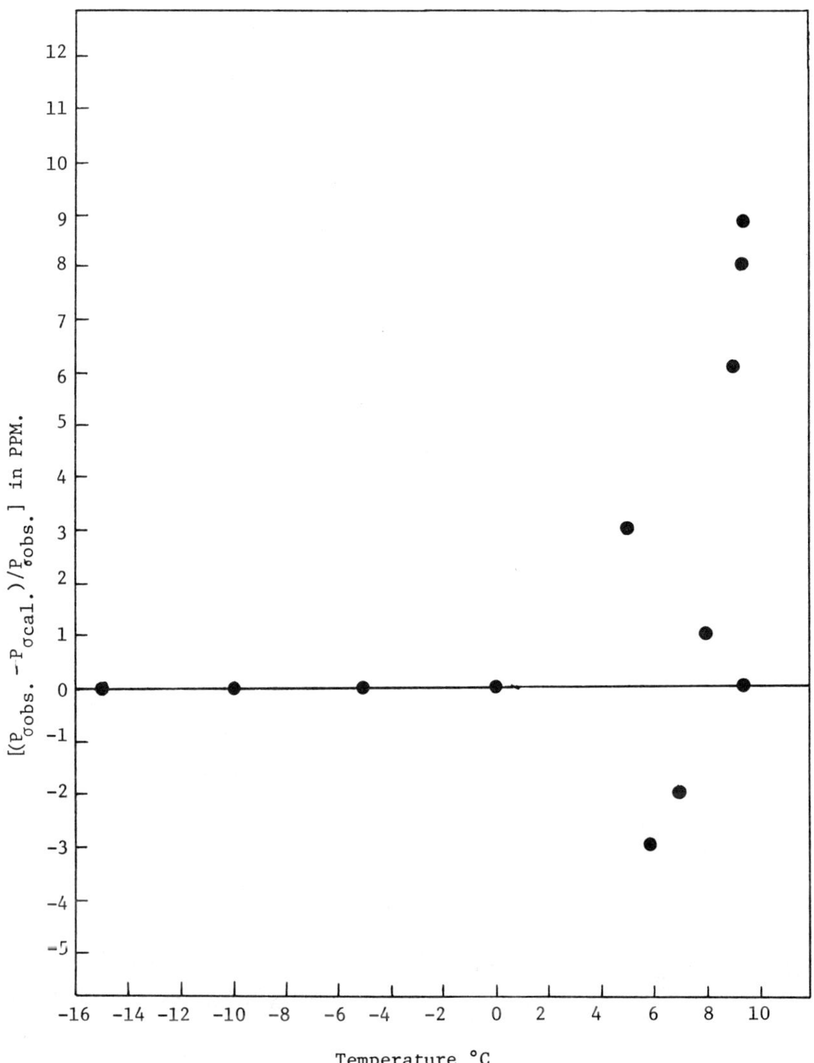

Figure 2. Relative deviations of the calculated vapor pressure for ethylene from the observed values of Douslin and Harrison (17) in parts per million

The isochoric slopes, Ψ_σ, obey a reasonably simple relationship in the critical region:

$$\Psi_\sigma = \Psi_c + k_5\tau \mp k_6\tau^\lambda \tag{6}$$

in which the minus sign denotes vapor values and the plus sign denotes liquid values for the last term. Equation 6 results from observations made by Hall and Eubank (*12*) regarding rectilinear behavior for the mean isochoric slopes and power law behavior for the difference in isochoric slopes. Figures 3 and 4 demonstrate these observations for ethylene.

The second derivative term, Φ_σ, which we will refer to as "curvature" for shorthand notation, has rather exotic behavior. Figure 5 is a qualitative sketch of the function plotted against density. An empirical relationship which can predict such behavior is

$$\Phi_\sigma = \Phi_c(1 - \xi^4) + k_9\xi(1 - \xi^2)\exp(k_{10}\xi^2), \tag{7}$$

where $\xi \equiv (\rho_\sigma - \rho_c)/\rho_c$. This expression is similar to the last term in the BWR equation. Figure 6 illustrates the fit of Equation 7 to ethylene data.

The saturation temperature, T_σ, is an implicit function of the saturation density:

$$(\rho_\sigma/\rho_c) = 1 + k_1\tau \mp k_2\tau^\beta \tag{8}$$

where β is a critical exponent and the minus sign again denotes a vapor value while the plus sign denotes a liquid value. This completes the equation of state which requires values for T_c, P_c, ρ_c, Ψ_c, Φ_c, θ, η, λ, β, k_1, k_2, k_5, k_6, k_7, k_8, k_9, and k_{10}, (five critical parameters, four critical exponents, and eight adjustable constants). Actually, relationships exist among θ, λ, and β which eliminate one of them from the parameter list. It is also likely that the critical exponents are universal.

Comparison with Data

We have chosen ethylene and water as substances to correlate with our equation of state. The reasons for picking these substances are their considerable practical and theoretical importance, and because of the apparently poor state of correlation in the critical region for ethylene in the IUPAC tables (*15*) and for water in the steam tables (*16*). Tables I and II present the results of the correlation. The agreement between observed and calculated pressures is excellent. The maximum percentage errors are 0.005% for ethylene and 0.09% for water. Tables III and IV present the values of the various parameters used in the correlation. For

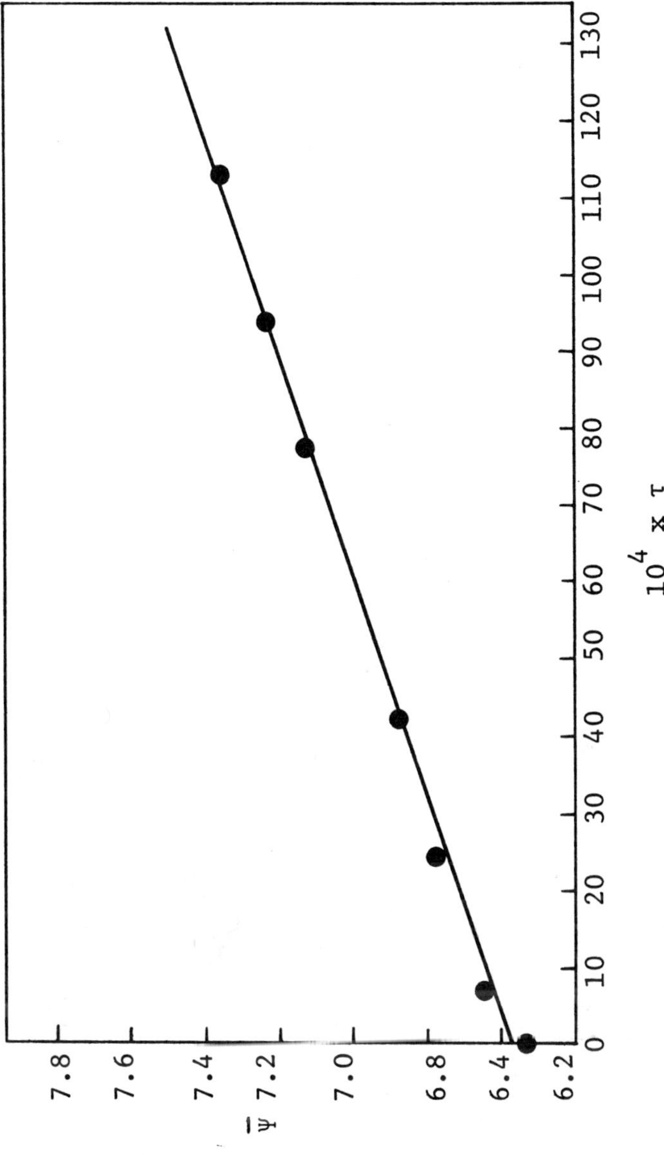

Figure 3. Mean reduced isochoric slopes at the coexistence curve for ethylene as a function of the reduced temperature difference, τ. The intercept is Ψ_c.

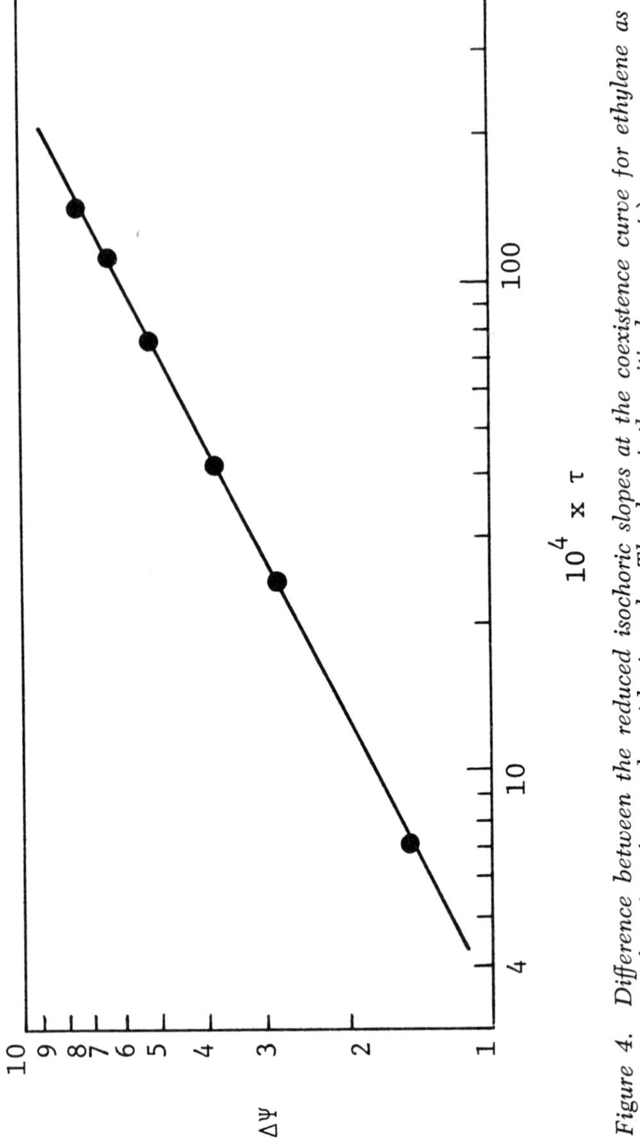

Figure 4. Difference between the reduced isochoric slopes at the coexistence curve for ethylene as a function of τ on a logarithmic scale. The slope is the critical exponent λ.

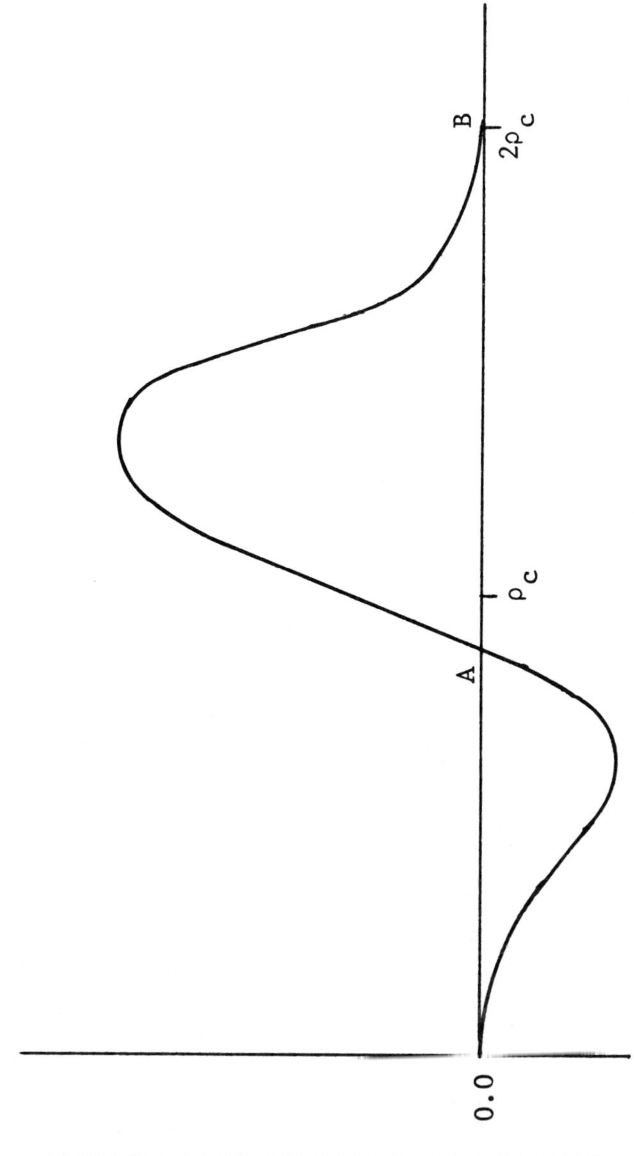

Figure 5. Curvature of the isochores at the coexistence curve vs. density. Points **A** *and* **B** *correspond to those in Figure 1.*

Table I. Comparison of Calculated Pressures with Those of Douslin and Harrison (*17*)

T_{68} *(K)*	$P_{observed}$ *MPa*	$P_{calculated}$ *MPa*	ϵ *(%)*
	$\rho = 5.0\ g\ mol/dm^3$		
280.15	4.79381	4.79383	+0.0004
281.15	4.86833	4.86824	−0.0018
282.15	4.94195	4.94195	0.0000
282.35	4.95656	4.95661	+0.0010
283.15	5.01491	5.01496	+0.0010
	$\rho = 6.0\ g\ mol/dm^3$		
282.15	5.01181	5.01184	+0.0006
282.25	5.02118	5.02107	−0.0022
282.35	5.03051	5.03028	−0.0046
283.15	5.10345	5.10353	+0.0016
	$\rho_c = 7.635\ g\ mol/dm^3$		
282.35	5.04197	5.04197	0.0000
283.15	5.13277	5.13298	+0.0042
	$\rho = 8.0\ g\ mol/dm^3$		
282.35	5.04200	5.04201	+0.0002
283.15	5.13594	5.13593	−0.0002
	$\rho = 9.0\ g\ mol/dm^3$		
282.15	5.02198	5.02207	+0.0018
282.25	5.03540	5.03547	+0.0014
282.35	5.04902	5.04889	−0.0026
283.15	5.15732	5.15731	−0.0002
	$\rho = 10.0\ g\ mol/dm^3$		
281.15	4.92924	4.92922	−0.0004
282.15	5.09211	5.09228	+0.0034
282.35	5.12506	5.12519	−0.0026
283.15	5.25792	5.25792	+0.0020

Table II. Comparison of Calculated Pressures with Those of Rivkin et al. (*18*)

T_{68} *(°C)*	$P_{observed}$ *bar*	$P_{calculated}$ *bar*	ϵ *(%)*
	$v = 2.7634\ cm^3\ g^{-1}$		
374.127	221.04	221.05	+0.01
374.997	223.62	223.56	−0.03
377.007	229.55	229.47	−0.03

Table II. Continued

T_{68} (°C)	$P_{observed}$ bar	$P_{calculated}$ bar	ε (%)
379.017	235.56	235.51	−0.02
382.027	244.73	244.80	+0.03
384.897	253.45	253.59	+0.05
	$v = 3.0128\ cm^3\ g^{-1}$		
375.237	224.07	224.02	−0.02
376.107	226.47	226.41	−0.03
377.067	229.14	229.05	−0.04
379.067	234.67	234.63	−0.02
380.977	240.05	240.03	−0.01
383.007	245.72	245.85	+0.05
	$v = 3.2707\ cm^3\ g^{-1}$		
374.147	221.01	221.05	+0.02
375.317	224.12	224.14	+0.01
376.017	225.96	225.99	+0.01
377.037	228.64	228.55	−0.04
379.177	234.28	234.43	+0.06
381.127	239.47	239.69	+0.09
383.027	244.36	244.55	+0.08
	$v = 3.4171\ cm^3\ g^{-1}$		
375.067	223.40	223.41	+0.01
376.117	226.09	226.11	+0.01
377.057	228.49	228.54	+0.02
379.007	233.47	233.59	+0.05
381.007	238.53	238.73	+0.08
383.167	244.01	244.16	+0.06
	$v = 3.6558\ cm^3\ g^{-1}$		
375.077	223.27	223.25	−0.01
376.007	225.54	225.52	−0.01
377.997	230.34	230.38	+0.02
380.997	237.51	237.67	+0.06
385.007	247.11	247.30	+0.08
	$v = 3.8619\ cm^3\ g^{-1}$		
374.907	222.65	222.61	−0.01
376.037	225.24	225.24	+0.00
377.977	229.71	229.72	+0.01
381.007	236.71	236.61	+0.04
384.967	246.08	246.01	−0.03

Table III. Ethylene Parameters and Standard Deviations

Parameter	*Value*	*Standard Deviation*
T_c	282.3502 K	0.000 K
P_c	5.04197 MPa	0.00014 MPa
ρ_c	7.635 g mol/dm^3	0.001 g mol/dm^3
β	0.3523	0.0042
λ	0.5304	0.0031
θ	0.2082	0.0006
η	1.5476	0.2142
Ψ_c	6.3696	0.0022
Φ_c	11.1017	0.3184
k_1	0.7649	0.0007
k_2	1.8902	0.0073
k_5	69.4549	0.4112
k_6	34.9071	0.0147
k_7	9.5355	0.0116
k_8	−6.4275	0.7123
k_9	282.9170	1.7139
k_{10}	−10.4318	0.4656

Table IV. Water Parameters and Their Standard Deviations

Parameter	*Value*	*Standard Deviation*
T_c (IPTS-68)	647.2262 K	0.0231 K
P_c	220.867 bar	0.0113 bar
ρ_c	0.32165 g/cm^3	0.00124 g/cm^3
β	0.3561	0.0062
λ	0.5288	0.0062
Θ	0.2308	0.0019
η	—	—
Ψ_c	7.8579	0.0003
Φ_c	6.3115	0.0983
k_1	2.2051	0.0910
k_2	2.3471	0.0142
k_5	75.529	0.0198
k_6	41.7893	0.1019
k_7	13.5956	0.0801
k_8	0	—
k_9	50.2477	0.6467

the water correlation, the expression of Φ_σ was considerably simpler than Equation 7 because the critical range essentially covered only the linear region of Figure 6. Therefore, for water we used

$$\Phi_\sigma = \Phi_c + k_9\xi \quad (9)$$

which is adequate as seen from Figure 7.

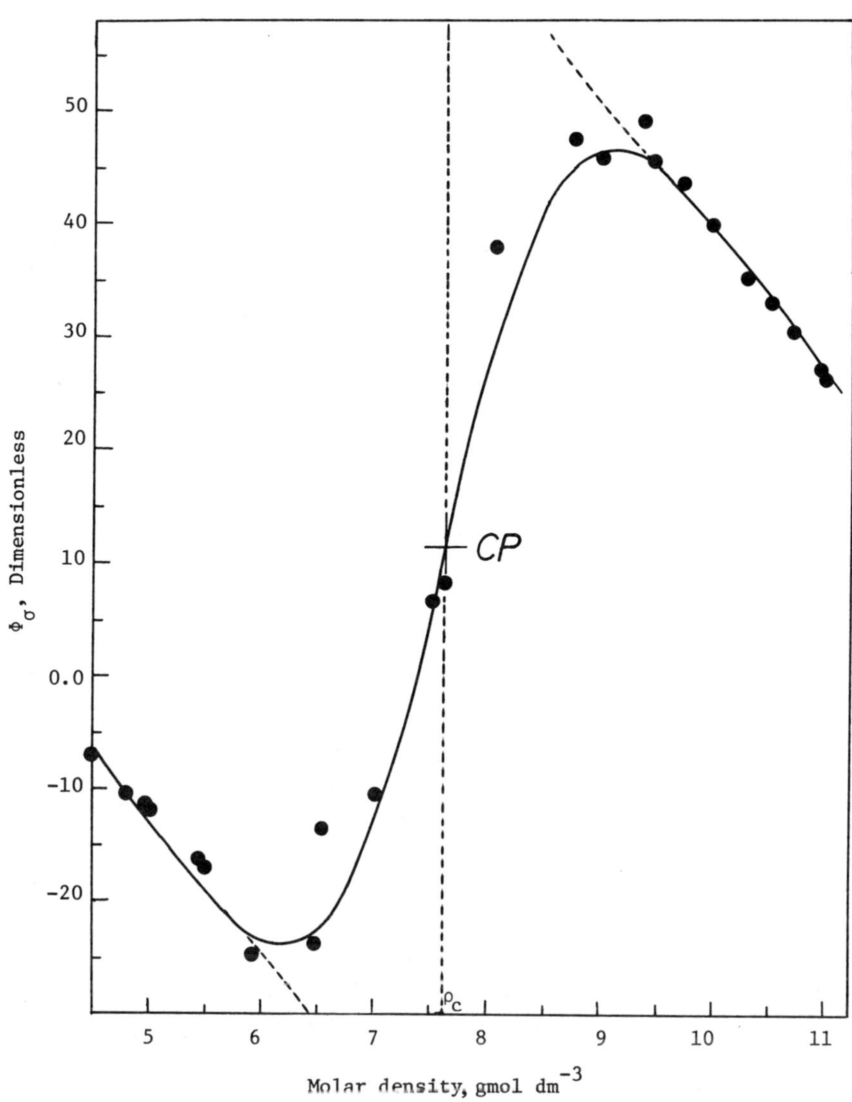

Figure 6. The reduced curvatures of isochores at the coexistence curve of ethylene vs. the molar density. CP indicates the critical point while the dotted curves show the trend of Φ_σ as predicted by the revised scaling theory.

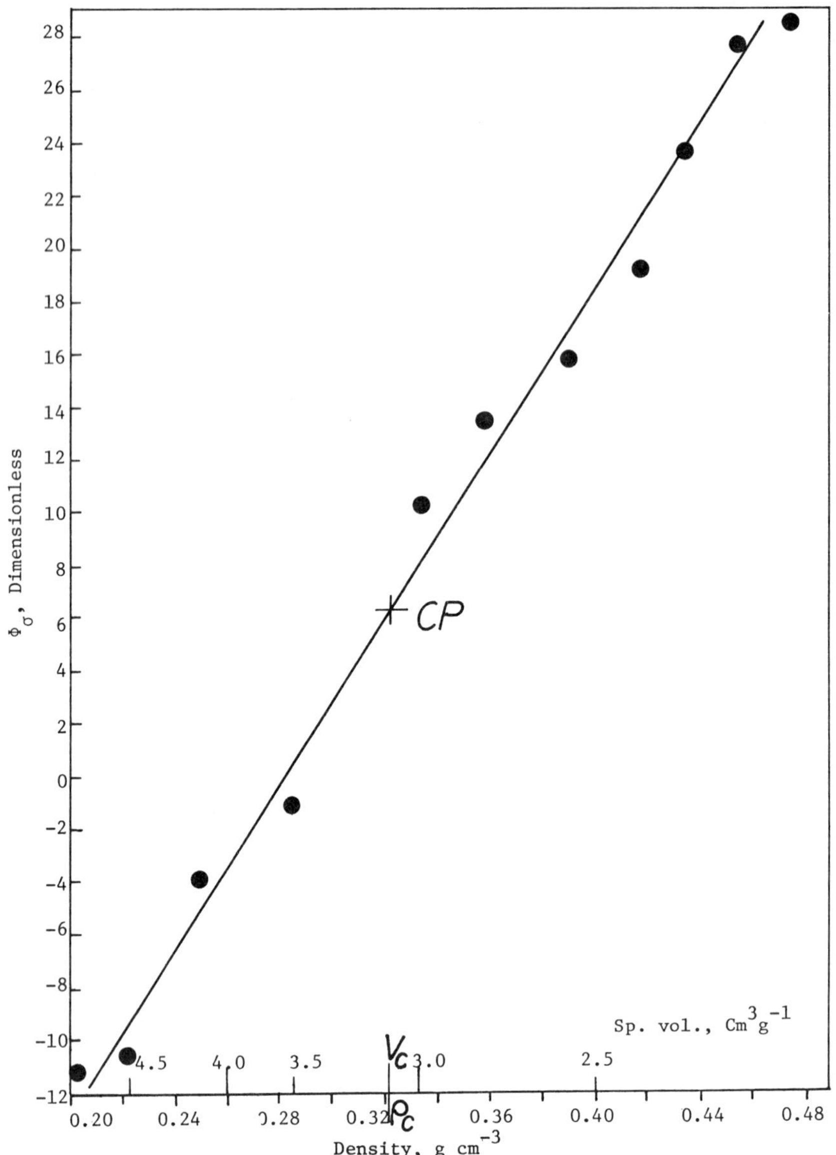

Figure 7. Reduced curvature of steam isochores at the coexistence curve vs. the density in the critical region based on the data of Rivkin et al. (18)

Conclusions

A relatively simple, isochoric equation of state can describe the critical region for fluids such as ethylene and water using five critical parameters, four critical exponents, and eight adjustable constants. Agreement between observed and calculated pressures is excellent and the current values are much better than those in standard reference tables.

Acknowledgment

The authors gratefully acknowledge financial support from National Science Foundation Grants ENG76-00692 and ENG77-01070.

Glossary of Symbols

c_1, c_2 = dimensionless constants in Equation 4
k_1, k_2, k_5–k_{10} = dimensionless constants in Equations 5, 6, 7, and 8
P = pressure, bar or MPa.
T = temperature, K or °C
V = specific volume, $cm^3\ g^{-1}$

Greek Letters

β = critical exponent of the orthobaric density difference
η = second order critical exponent of the vapor pressure curvature
θ = primary critical exponent of the vapor pressure curvature
λ = critical exponent of the difference of isochoric slopes issuing from the vapor pressure curve
ξ = the dimensionless density, $(\mathrm{P} - \rho_c)/\rho_c$
ρ = density, g mol/dm^3
τ = the dimensionless temperature, $(T_c - T)/T_c$
Φ = reduced isochoric second derivative, $(\partial^2 P_R/\partial T_R{}^2)_{\rho_R}|_\sigma$
Ψ = reduced isochoric first derivative, $(\partial P_R/\partial T_R)_{\rho_R}|_\sigma$

Subscripts

c = critical point
R = value divided by its critical value
σ = value at the coexistence curve
ρ = value along an isochore

Literature Cited

1. Andrews, T. *Proc. R. Soc. London, Ser A* **1869,** *159,* 575.
2. Van der Waals, J. D. Ph.D. Dissertation, Sijthoff, Leiden, 1873.
3. Levelt Sengers, J. M. H. *Physica (Utrecht)* **1976,** *82A,* 319.

4. Widom, B. *J. Chem. Phys.* **1965**, *43*, 3898.
5. Stanley, H. E. "Introduction to Phase Transitions and Critical Phenomena"; Oxford University Press: New York, 1971.
6. Ma, S-K. "Modern Theory of Critical Phenomena"; W. A. Benjamin, Inc.: Reading, MA, 1976.
7. Levelt Sengers, J. M. H.; Greer, W. L.; Sengers, J. V. *J. Phys. Chem. Ref. Data* **1976**, *5*, 1.
8. Levelt Sengers, J. M. H. *Proc. Symp. Thermophys. Prop.* **1977**, *7*, 774.
9. Chapela, G. A.; Rowlinson, J. S. *J. Chem. Soc., Faraday Trans. I* **1974**, *70*, 584.
10. Goodwin, R. D. *J. Res. Natl. Bur. Stand., Sect. A* **1969**, *73*, 487.
11. Verbeke, O. B. *J. Chem. Phys.* **1973**, *59*, 4763.
12. Hall, K. R.; Eubank, P. T. *Ind. Eng. Chem., Fundam.* **1976**, *15*, 80.
13. Ibid., p. 323.
14. Walton, C. W.; Mullins, J. C.; Holste, J. C.; Hall, K. R.; Eubank, P. T. *AIChE J.* **1978**, *24*, 1000.
15. Angus, S.; Armstrong, B.; de Reuck, D. M. "International Thermodynamic Tables of the Fluid State, Ethylene, 1972"; Butterworth: London, 1972.
16. Keenan, J. H.; Keyes, F. G.; Hill, P. G.; Moore, J. G. "Steam Tables"; Wiley: New York, 1969.
17. Douslin, D. R.; Harrison, R. H. *J. Chem. Eng. Data* **1977**, *22*, 74.
18. Rivkin, S. L.; Troyanovska, G. V.; Akhundov, T. S. *Teplofizika Vysokikh Temperatur* **1964**, 2(2), 219.

RECEIVED September 20, 1978.

7

Three-Parameter, Corresponding-States Conformal Solution Mixing Rules for Mixtures of Heavy and Light Hydrocarbons

T. J. LEE, L. L. LEE, and K. E. STARLING

School of Chemical Engineering and Materials Science,
University of Oklahoma, Norman, OK 73019

The conformal solution method is used as the basis for developing mixing rules for the characteristic parameters appearing in a three-parameter, corresponding-states correlation of thermodynamic properties. A logical extension of the van der Waals one-fluid mixing rules from two to three parameters is shown to yield poor predictions of vapor–liquid equilibrium for mixtures of paraffin hydrocarbons with highly dissimilar molecular sizes. Therefore, semiempirical rules were developed with improved predictive capability. The average absolute deviations of predicted methane K-values from experimental binary mixture data for methane with heavier normal paraffin hydrocarbons ranging from ethane through normal decane were 11.0% and 4.6%, respectively, using the modified van der Waals one-fluid mixing rules and the semiempirical mixing rules.

Using conformal solution theory models for the prediction of mixture thermodynamic behavior is becoming increasingly popular for industrial calculations. The attractiveness of the conformal solution approach stems largely from the fact that it is faster computationally than purely theoretical methods and yet has a sufficiently good basis in theory to allow extension to complex molecular interactions (e.g., multipole, dispersion, and steric effects), which would be difficult using purely empirical methods.

The formulation of conformal solution theory which has received the widest use to date is the so-called VDW one-fluid theory (*1*). Strictly, the VDW one-fluid theory applies to mixtures of similar size molecules

0-8412-0500-0/79/33-182-125$05.00/1

for which all pair potentials can be expressed in the form $u_{ij} = \epsilon_{ij} \phi(r_{ij}/\sigma_{ij})$. Unfortunately, for many industrial mixtures molecular size differences can be large and orientation effects make important contributions to the pair potentials. Thus, aside from the approximations inherent to conformal solution theory, factors which adversely affect the accuracy of the VDW one-fluid theory for the complex molecular systems encountered industrially include using the two parameter (ϵ_{ij} and σ_{ij}) pair potential and requiring of similar molecular sizes for the mixture components.

Efforts are in progress at the University of Oklahoma to develop a multiparameter, corresponding-states framework for correlation of thermodynamic properties, taking into account the various orientation contributions to pair interactions (e.g., dipole–dipole, quadrupole–quadrupole, dipole–quadrupole, and higher multipole effects as well as dispersion and steric effects). Preliminary research (*2*) in this direction has involved lumping the collective effects of orientation contributions into a single term in the pair potential and the resultant expressions for the thermodynamic properties from the Pople perturbation theory (*3*). This approach leads to the three-parameter, corresponding-states correlation framework reported in recent work (*2*) and used herein. The three characterization parameters in this correlation framework are the characteristic molecular size/separation parameter, σ, the characteristic molecular energy parameter, ϵ, and the characteristic orientation parameter, γ. Within this three-parameter, corresponding-states framework it is possible to derive, along the lines of the method used by Smith (*4*), a three-parameter, conformal solution model, which is presented in Section 2. In the derivation of the three-parameter conformal solution model, certain parameters (exponents) in the mixing rules for the three characterization parameters are arbitrary. Using the VDW one-fluid mixing rules for the energy and separation parameters, along with a mixing rule for the orientation parameter derived along the lines of the VDW one-fluid theory yields the so-called modified VDW one-fluid mixing rules in Section 3. The methodology for the thermodynamic properties calculations presented herein is presented in Section 4. It is shown in Section 5 that using the modified VDW one-fluid mixing rules yields accurate predictions of mixture thermodynamic behavior for mixtures of molecules with dissimilarities as great as methane and propane, but that the accuracy of prediction decays for larger molecular dissimilarities. In Section 6, the exponents in the modified VDW one-fluid mixing rules are varied empirically; the resultant mixing rules, referred to herein as semiempirical mixing rules, yield significantly improved predictions for mixtures with components as dissimilar as methane and normal decane. The implications of these results are discussed in Section 7.

Anisotropic Fluid Conformal Solution Model

The method used here for considering conformal solution models for fluids with molecular anisotropies is based on the method used by Smith (*4*) for treating isotropic one-fluid conformal solution methods as a class of perturbation methods. The objective of the method is to closely approximate the properties of a mixture by calculating the properties of a hypothetical pure reference fluid. The characterization parameters (in this case, intermolecular potential parameters) of the reference fluid are chosen to be functions of composition (i.e., mole fractions) and the characterization parameters for the various possible molecular pair interactions (like–like and unlike–unlike). In principle, all molecular anisotropies (dipole–dipole, quadrupole–quadrupole, dipole–quadrupole, and higher multipole interactions, as well as overlap and dispersion interactions) can be included in the method. Here, the various molecular anisotropies are lumped into a single term, so that the intermolecular potential energy $u_{ij}(\mathbf{r}_{12}, \omega_1, \omega_2)$ between Molecules 1 and 2 of Species i and j can be written in the form

$$u_{ij}(\mathbf{r}_{12}, \omega_1, \omega_2) = \epsilon_{ij}\,\phi^{\circ}\left(\frac{r_{12}}{\sigma_{ij}}\right) + \delta_{ij}\epsilon_{ij}\,\phi^{p}\left(\frac{\mathbf{r}_{12}}{\sigma_{ij}}, \omega_1, \omega_2\right) \tag{1}$$

In Equation 1, $\mathbf{r}_{12}$ is the vector displacement of the molecular centers of Molecules 1 and 2, r_{12} is the scaler separation, $r_{12} = |\mathbf{r}_{12}|$, and ω_1 and ω_2 are the Euler angles describing the orientations of Molecules 1 and 2. The first term on the right-hand side of Equation 1 involving ϕ° is recognized as an isotropic potential form, so that the term involving ϕ^{p} describes anisotropic effects. The characterization parameters σ_{ij}, ϵ_{ij}, and δ_{ij}, respectively, are characteristic distance, energy, and anisotropic strength parameters for the interaction between molecules of Species i and j.

The extension of the isotropic mixture conformal solution method of Smith (*4*) to the case of anisotropic molecular systems can be made easily in the following manner. The quantities a_{ij}, b_{ij}, and c_{ij} are defined by the relations $a_{ij} = \delta_{ij}^{\mathrm{k}}\epsilon_{ij}^{\mathrm{l}}\sigma_{ij}^{\mathrm{m}}$. $b_{ij} = \delta_{ij}^{\mathrm{p}}\epsilon_{ij}^{\mathrm{q}}\sigma_{ij}^{\mathrm{r}}$, $c_{ij} = \delta_{ij}^{\mathrm{u}}\epsilon_{ij}^{\mathrm{v}}\sigma_{ij}^{\mathrm{w}}$, where the exponents k, l, m, p, q, r, u, v, and w are left unspecified at this point in the development. The configurational Helmholtz free energy A for an anisotropic mixture then can be expanded about the configurational Helmholtz tree energy of a hypothetical pure reference fluid, A_x, with characterization parameters δ_x, ϵ_x, and σ_x (or a_x, b_x, c_x),

$$A = A_x + \frac{\partial A_x}{\partial a_x}\sum_i\sum_j x_i x_j (a_{ij} - a_x) + \frac{\partial A_x}{\partial b_x}\sum_i\sum_j x_i x_j (b_{ij} - b_x) + \frac{\partial A_x}{\partial c_x}\sum_i\sum_j x_i x_j (c_{ij} - c_x) + \text{higher order terms} \tag{2}$$

where x_i is the mole fraction of the ith component in the mixture. The following mixing rules annul the first order terms in the expansion in Equation 2,

$$\delta_x^{\mathrm{k}}\epsilon_x^{\mathrm{l}}\sigma_x^{\mathrm{m}} = \sum_i\sum_j x_i x_j \delta_{ij}^{\mathrm{k}}\epsilon_{ij}^{\mathrm{l}}\sigma_{ij}^{\mathrm{m}} \tag{3}$$

$$\delta_x^{\mathrm{p}}\epsilon_x^{\mathrm{q}}\sigma_x^{\mathrm{r}} = \sum_i\sum_j x_i x_j \delta_{ij}^{\mathrm{p}}\epsilon_{ij}^{\mathrm{q}}\sigma_{ij}^{\mathrm{r}} \tag{4}$$

$$\delta_x^{\mathrm{u}}\epsilon_x^{\mathrm{v}}\sigma_x^{w} = \sum_i\sum_j x_i x_j \delta_{ij}^{\mathrm{u}}\epsilon_{ij}^{\mathrm{v}}\sigma_{ij}^{\mathrm{w}} \tag{5}$$

The application of the conformal solution method in industrial calculations requires the use of the approximation $A = A_x$ to avoid the lengthy computation required to calculate the higher order terms in Equation 2. Thus, a practical strategy for choosing the exponents k, l, m, p, q, r, u, v, and w in Equations 3, 4, and 5 would be through minimization of the difference $A - A_x$ (actually, data for all available mixture thermodynamic properties can be used simultaneously to determine the exponents by regression). However, most applications of the conformal solution method have involved the use of exponents based on molecular theory and so this approach was used in the initial phases of the present work.

Modified van der Waals One-Fluid Mixing Rules

The well known VDW one-fluid mixing rules for the characterization parameters σ_x and ϵ_x for isotropic fluids are

$$\sigma_x^3 = \sum_i\sum_j x_i x_j \sigma_{ij}^3 \tag{6}$$

$$\epsilon_x\sigma_x^3 = \sum_i\sum_j x_i x_j \epsilon_{ij}\sigma_{ij}^3 \tag{7}$$

Thus, the VDW one-fluid rules correspond to the use of the following values of the exponents in Equations 3 and 4, k = 0, l = 0, m = 3, p = 0, q = 1, r = 3. Smith (5) has discussed the fact that for hard-sphere mixtures, Equation 6 is the most reasonable theoretical choice for specifying σ_x (although other mixing rules have been used). Also, Smith (5) has shown for hard-sphere binary mixtures that using the arithmetic mean rule, $\sigma_{12} = 1/2(\sigma_{11} + \sigma_{22})$, the second-order terms in Equation 2 for the Helmholtz free energy probably can be neglected only when σ_{11} and σ_{22} differ by less than about 10%. For isotropic fluids, the perturbation expansion of the Helmholtz free energy about that of a hard-sphere system leads to Equation 7 when the mean density approximation is used for the hard-sphere pair distribution function (5).

Although the VDW one-fluid mixing rules yield reasonably accurate predictions of mixture behavior for molecules which are not greatly dissimilar, the cases of evaluation of the unlike interaction parameters, σ_{ij} and ϵ_{ij}, where $i \neq j$, from the data may be compensatory in an empirical way.

For the derivation of a mixing rule for the anisotropic strength parameter, δ_x, consider the Pople expansion (*3*) of the Helmholtz free energy, A, about the free energy, A_o, of an isotropic fluid reference system

$$A = A_o + A_1 + A_2 + \ldots \tag{8}$$

where A_i are the ith order terms in the expansion. The isotropic reference system pair potential is defined as the unweighted average of the anisotropic pair potential in Equation 1, that is,

$$\epsilon_{ij}\, \phi^{\circ} \left(\frac{r_{12}}{\sigma_{ij}}\right) = \langle u_{ij}(\mathbf{r}_{12}, \omega_1, \omega_2) \rangle_\omega \tag{9}$$

where the brackets $\langle\ \rangle_\omega$ denote the angle average. Thus, $A_1 = 0$ and Equation 8 is a perturbation expansion for A if higher-order terms are small. For small anisotropies, truncation at A_2 is accurate, while for large anisotropies the use of the Padé approximant used by Stell (*6*)

$$A = A_o + \frac{A_2}{(1 - A_3/A_2)}$$

yields good results. Herein the truncation at A_2 will be used. The second-order term A_2 is given by the relation

$$A_2 = \frac{-\rho^2}{4kT} \sum_i \sum_j x_i x_j \delta_{ij}^2 \epsilon_{ij}^2 \int d\mathbf{r}_1 d\mathbf{r}_2 < (\phi_{ij}^{\mathrm{p}})^2 >_\omega g_{ij}^{\circ} \tag{10}$$

where ρ is the molecule number density, T is absolute temperature, k is Boltzmann's constant, $\mathbf{r}_1$ and $\mathbf{r}_2$ are the position vectors of Molecules 1 and2, and g_{ij}° is the isotropic pair distribution function. For the case in which ϕ_{ij}^{p} can be written as the product function

$$\phi_{ij}^{\mathrm{p}} = F_{ij} \left(\frac{r_{12}}{\sigma_{ij}}\right) D(\omega_1, \omega_2) \tag{11}$$

A_2 becomes

$$A_2 = -\frac{\pi\rho}{(kT)^2} \sum_i \sum_j x_i x_j \delta_{ij}^2 \epsilon_{ij}^2 \sigma_{ij}^3 \int dr_{12}^* r_{12}^{*2} F_{ij}^2 g_{ij}^{\circ} \langle D^2 \rangle_\omega \tag{12}$$

where $r_{12}^* = r_{12}/\sigma_{12}$. For example, if the perturbation contribution to the pair potential were the overlap potential for linear molecules, the perturbation contribution could be approximated by the following expression, owing to Pople (3),

$$\delta_{ij}\epsilon_{ij}\phi_{ij}^{\mathrm{p}} = \delta_{ij}\epsilon_{ij}\left(\frac{\sigma_{ij}}{r_{12}}\right)^{12} [3\cos^2\theta_1 - 3\cos^2\theta_2 - 2] \tag{13}$$

so that $F_{ij} = (\sigma_{ij}/r_{12})^{12}$ and $D = [3\cos^2\theta_1 - 3\cos^2\theta_2 - 2]$, where θ_1 and θ_2 are the polar angles of orientation of Molecules 1 and 2. To obtain the expression for δ_x the following approximation is introduced,

$$g_{ij}^{\circ}\left(\frac{r_{12}}{\sigma_{ij}}, \frac{\epsilon}{kT}, \rho\sigma_{ij}^3, x_1, \ldots\right) = g_x^{\circ}(r^*, \rho^*, T^*) \tag{14}$$

where $r^* = r/\sigma_x$, $\rho^* = \rho\sigma_x^3$, $T^* = kT/\epsilon_x$. The approximation in Equation 14 is similar to, but more stringent than, the mean density approximation. With the assumption in Equation 14, A_2 becomes

$$A_2 = \frac{-\pi\rho\langle D^2\rangle_\omega}{(kT)^2}\sum_i\sum_j x_i x_j \delta_{ij}^2\epsilon_{ij}^2\sigma_{ij}^3 \int \mathrm{d}r^* r^{*2}F^2 g_x^{\circ} \tag{15}$$

It is then logical to choose the following mixing rule for the anisotropic strength parameter (overlap parameter in the specific example) δ_x,

$$\delta_x^2\epsilon_x^2\sigma_x^2 = \sum_i\sum_j x_i x_j \delta_{ij}^2\epsilon_{ij}^2\sigma_{ij}^3 \tag{16}$$

This mixing rule corresponds to the use of the following values of the exponents in Equation 5, u = 2, v = 2, w = 3. The reduced Helmholtz free energy, $A^* = A/NkT$, where N is the number of molecules then takes the form

$$A^* = A_o^* - \delta_x^2\,\pi\,\langle D^2\rangle\,\rho^* J_x/(T^*)^2 \tag{17}$$

where $\rho^* = \rho\sigma_x^3$, $T^* = kT/\epsilon_x$, and J_x is the integral

$$J_x = \int \mathrm{d}r^* r^{*2}F^2 g_x^{\circ} \tag{18}$$

Note that A^* is of the form

$$A^* = A_o^* + \delta_x^2 f^*(T^*, \rho^*) \tag{19}$$

This result is identical to the expression which is obtained from the perturbation expansion of A for a pure fluid. Thus, referring to Equation 2, the first-order conformal solution relation for anisotropic fluids is

$$A^*(T, \rho, \{\sigma_{ij}\}, \{\epsilon_{ij}\}, \{\delta_{ij}\}, \{x_k\}) = A_x(T^*, \rho^*, \delta_x) \tag{20}$$

where $\{\sigma_{ij}\}$, $\{\epsilon_{ij}\}$, and $\{\delta_{ij}\}$ denote the sets of characterization parameters for the mixture constituent binary pairs, $\{x_k\}$ denotes the set of mole fractions of the mixture components, and

$$A_x^*(T^*, \rho^*, \delta_x) = A_o^*(T^*, \rho^*) + \delta_x^2 f^*(T^*, \rho^*) \tag{21}$$

with the modified VDW mixing rules for σ_x, ϵ_x, and δ_x given in Equations 6, 7, and 16.

The equation-of-state expression for the absolute pressure P is obtained from Equation 19 using the thermodynamic relation

$$\frac{P}{\rho kT} = \rho^* \left(\frac{\partial A^*}{\partial \rho^*}\right)_{N,\,T} \tag{22}$$

the resultant expression for the compressibility factor $Z = P/\rho kT$ is

$$Z = Z_o + \delta_x^2 Z_1 \tag{23}$$

where n

$$Z_o = \rho^* \left(\frac{\partial A_o^*}{\partial \rho^*}\right)_{N,\,T} \tag{24}$$

$$Z_1 = \rho^* \left(\frac{\partial f^*}{\partial \rho^*}\right)_{N,\,T} = \rho^* \left[\frac{\partial}{\partial \rho^*}\left(\frac{-\pi \langle D\rangle^2 \rho^* J_x}{T^{*2}}\right)\right]_{N,\,T} \tag{25}$$

Calculation of Thermodynamic Properties

For the calculation of thermodynamic properties, Equation 23 was used in an empirical manner. Only data for nonpolar normal paraffin hydrocarbon systems were used in the correlation development so that as an approximation, the Pitzer acentric factor, ω, could be taken as an estimate of the collective strength of molecular anisotropies (i.e., $\delta^2 = \omega$). Because the use of the resultant correlation for polar systems was anticipated, the parameter γ ($\gamma = \delta^2$), referred to herein as the orientation parameter, was used instead of the acentric factor ($\gamma \neq \omega$ for other fluids). The equation of state in Equation 23 then takes the form

$$Z(T^*, \rho^*, \gamma) = Z_o(T^*, \rho^*) + \gamma Z_1(T^*, \rho^*) \tag{26}$$

where Z is the compressibility factor and Z_o and Z_1 are functions of the reduced temperature $T^* = kT/\epsilon$ and reduced density $\rho^* = \rho\sigma^3$.

The equation-of-state form used herein is the modified Benedict–Webb–Rubin (MBWR) equation as given by Han and Starling (7). It is reformulated into the form of Equation 26 by expressing the constants appearing linearly in the equation into two parts—one isotropic part and one anisotropic part,

$$B_i = a_i + \gamma b_i \tag{27}$$

a_i being the isotropic part and b_i being the anisotropic part, where as noted above, $\gamma \simeq \delta^2$ is an orientation parameter accounting for the nonsphericity of the molecule-pair potentials under consideration. Therefore, the MBWR equation corresponding to Equation 26 assumes the form

$$Z = 1 + \rho^*[B_1 - B_2T^{*-1} - B_3T^{*-3} + B_9T^{*-4} - B_{11}T^{*-5}] + \rho^{*2}[B_5 - B_6T^{*-1} - B_{10}T^{*-2}] + \rho^{*5}[B_7T^{*-1} + B_{12}T^{*-2}] + B_8\rho^{*2}T^{*-3}[(1 + B_4\rho^{*2})\exp(-B_4\rho^{*2})] \quad (28)$$

where b_4 in Equation 27 is zero to insure linearity of Z in γ, ρ^* is the reduced density, $\rho^* = \rho\sigma^3$, and T* is the reduced temperature, $T^* = kT/\epsilon$. The characteristic molecular distance parameter, σ, and energy parameter, ϵ, were estimated from the critical constants using the relations

$$\sigma^3 = \frac{0.3189}{\rho_c} \quad (29)$$

$$\epsilon = \frac{kT_c}{1.2593} \quad (30)$$

where k is the Boltzmann constant. Pertinent relations for other thermodynamic properties have been presented elsewhere (*2*). Equations 29 and 30 are based on the relationships of the Lennard–Jones (12–6) potential parameters for argon to the argon critical constants. The use of Equations 29 and 30 in the MBWR equation of state given in Equations 27 and 28 works well for pure normal paraffin hydrocarbons. The universal constants a_i and b_i, $i = 1, \ldots 12$ ($b_4 = 0$) were determined by simultaneously using density, vapor pressure, and enthalpy departure data for methane through normal decane in multiproperty analysis. Average absolute deviations of predicted from experimental properties were 1.00% for density, 1.13 Btu/lb for enthalpy and 0.85% for vapor

Table I. Generalization Parameters of Pure Materials to Be Used with Generalized Equation of State

	Critical Temp. (°F)	*Critical Density (lb–mol/cu ft)*	*Molecular Weight*	*Orientation Parameter (γ)*
Methane	−116.43	0.6274	16.042	0.01289
Ethane	90.03	0.4218	30.068	0.09623
Propane	206.13	0.3121	44.094	0.1538
n-Butane	305.67	0.2448	58.12	0.1991
n-Pentane	385.42	0.2007	72.146	0.2530
n-Hexane	453.45	0.1696	86.172	0.3054
n-Heptane	512.85	0.1465	100.198	0.3499
n-Octane	563.79	0.1284	114.224	0.4004
n-Nonane	610.50	0.1150	128.24	0.4463
n-Decane	651.90	0.1037	142.276	0.4880

pressure. Thus, the multiparameter, corresponding-states correlation framework provided by the perturbation equation form in Equation 26 and the resultant generalized MBWR equation in Equation 28 yields good results for the pure normal paraffin hydrocarbons. Values of the critical constants and orientation parameters used in this work are given in Table I, while the values of the constants a_i and b_i in Equation 27 are given in Table II.

Table II. Generalized Parameters Used in the MBWR Equation

Parameter i	$B_i = a_i + \gamma b_i$	
	a_i	b_i
1	1.45907	0.32872
2	4.98813	−2.64399
3	2.20704	11.3293
4	4.86121	
5	4.59311	2.79979
6	5.06707	10.3901
7	11.4871	10.3730
8	9.22469	20.5388
9	0.094624	2.76010
10	1.48858	−3.11349
11	0.015273	0.18915
12	3.51486	0.94260

Use of the Modified van der Waals One-Fluid Rules

The modified VDW one-fluid mixing rules for σ_x, ϵ_x and δ_x in Equations 6, 7, and 16 were used to determine the ability of this formulation of the conformal solution model for predicting mixture behavior. The following relations were used for σ_{ij}, ϵ_{ij}, and δ_{ij} where $i \neq j$,

$$\sigma_{ij} = \xi_{ij}(\sigma_{ii}\sigma_{jj})^{1/2} \quad (31)$$

$$\epsilon_{ij} = \zeta_{ij}(\epsilon_{ii}\epsilon_{jj})^{1/2} \quad (32)$$

$$\gamma_{ij} = \phi_{ij}(\gamma_i + \gamma_j)^{1/2} \quad (33)$$

where ξ_{ij}, ζ_{ij}, and ϕ_{ij} are binary interaction parameters to be determined from binary mixture thermodynamic property data. There was little loss in accuracy of prediction when ϕ_{ij} was fixed at unity; therefore $\phi_{ij} = 1$ was used for the calculations discussed herein. Values of the parameters ξ_{ij} and ζ_{ij} determined from available binary density, enthalpy, and vapor–liquid equilibrium data for methane with heavier hydrocarbons are given in Table III. Table IV presents a summary of the deviations of predicted densities and methane *K*-values (equilibrium ratio of vapor-to-liquid mole fractions). Deviations of predicted heavy component *K*-values from experimental data were not used to evaluate the accuracy of prediction

Table III. Binary Interaction Parameters for Methane (First Component) with Heavier Hydrocarbons

Second Component	*Data Ref.*	*Modified VDW One-Fluid Rules*		*Semiempirical Exponent Rules*	
		ξ_{12}	ξ_{12}	ζ_{12}	ζ_{12}
Ethane	*8*	0.999079	0.996810	0.999925	0.979586
Propane	*8*	1.02116	0.974404	1.01188	0.936840
n-Butane	*9*	1.03946	0.958079	1.02559	0.899345
n-Pentane	*10*	1.05214	0.936798	1.03220	0.860984
n-Hexane	*11*	1.07738	0.920368	1.05049	0.832570
n-Heptane	*12*	1.08744	0.921744	1.06234	0.816646
n-Nonane	*13*	1.09674	0.937876	1.07753	0.799090
n-Decane	*14*	1.11940	0.978290	1.08519	0.790355

because the vapor-phase mole fraction of the heavy component often is so small that the measurement error is extremely large on a percentage basis. The trend which can be noted in Table IV occurs because properties are predicted with reasonable accuracy for the methane–ethane and methane–propane systems, but there is a decay in the accuracy of prediction for the mixtures of methane with normal butane and heavier components. This trend would be anticipated by virtue of the approximations made herein to develop the multiparameter, corresponding-states–conformal solution formulas. The major approximations of concern are (1) the second-order truncation of the Pople expansion; (2) the lumping of the collective effects of molecular anisotropies into a single term, characterized by a single orientation parameter, γ; (3) the first-order truncation of the conformal-solution expansion of the Helmholtz free energy; and (4) the choices made for the exponents in the mixing rules for the reference system characterization parameters σ_x, ϵ_x, and δ_x. Because of the success of the formulation for predicting pure-fluid properties, even

Table IV. Summary of Deviations of Predicted Binary Mixture Densities and Methane *K*-Values from Experimental Data

Second Component with Methane	*Average Absolute Deviations, %*			
	Modified VDW One-Fluid Rules		*Semiempirical Exponent Rules*	
	Densities	*Densities*	*K-values*	*K-values*
Ethane	2.20	1.14	1.96	1.00
Propane	0.94	1.14	1.06	0.90
n-Butane	2.65	8.10	2.17	4.04
n-Pentane	2.12	9.61	1.54	4.62
n-Hexane	—	17.9	—	7.31
n-Heptane	3.57	13.6	3.31	11.2
n-Nonane	1.41	16.1	2.73	2.15
n-Decane	4.34	20.5	5.68	5.55

as heavy as normal decane, the first two approximations appear adequate for practical, industrially oriented correlations such as that used herein. Although the third approximation has been shown to be poor for binary mixtures of hard-sphere molecules with large size differences, the use of second-order conformal solution method introduces additional computational requirements which would slow practical calculations, especially multicomponent vapor–liquid equilibrium predictions. For these reasons, the fourth approximation was focused on and a first alternative to the modified VDW one-fluid rules used above was considered.

Semiempirical Exponent Mixing Rules

To determine if a significant level of improvement in predicted mixture properties over the VDW one-fluid mixing rules is possible, the nine exponents in the general mixing rules for σ_x, ϵ_x, and δ_x in Equations 3, 4, and 5 could be determined empirically. However, all contact with the VDW one-fluid formulas might be lost by such an approach. Therefore, the exponents k, l, and p were fixed at zero and nonlinear regression was performed to determine the remaining exponents, starting the nonlinear regression with the VDW one-fluid values for the remaining exponents, i.e. m = 3, q = 1, r = 3, u = 2, v = 2, and w = 3. Vapor–liquid equilibrium data for the eight binary systems in Table IV were used to determine the revised exponent values. The optimal values of the exponents are m = 4.5255, q = 1.0, r = 4.4271, u = 2.0, v = 0.0, and w = 3.4959. Rounding off these exponents yields the following semiempirical mixing rules,

$$\sigma_x^{4.5} = \sum_i \sum_j x_i x_j \sigma_{ij}^{4.5} \quad (34)$$

$$\epsilon_x \sigma_x^{4.5} = \sum_i \sum_j x_i x_j \epsilon_{ij} \sigma_{ij}^{4.5} \quad (35)$$

$$\delta_x^2 \sigma_x^{3.5} = \sum_i \sum_j x_i x_j \delta_{ij}^2 \sigma_{ij}^{3.5} \quad (36)$$

Table V. Summary of Deviations of Predicted Vapor–Liquid-Phase Compositions with Experimental Data (*8*) for the System Methane–Ethane–Propane (Subscripts 1, 2, and 3, Respectively)

No. Data Points: 33
Temperature Range: −176– −76°F
Pressure Range: 32–800 psia

Mixing Rules	*Average Absolute Deviations in Mole Fractions (%)*					
	x_1	x_2	x_3	y_1	y_2	y_3
[a]	2.92	3.34	3.77	1.09	7.16	13.67
[b]	2.72	3.67	4.01	1.29	8.61	16.77

[a] Modified VDW one-fluid mixing rules.
[b] Semiempirical mixing rules.

Table VI. Comparison of Predicted and Experimental (*15*) Methane–Ethane–Propane-*n*-Butane (Subscripts

T (°F)	P (psia)		x_1	x_2
−60	204	Expt.	0.16	0.188
		Calc.[a]	0.145	0.188
		Calc.[b]	0.152	0.189
−60	288	Expt.	0.234	0.178
		Calc.[a]	0.212	0.178
		Calc.[b]	0.221	0.178

[a] Modified VDW one-fluid mixing rules.

The binary interaction parameters for use with the semiempirical mixing rules are given in Table III. Summaries of deviations of predicted properties from experimental values using these semiempirical mixing rules are given for binary systems in Table IV. The improvement in vapor–liquid predictions is significant. The average absolute deviation of predicted methane *K*-values from experimental data for the semiempirical mixing rules is 4.6% compared with 11.0% for the modified

Table VII. Comparison of Predicted and Experimental (*16*) Methane–Ethane–Propane–*n*-Pentane–*n*-Hexane–*n*-Decane

	x_1	x_2	x_3	x_4	x_5
$T = 609.67$°R[a]; $P = 100$ psia					
Expt.	0.0266	0.0025	0.0019	0.2023	0.2004
Calc.[b]	0.0206	0.0059	0.0031	0.2040	0.1983
Calc.[c]	0.0225	0.0086	0.0038	0.2076	0.1973
$T = 609.67$°R; $P = 3000$ psia					
Expt.	0.5753	0.0409	0.0272	0.0509	0.0616
Calc.[b]	0.5547	0.0490	0.0297	0.0526	0.0626
Calc.[c]	0.5264	0.0561	0.0328	0.0561	0.0664
$T = 709.$°R; $P = 100$ psia					
Expt.	0.0177	0.0016	0.0014	0.1694	0.1785
Calc.[b]	0.0122	0.0033	0.0022	0.1800	0.1869
Calc.[c]	0.0161	0.0053	0.0029	0.1900	0.1882
$T = 709.7$°R; $P = 1000$ psia					
Expt.	0.2157	0.0232	0.0182	0.1265	0.1428
Calc.[b]	0.1603	0.0302	0.0212	0.1359	0.1511
Calc.[c]	0.1934	0.0366	0.0230	0.1323	0.1443
$T = 709.7$°R; $P = 2000$ psia					
Expt.	0.4122	0.0342	0.0235	0.0947	0.0993
Calc.[b]	0.3123	0.0428	0.0286	0.1138	0.1170
Calc.[c]	0.3539	0.0481	0.0297	0.1084	0.1094

[a] °R stands for degrees Rankine.
[b] Modified VDW one-fluid mixing rules.

Vapor–Liquid Equilibrium Mole Fractions for the System 1, 2, 3, and 4, Respectively)

x_3	x_4	y_1	y_2	y_3	y_4
0.583	0.069	0.852	0.08	0.063	0.005
0.594	0.072	0.880	0.077	0.041	0.0008
0.586	0.071	0.884	0.074	0.040	0.0007
0.527	0.061	0.874	0.060	0.059	0.007
0.543	0.066	0.910	0.057	0.032	0.006
0.535	0.065	0.913	0.055	0.031	0.006

[b] Semiempirical mixing rules.

VDW one-fluid rules. Comparisons of predicted and experimental vapor–liquid equilibrium for ternary and multicomponent systems are given in Tables V, VI, and VII, for both the semiempirical and VDW one-fluid mixing rules. In these calculations, the unlike interaction parameters for interactions of ethane and heavier components with each other were taken to be unity. This is a reasonable approximation for the unlike interaction parameters for the heavier components; for the interaction of ethane and

Vapor–Liquid Equilibrium Mole Fractions for the System (Subscripts 1, 2, 3, 4, 5, and 6, Respectively)

x_6	y_1	y_2	y_3	y_4	y_5	y_6
0.5703	0.8712	0.0216	0.0062	0.0701	0.0269	0.004
0.5680	0.8802	0.0182	0.0049	0.0670	0.0275	0.0019
0.5599	0.8910	0.0155	0.0042	0.0612	0.0260	0.0018
0.2441	0.8585	0.0414	0.0183	0.0179	0.0172	0.0466
0.2511	0.9148	0.0315	0.0142	0.0121	0.0110	0.0162
0.2620	0.9272	0.0246	0.0116	0.0104	0.00955	0.0165
0.6314	0.5986	0.0231	0.0090	0.2103	0.1207	0.0382
0.6152	0.6374	0.0223	0.0085	0.2008	0.1080	0.0227
0.5972	0.6566	0.0207	0.0079	0.1896	0.1034	0.0216
0.4737	0.8797	0.0383	0.0165	0.0337	0.0211	0.0105
0.5010	0.9011	0.0311	0.0136	0.0291	0.0189	0.00598
0.4701	0.9148	0.0247	0.0112	0.0259	0.0173	0.00592
0.3362	0.8510	0.0415	0.0192	0.0363	0.0262	0.0259
0.3853	0.9011	0.0336	0.0152	0.0246	0.0168	0.0084
0.3501	0.9146	0.0273	0.0127	0.0216	0.0150	0.0085

[c] Semiempirical mixing rules.

perhaps propane with heavier components, the determination of unlike interaction parameters from binary data should yield improved results. From inspection of Tables IV, V, VI, and VII, it is obvious that for systems containing components heavier than propane, vapor–liquid equilibrium predictions are more accurate using the semiempirical mixing rules rather than the VDW one-fluid mixing rules. Thus, from the point of view of practical industrial computations, the semiempirical mixing rules are recommended.

It is difficult to ascertain the reasons for the magnitude of improvement in vapor–liquid predictions using the semiempirical mixing rules instead of the modified VDW one-fluid rules. The semiempirical rules probably offset the truncation error in the approximation $A = A_x$ to some extent. It certainly is interesting that the orientation parameter $\gamma_x = \delta_x^2$ is independent of the set of characteristic energy parameters, $\{\epsilon_{ij}\}$, for the mixture. This possibly indicates for normal paraffin hydrocarbons (and perhaps similar nonpolar fluids) that contributions of molecular attraction and orientation effects are essentially independent, but that steric and size effects are dependent.

Conclusion

In this study the modified VDW conformal solution method based on the three-parameter, corresponding-states correlation of pure fluid thermodynamic properties yields accurate mixture property predictions if the components are not greatly dissimilar. However, there is a progressive decay in prediction accuracy as molecular dissimilarities increase. This study found that if the VDW one-fluid mixing rule exponents are modified empirically (i.e., to noninteger values), the resulting semiempirical mixing rules yield significant improvement in vapor–liquid equilibrium predictions for mixtures of molecules as dissimilar as methane and normal decane.

The study presented herein has a number of implications. First, this study implies that it is possible to obtain accurate predictions of the thermodynamic behavior of mixtures within a multiparameter, corresponding-states framework using empirically determined exponents for the characterization parameters of the reference system in a first-order truncation of the conformal solution method expression for the Helmholtz free energy. This result is important to the continuing effort to develop a highly accurate multiparameter, corresponding-states framework for correlation of fluid properties, and to the industrial use of such a correlation. Second, this study demonstrates that there is a need to study separately rather than collectively (as herein) the errors introduced by the various major approximations introduced into the correlation methodology. These approximations include: (1) the choice for the form of the pair potential; (2) the method for estimation of the pure-fluid pair potential parameters; (3) the order and method (e.g., use of the Padé approximant) of truncation of the Pople expansion of the thermodynamic properties; (4) the order of truncation of the expansion of mixture properties about the properties of the pure-fluid reference system in the conformal solution methodology; (5) the method for choosing the mixing rules for the reference system characterization parameters as functions of composition and the pair parameters for the molecular interactions of the components; and (6) the method for determination of the unlike pair parameters.

With a better understanding of the errors introduced by these approximations, the development of a more truly comprehensive correlation, capable of describing fluid systems with wide ranges of charcateristics over wide ranges of conditions, should be possible.

Glossary of Symbols

A = Helmholtz free energy
A_o = reference system Helmholtz free energy
A_i = the ith order term in perturbation of A
A_x = Helmholtz free energy of a hypothetical pure reference fluid
A^* = reduced Helmholtz free energy = A/NkT
a_i = constant in expression for B_i, $i = 1, \ldots 12$
a_x, b_x, c_x = parameters in Equation 2
a_{ij} = $\delta_{ij}^{k} \epsilon_{ij}^{l} \sigma_{ij}^{m}$
B_i = coefficients of MBWR equation, $B_i \equiv a_i + \gamma b_i$, $i = 1, \ldots 12$
b_i = constant in expression for B_i, $i = 1, \ldots 12$
b_{ij} = $\delta_{ij}^{p} \epsilon_{ij}^{q} \sigma_{ij}^{r}$
c_{ij} = $\delta_{ij}^{u} \epsilon_{ij}^{v} \sigma_{ij}^{w}$
D = angle dependent part of an isotropic potential in Equation 11
F_{ij} = distance dependent part of an isotropic potential in Equation 11
f^* = function (*see* Equation 21)
g_{ij}° = radial distribution function of the reference system with pair potential U_{ij}°
g_x° = radial distribution function of a hypothetical pure reference fluid with pair potential U_{ij}°
J_x = integral defined in Equation 18
k = Boltzmann constant

k, l, m, p, q, r, u, v, w = exponents in Equations 3, 4, and 5
N = number of molecules in system
P = absolute pressure
P_c = critical pressure
$\mathbf{r}_1$ = position vector of Molecule 1
r_{12} = distance between molecular centers
r^* = reduced distance, $r^* \equiv r/\sigma_x$
T = absolute temperature
T_c = critical temperature
T^* = reduced temperature, $T^* \equiv kT/\epsilon_x$
u_{ij} = intermolecular pair potential
u_{ij}° = spherically symmetric part of the interaction potential
x_i = mole fraction of Component i
Z = compressibility factor of fluid
Z_0 = reference fluid compressibility factor
Z_1 = perturbation contribution to compressibility factor

Greek Symbols

γ = orientation parameter
γ_{ii} = orientation parameter for Component i
γ_{ij} = interaction orientation parameter species for i and j
γ_x = orientation parameter obtained from conformal solution theory mixing rule
δ = overlap potential parameter
δ_{ij} = unlike interaction overlap potential parameter
δ_x = mixture reference overlap parameter
ϵ = characteristic molecular energy parameter
ϵ_{ii} = ϵ for Component i
ϵ_{ij} = interaction parameter for characteristic molecular energy parameter ϵ between species i and j
ϵ_x = mixture reference systems energy parameter
ζ_{ij} = unlike-pair separation parameter coefficient for Species i and j
θ_1 = polar angles of orientation for Molecule 1
ξ_{ij} = unlike-pair energy parameter coefficient for i and j species
ρ = molecule number density
ρ^* = reduced number density $\rho^* \equiv \rho\sigma_x^3$
ρ_c = critical density
σ = characteristic molecular distance parameter

σ_{ii} = σ for Component i

σ_{ij} = interaction parameter for characteristic molecular distance parameter, σ between substance i and j

σ_x = mixture reference system molecular distance parameter

ϕ° = isotropic part of potential

ϕ^{p} = anisotropic part of potential

ω = acentric factor

ω_i = orientation of Molecule i

Superscript

$*$ = reduced form

Acknowledgment

The support of the Gas Research Institute and the American Gas Association through Project BR-111-1 is acknowledged. Important phases of this work were supported by the University of Oklahoma; in particular, the extensive use of the computer facilities is acknowledged.

Literature Cited

1. Leland, T. W.; Rowlinson, J. S.; Sather, G. A. *Trans. Faraday Soc.* **1968,** *64,* 1447.
2. Lee, L. L.; Starling, K. E.; Mo, K. C. "Proceedings of the Seventh Symposium on Thermophysical Properties," American Society of Mechanical Engineers, New York, 1977, p. 451.
3. Pople, J. A. *Proc. R. Soc. London, Ser. A* **1954,** *221,* 498.
4. Smith, W. R. *Can. J. Chem. Eng.* **1972,** *50,* 271.
5. Smith, W. R. *Mol. Phys.* **1971,** *21,* 105.
6. Stell, G.; Rasaiah, J. C.; Narang, H. *Mol. Phys.* **1972,** *23,* 393.
7. Starling, K. E. "Fluid Thermodynamic Properties for Light Petroleum Systems"; Gulf Publication Company: Houston, 1973.
8. Wichterle, I.; Kobayashi, R. "Low Temperature Vapor–Liquid Equilibria in the Methane–Ethane–Propane Ternary and Associated Binary Methane Systems with Special Consideration of the Equilibria in the Vicinity of the Critical Temperature of Methane"; Rice University Monograph: Houston, 1970.
9. Kahre, L. C. *J. Chem. Eng. Data* **1973,** *18,* 267.
10. Sage, B. H.; Lacey, W. N. "Thermodynamic Properties of the Lighter Paraffin Hydrocarbons and Nitrogen"; American Petroleum Institute: New York, 1950.
11. Chu, T. C.; Chen, R. J. J.; Chappelear, P. S.; Kobayashi, R. *J. Chem. Eng. Data* **1976,** *21,* 41.
12. Chang, H. L.; Lewis, J. H.; Kobayashi, R. *AIChE J.* **1966,** *12,* 1212.
13. Shipman, L. M.; Kohn, J. *J. Chem. Eng. Data* **1966,** *11,* 176.
14. Reamer, H. H.; Olds, R. H.; Sage, B. H.; Lacey, W. N. *Ind. Eng. Chem.* **1942,** *34,* 1526.
15. DePriester, C. L. *Chem. Eng. Prog. Symp. Ser.* **1953,** *49*(7), 1–43.
16. Vairogs, J.; Klekers, J.; Edmister, W. C. *AIChE J.* **1971,** *17,* 308.

RECEIVED August 14, 1978.

8

Second Virial Cross-Coefficients: Correlation and Prediction of k_{ij}

CONSTANTINE TSONOPOULOS

Exxon Research and Engineering Company, Florham Park, NJ 07932

The prediction of the second virial cross-coefficient B_{ij} *depends most sensitively on the mixing rule for the characteristic critical constant* T_{cij} *(Equation 5). The characteristic binary constant,* k_{ij}*, can be determined by fitting* B_{ij} *and other data. The prediction of* B_{ij}*, therefore, requires knowledge of* k_{ij}*. The* k_{ij} *values are available for binaries of inorganic gases and hydrocarbons with up to 8 carbon atoms. Extensions in the data base and correlations for* k_{ij}*'s are presented here in two areas. First, recent* B_{ij} *data for binaries involving* C_{10}*–*C_{30} *hydrocarbons are reduced to* k_{ij}*'s, which are shown to be both meaningful and satisfactory in the calculation of* B_{ij} *over the entire carbon-number range. A similar approach is taken in the second area investigated, binaries involving polar compounds. New* k_{ij} *data are used to establish trends and develop preliminary* k_{ij} *correlations so that* B_{ij} *and, more importantly, fugacity coefficients can be predicted reliably.*

In the analysis and correlation of vapor–liquid equilibrium (VLE) data it is essential, especially at superatmospheric pressures, to take into account the effect of vapor-phase nonideality. This is expressed by the fugacity coefficient which, as long as the density of the mixture is not greater than one fourth of its critical value, can be calculated reliably with the following equation (for a binary mixture):

$$\ln \phi_i = \frac{2}{V_m} (y_i B_{ii} + y_j B_{ij}) - \ln z_m \tag{1}$$

where ϕ_i, y_i, and B_{ii} are the fugacity coefficient, vapor mole fraction, and second virial coefficient of Component i, while v_m and z_m are the molar

0-8412-0500-0/79/33-182-143$05.00/1

volume and compressibility factor of the vapor mixture. It is the prediction of B_{ij}, the second virial cross-coefficient, that will primarily concern us.

B_{ij} will be predicted with an empirical correlation that was proposed in 1974 (*1*). The original correlation was developed primarily for oxygenated polar compounds, but it has been extended to haloalkanes (*2*) and water pollutants (*3*). Refs. *1*, *2*, and *3* can be consulted for details on the development of the correlation and the results.

Nonpolar Mixtures

For nonpolar compounds and mixtures, the correlation is a modified form of the Pitzer–Curl relationship (*see* Ref. *1*):

$$\frac{BP_c}{RT_c} = f^{(0)}(T_R) + \omega f^{(1)}(T_R) \tag{2}$$

$$f^{(0)}(T_R) = 0.1445 - 0.330/T_R - 0.1385/T_R^2 - 0.0121/T_R^3 - 0.000607/T_R^8 \tag{3}$$

$$f^{(1)}(T_R) = 0.0637 + 0.331/T_R^2 - 0.423/T_R^3 - 0.008/T_R^8 \tag{4}$$

The same equations are used for B_{ij}, but with characteristic parameters $P_{c_{ij}}$, $T_{c_{ij}}$, and ω_{ij} in place of the pure-component constants. The former are related to the latter through the following mixing rules:

$$T_{c_{ij}} = (T_{c_i} T_{c_j})^{1/2}(1 - k_{ij}) \tag{5}$$

$$P_{c_{ij}} = \frac{4\,T_{c_{ij}}(P_{c_i}V_{c_i}/T_{c_i} + P_{c_j}V_{c_j}/T_{c_j})}{(V_{c_i}^{1/3} + V_{c_j}^{1/3})^3} \tag{6}$$

$$\omega_{ij} = 0.5(\omega_i + \omega_j) \tag{7}$$

The most sensitive mixing rule is Equation 5. A characteristic constant for each binary, k_{ij}, expresses the deviation from the geometric mean for $T_{c_{ij}}$, which applies (that is, $k_{ij} = 0$) only when i and j are very similar in size and chemical nature. Otherwise, in the absence of any strong specific chemical interaction between i and j, k_{ij} should be positive and thus $T_{c_{ij}}$ would be less than the geometric mean.

An extensive tabulation of k_{ij}'s for nonpolar systems was reported by Chueh and Prausnitz in 1967 (*4*) and, with a few additions, in 1968 (*5*). The k_{ij}'s presented by Chueh and Prausnitz are primarily for binaries of hydrocarbons with up to 8 carbon atoms—along with a few naphthalene systems. There are also data for binaries of CO_2, H_2S, N_2, Ar, etc. Most of them were obtained from experimental information on the second virial coefficient or the saturated liquid volume of binary systems, although the resulting k_{ij}'s then were used by Chueh and Prausnitz in their calculations with a modified form of the Redlich–Kwong (RK) equation of state. What this says (and it is an important point) is that once the value of k_{ij} is determined for an i/j binary, by fitting (reliable) B_{ij} or other mixture data, it can be used in the prediction of any other mixture property. That is, under ideal conditions, k_{ij} is a true constant that can be used with any equation of state or correlation that uses $T_{c_{ij}}$.

Another benefit drawn from using k_{ij} to correct for the deviation of $T_{c_{ij}}$ from the geometric mean is that it also can be used to predict ϵ_{ij}, the characteristic energy parameter for a binary molecular interaction. Since ϵ is proportional to T_c, it follows that

$$\epsilon_{ij} = (\epsilon_i \, \epsilon_j)^{1/2} \, (1 - k_{ij}) \tag{8}$$

The more theoretically oriented investigators probably consider the k_{ij} in Equation 8 to be physically more significant than that in Equation 5. However, the two should be very similar numerically—especially if we consider that the uncertainty in the determination of k_{ij} is at least $\pm$ 0.01 units.

Hiza and Duncan (*6*) have examined the deviation of ϵ_{ij} from the geometric mean primarily for cryogenic systems. They also developed a correlation for the prediction of k_{ij} for such systems that will be discussed later.

The k_{ij} tabulations of Hiza and Duncan and of Chueh and Prausnitz are adequate for nonpolar systems involving inorganic gases and hydrocarbons with perhaps up to 8 carbon atoms. However, for process calculations on heavy hydrocarbons, coal liquids, etc., k_{ij} must be known or predicted for binaries of hydrocarbons with carbon number (much) greater than 10.

Available B_{ij} data on binaries involving hydrocarbons with up to 30 carbon atoms were fitted with Equations 2–7 to determine optimum values for the characteristic constant k_{ij}. These results, which will be examined below, have also helped establish trends and even suggest ways of correlating and predicting k_{ij}'s for hydrocarbon binaries. Before the new information is reviewed, therefore, a few comments are appropriate on what is known and what has been done in attempting to correlate k_{ij} with pure-component properties.

A Review of Correlations for k$_{ij}$. Justification for using the geometric mean for ϵ_{ij}—and for $T_{c_{ij}}$—is provided by London's theory of dispersion forces. When London's expression for the intermolecular energy is equated to the attractive part of the Lennard–Jones (LJ) 12:6 potential, the following relationship obtains:

$$k_{ij} = 1 - \left[\frac{2\,(I_i I_j)^{1/2}}{I_i + I_j}\right]\left[\frac{2\,(\sigma_i \sigma_j)^{1/2}}{\sigma_i + \sigma_j}\right]^6 \tag{9a}$$

where I_i is the first ionization potential and σ_i is the collision diameter of i. If the latter is replaced by the critical volume, Equation 9a assumes the form

$$k_{ij} = 1 - \left[\frac{2\,(I_i I_j)^{1/2}}{I_i + I_j}\right]\left[\frac{2\,(V_{c_i} V_{c_j})^{1/6}}{V_{c_i}{}^{1/3} + V_{c_j}{}^{1/3}}\right]^6 \tag{9b}$$

Equation 9 may have some theoretical basis, but at best it provides only rough estimates for the k_{ij} parameter. Indeed, better results are sometimes obtained when the exponent of the second bracketed quantity is changed to 3. For paraffin–paraffin mixtures only, Chueh and Prausnitz (*7*) found that

$$k_{ij} = 1 - \left[\frac{2\,(V_{c_i} V_{c_j})^{1/6}}{V_{c_i}{}^{1/3} + V_{c_j}{}^{1/3}}\right]^3 \tag{10}$$

gives satisfactory results. (Even for the methane–*n*-decane binary, the first bracketed quantity in Equation 9 is equal to 0.992 and hence can be ignored.) As Teja (*8*) has shown, Equation 10 follows from Berthelot's mixing rule for the van der Waals (VDW) a_{ij} parameter: $a_{ij} = (a_i a_j)^{1/2}$.

The inadequacy of Equation 9 may result from the fact that its derivation ignored the contribution of the repulsive forces. Taking these forces into account, however, would make the theoretical treatment of the problem formidable, and therefore Hiza and Duncan (*6*) opted for an empirical approach. Their analysis of data for binaries of H_2, He, and Ne with light hydrocarbons and Ar suggested the use of I as the correlating parameter for k_{ij} in the following form:

$$k_{ij} = 0.17\,(I_i - I_j)^{1/2} \ln \frac{I_i}{I_j} \tag{11}$$

where i is the component with the larger ionization potential. With the exception of the O_2 binaries, Equation 11 satisfies binaries of inorganic gases, methane, ethane, and ethylene. Agreement with experimental values was within ±0.04. More recently, it was shown by H.–M. Lin that Equation 11 (with 0.18 in place of 0.17) gives a good representation of the data for rare-gas mixtures (*9*).

Equations 10 (for paraffin–paraffin mixtures) and 11 (for mixtures of inorganic gases, CH_4, C_2H_6, and C_2H_4) are probably the only reasonably reliable correlations available in the literature for the prediction of k_{ij} for nonpolar mixtures.

Finally, Fender and Halsey (*10*) proposed a geometric–arithmetic mean that follows, by drastic simplification, from the Kirkwood–Müller formula for ϵ_{ij}:

$$\epsilon_{ij} \cong \frac{2\,\epsilon_i\,\epsilon_j}{\epsilon_i + \epsilon_j}$$

This relationship, which gives very good results for Ar–Kr, leads to the following expression for k_{ij}:

$$k_{ij} = 1 - \frac{2\,(\epsilon_i\,\epsilon_j)^{1/2}}{\epsilon_i + \epsilon_j} \qquad (12a)$$

$$= 1 - \frac{2\,(T_{c_i}\,T_{c_j})^{1/2}}{T_{c_i} + T_{c_j}} \qquad (12b)$$

Equation 12 will be examined in the next section.

The k_{ij}'s for Methane–Hydrocarbon Binaries. As noted earlier, most of the k_{ij} values available in the literature are for binaries of inorganic gases and hydrocarbons with up to 8 carbon atoms. New determinations of optimum k_{ij} values for methane–hydrocarbon binaries are listed in Table I. These and Chueh's results make a total of 26 k_{ij} values (for 25 binaries; two different k_{ij}'s were included for the naphthalene binary) over the carbon-number range of 2 to 30 for *j*. This is the most

Table I. Optimum k_{ij} Values for Methane–Hydrocarbon Binaries

j	*Average Deviation of* B_{ij} *(cm*^{3}*/gmol)*	k_{ij}	t *Range (°C) (No. Points)*	*References*
Benzene	0.8	0.09	50 (1)	*11*
2,2,5-Trimethylhexane	3	0.19	25–100 (4)	*12*
n-Decane	13	0.16	50–125 (4)	*12*
tert-Butylbenzene	6	0.21	50–125 (4)	*12*
Naphthalene	12	0.13	21– 68 (6)	*13*
1-Methylnaphthalene	6	0.24	75–175 (3)	*14*
n-Dodecane	4	0.15	75–150 (4)	*12*
Bicyclohexyl	27	0.20	50–170 (4)	*14*
Diphenylmethane	17	0.12	65–170 (4)	*14*
Phenanthrene	11	0.18	40–138 (7)	*15*
Anthracene	8	0.19	66–185 (7)	*16*
n-Hexadecane	21	0.22	75–175 (5)	*14*
n-Eicosane	8	0.21	165–270 (4)	*14*
Squalane[a]	22–25	0.33 ± 0.02	230, 272 (2)	*14*

[a] 2,6,10,15,19,23-Hexamethyltetracosane.

extensive set of k_{ij}'s available for a given Component i, and therefore was selected to check the various relationships presented in the previous section for correlating k_{ij}.

As expected, Equation 10 worked well for the 14 paraffin binaries; the average deviation of the predicted k_{ij} from the optimum values was only 0.02. However, when all 26 hydrocarbon binaries were included, the average deviation increased to 0.05.

Equation 11 was much less satisfactory, even when the coefficient was adjusted to minimize the deviation. The optimum value of the coefficient was 0.18, in agreement with Lin's conclusion for rare-gas mixtures, but the root-mean-square deviation of k_{ij} was 0.08. Any relationship involving I, the first ionization potential, is doomed to failure when applied to heavy hydrocarbon mixtures because I, for a given homologous series, is very weakly dependent on carbon number. For example, the I of n-decane is 10.19 eV, only 0.24 eV less than that of n-hexane, while that of n-eicosane should be about 10.04 eV. Thus, a correlation of k_{ij}'s for methane–paraffin binaries based solely on ionization potentials would give the same result for all $C_{10}+$ paraffins.

Equation 12b works surprisingly well. Although it is inferior to Equation 11 when applied to the paraffin binaries (average deviation of 0.03 vs. 0.02), it comes out better when all hydrocarbon binaries are considered (average deviation of 0.04 vs. 0.05). Equation 12b is also satisfactory for the binaries of methane with Ar, Kr, N_2, and H_2S (average deviation of 0.01 vs. 0.03 with Equation 10), but is very poor for methane–H_2.

Both the critical volume and the critical temperature are therefore satisfactory correlating parameters for k_{ij}. However, the best correlating parameter for all 26 methane–hydrocarbon k_{ij} values was the carbon number of the hydrocarbon. This is shown in Figure 1, where all the data have been plotted along with the simple relationship

$$k_{ij} = 0.0279\,[\ln\,(n_{c_j})\,]^2 \tag{13}$$

where n_{c_j} is the carbon number of the hydrocarbon. The average deviation of the k_{ij} predicted with Equation 13 is only 0.02—for all 26 points. (The largest deviations from the correlation in Figure 1 are for both paraffins and aromatics.) Thus, Equation 13 provides the best correlation for the effect of the hydrocarbon carbon number on k_{ij}. The same approach will be used now with other binaries for which extensive data are available.

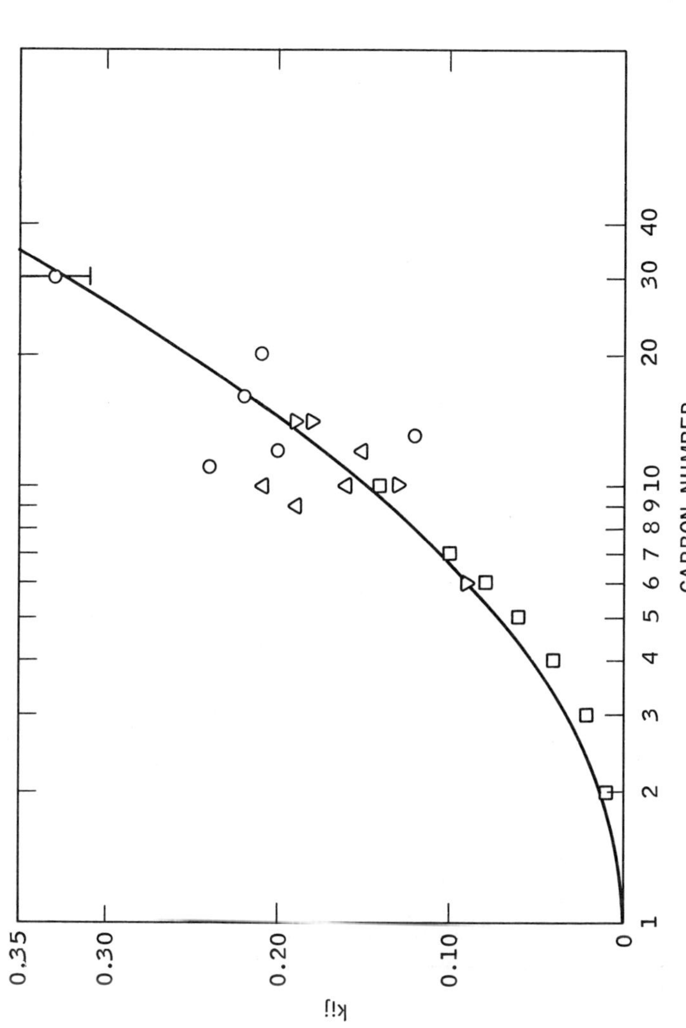

Figure 1. Optimum k_{ij} *values for methane–hydrocarbon binaries: (——), Equation 13; (○) Kaul and Prausnitz (1978); (△), D'Avila et al. (1976); (□), results of Chueh and Prausnitz (1967); (▽), King et al. (1966–1970).*

The k_{ij}'s for Other Hydrocarbon Binaries. Table II presents optimum k_{ij} values for heavy hydrocarbon binaries of ethylene and ethane. The two sets of k_{ij}'s are very similar, and this is shown more clearly in Figure 2, where the results of Chueh and Prausnitz (*4*) have been plotted also. Chueh and Prausnitz actually found that there was no significant difference in the k_{ij}'s between ethylene and ethane—or between n-hexane and benzene.

The line in Figure 2 was calculated with a more general form of Equation 13:

$$k_{ij} = m[ln\ (n_{c_j} - n_{c_i} + 1)]^2 \tag{14}$$

(This form ensures that $k_{ij} = 0$ when $n_{c_j} = n_{c_i}$; it is implied that $n_{c_j} \geq n_{c_i}$.) For the ethylene and ethane binaries, the value of m is 0.0202. A substantially better fit of the data is possible if k_{ij} is assumed to be linearly dependent on carbon number:

$$k_{ij} = 0.0117\ (n_{c_j} - n_{c_i} - 1);\quad n_{c_j} > n_{c_i} \tag{15}$$

Teja (*8*) recommended Equation 15 (with 0.01 as the coefficient) for the paraffin binaries of ethane.

Table II. Optimum k_{ij} Values for Ethylene and Ethane Binaries

j	*Average Deviation of* B_{ij} *(cm³/gmol)*	k_{ij}	t *Range (°C) (No. Points)*	*References*
i = Ethylene				
Naphthalene	8	0.09	23–69 (9)	*13, 17*
1-Methylnaphthalene	37	0.10	75–175 (3)	*14*
Bicyclohexyl	19	0.14	50–170 (3)	*14*
Diphenylmethane	21	0.12	65–175 (3)	*14*
Phenanthrene	26	0.15	37–142 (7)	*15*
Anthracene	11	0.15	65–180 (7)	*16*
n-Hexadecane	9	0.15	75–175 (3)	*14*
i = Ethane				
Naphthalene	8	0.12	26– 67 (7)	*18*
1-Methylnaphthalene	24	0.08	75–175 (3)	*14*
Bicyclohexyl	23	0.10	50–170 (3)	*14*
Diphenylmethane	18	0.12	65–175 (3)	*14*
Anthracene	8	0.16	63–175 (6)	*16*
n-Hexadecane	18	0.12	75–175 (3)	*14*
n-Eicosane	6	0.19	165–270 (4)	*14*
Squalane	21	0.29	230, 272 (2)	*14*

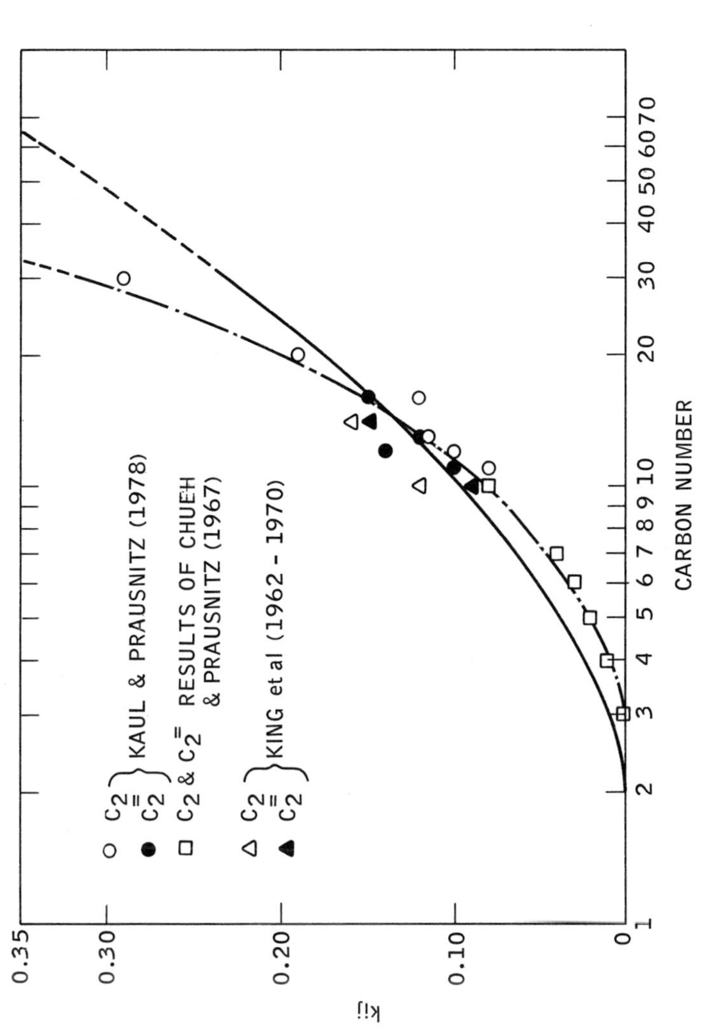

Figure 2. *Optimum* k_{ij} *values for ethylene– and ethane–hydrocarbon binaries:* (— · —), *Equation 15;* (———), *Equation 14 with* m = 0.0202.

The similarity between ethylene and ethane or even benzene and n-hexane no longer holds when some specific chemical interaction takes place between i and j. A good example of this is provided by the acetylene binaries, which have substantially higher k_{ij}'s than the ethylene or ethane binaries. Furthermore, the k_{ij} for acetylene–ethylene is 0.02 units lower than that for acetylene–ethane (*4*).

The k_{ij} values for N_2–hydrocarbon binaries are listed in Table III and plotted on Figure 3—along with the results of Chueh and Prausnitz (*4*). As shown in Figure 3, Equation 14 (with $m = 0.0364$) gives a satisfactory fit of the data. That is, Equation 14, with $n_{c_i} = 0$, applies to inorganic–hydrocarbon binaries as well. Indeed, it appears to fit reasonably well the k_{ij}'s for the hydrocarbon binaries of H_2—the most important inorganic gas in the heavy hydrocarbon and synthetic fuels processes—and even of Ar. The results of new k_{ij} determinations for the H_2 and Ar binaries are included in Table III, and all the values have been plotted on Figure 4.

The similarity of the k_{ij}'s for the binaries of N_2, H_2, and Ar naturally suggests that the three must have similar values for a property that

Table III. Optimum k_{ij} Values for Inorganic–Hydrocarbon Binaries

j	*Average Deviation of* B_{ij} *(cm³/gmol)*	k_{ij}	t *Range (°C) (No. Points)*	*References*
$i = N_2$				
Benzene	2	0.135 ± 0.015	35– 50 (5)	*11*[a]
2,2,5-Trimethylhexane	2	0.24	25–100 (4)	*12*
n-Decane	5	0.18	50–125 (4)	*12*
tert-Butylbenzene	3	0.25	50–125 (4)	*12*
Naphthalene	3	0.19	22, 72 (2)	*17*
n-Dodecane	4	0.20	75–150 (4)	*12*
$i = H_2$				
Benzene	—	0.175 ± 0.045	50 (2)	*11*[a]
Naphthalene	4	0.18	22, 70 (2)	*17*
$i =$ Ar				
Benzene	1	0.21	25 (1)	*19*
Benzene	6	0.14	22, 50 (2)	*11*
Naphthalene	7	0.23	24–74 (3)	*17, 19*
Anthracene	1	0.25	75 (1)	*19*

[a] Optimum k_{ij} value also reflects data of Gainey and Young (N_2) and of Everett et al. (N_2, H_2) (*11*).

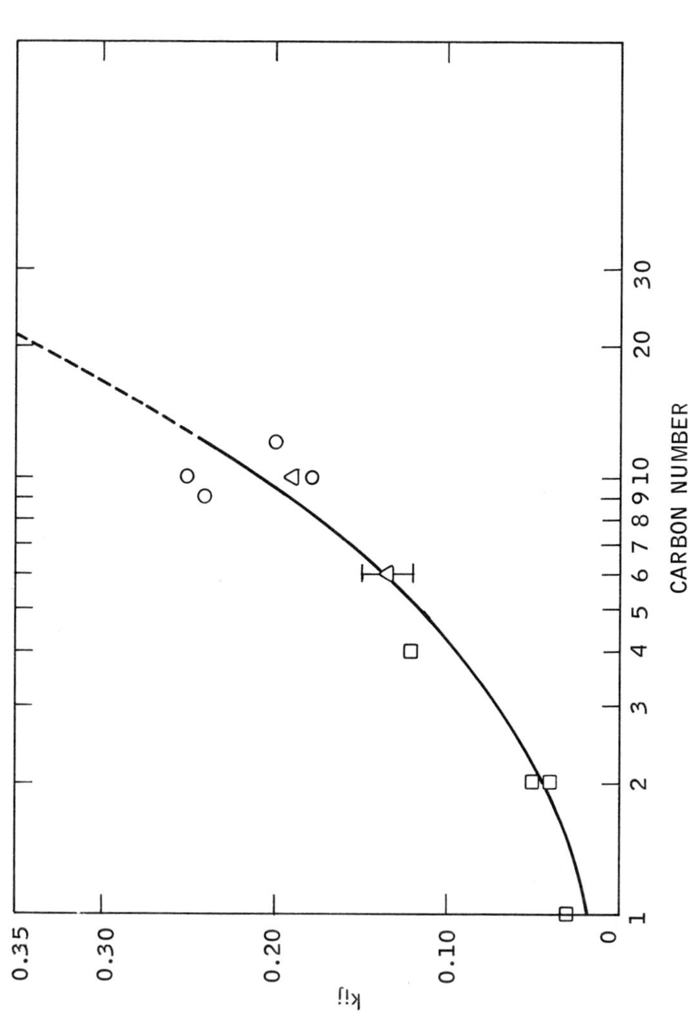

Figure 3. Optimum k_{ij} *values for* N_2*–hydrocarbon binaries:* (——), *Equation 14 with* m = 0.0364; (○), D'Avila *et al.* (1976); (□), *results of Chueh and Prausnitz* (1967); (△), King *et al.* (1962, 1969).

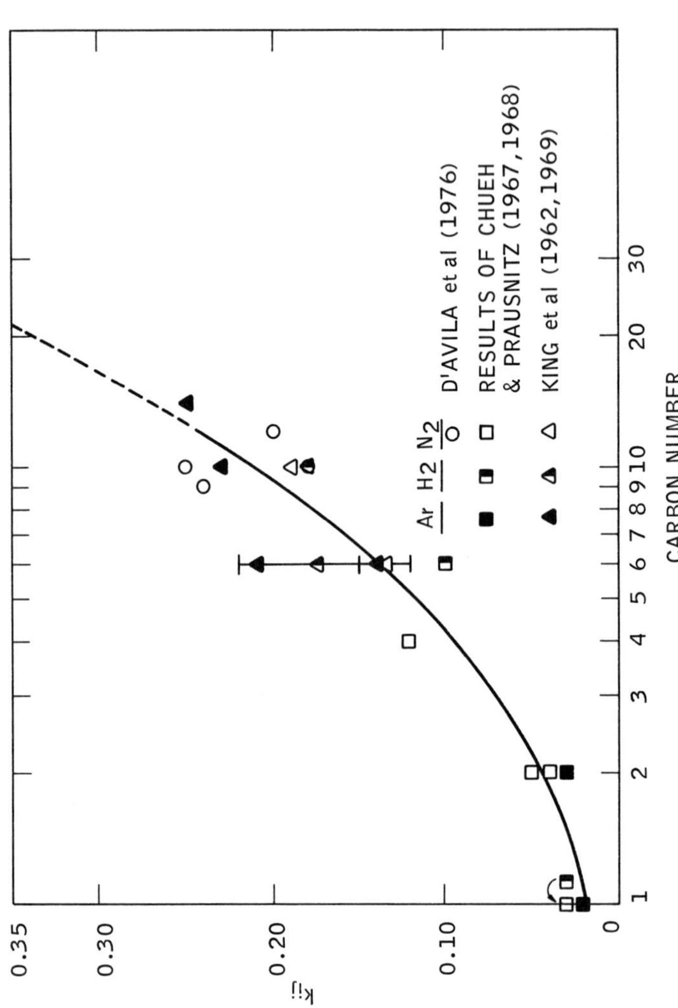

Figure 4. Optimum k_{ij} *values for* N_2*–,* H_2*–, and Ar–hydrocarbon binaries: (——), Equation 14 with* $m = 0.0364$.

affects k_{ij}. Their critical volumes are only roughly similar (51.5, is the "classical" constant for H_2 used in B_{ij} calculations (*3*)):

	V_c *(cm³/gmol)*
N_2	89
H_2	51.5
Ar	75

but their first ionization potentials are nearly exactly equal (*20*)

	I, *eV*
N_2	15.5
H_2	15.4
Ar	15.4

The first ionization potential may prove useful in sorting out and ordering the data for binaries of inorganic compoounds, but it is certainly of little use in the prediction of k_{ij}'s for hydrocarbon–hydrocarbon binaries. For example, the I of *n*-pentane (10.55 eV) is very similar to that of ethylene (10.51 eV), but the k_{ij}'s for the *n*-pentane binaries are markedly lower than those for the corresponding ethylene binaries (*4*). On the other hand, the ethylene and ethane binaries have similar k_{ij}'s (Figure 2), even though the I of ethane is 1.25 eV higher than for ethylene (*20*).

Concluding Remarks on Nonpolar Mixtures. Plotting k_{ij}, for a given Component i against the carbon number of the hydrocarbon j provides a convenient way for checking individual values and trends. However, this should not be taken to imply that the carbon number is a good correlating parameter for all cases. For hydrocarbon binaries of components not considered here (or in Refs. *4, 5*, and *6*), the k_{ij} still may be predicted with either Equation 10 (if both components are paraffins) or 12b (for all others). However, if m can be correlated with some pure-component property (perhaps V_c, although I works better for inorganics), Equation 14 probably will prove superior to Equations 10 and 12b for all hydrocarbon binaries.

Actually, Equation 15 may prove to be the best relationship for hydrocarbon–hydrocarbon binaries with $n_{c_i} \geq 2$ (and $n_{c_j} > n_{c_i}$). It certainly works very well for $n_{c_i} = 2$, but there are insufficient data for $n_{c_i} > 2$ to determine the optimum form of the relationship and the value(s) of the coefficient.

Figures 1–4 show that the extension of the k_{ij} approach to heavy hydrocarbons is reasonable and more or less straightforward. For the moderately dense vapor mixtures considered here, there is continuity between low- and high-molecular-weight systems.

Although k_{ij} apparently retains its usefulness up to very high carbon numbers, the description of the properties of liquid mixtures of $C_{10}+$ hydrocarbons is not completely satisfactory when only k_{ij} is used. An additional parameter is required, most likely as a correction for the deviation of σ_{ij} from the arithmetic mean (or $V_{c_{ij}}$ from the Lorentz or cube-root mean). Such a correction has been used even for cryogenic mixtures (*9*), but it should be more meaningful—and important—for mixtures of heavy hydrocarbons. If the value of k_{ij} has been fixed already by fitting B_{ij} or some other property of the vapor mixture, then the analysis of VLE data will permit the unique determination of the second parameter.

Polar Mixtures

The polar mixtures should be divided into polar–nonpolar and polar–polar systems. Very little can be said about the latter because the data are limited and the polar–polar interactions are generally too system-specific to allow any generalizations. Certainly, the various correlations already examined for the nonpolar mixtures cannot be expected to work for polar–polar binaries. Indeed, they are unsatisfactory even for most polar–nonpolar binaries.

The k_{ij}'s for a few polar–polar binaries have been presented in Refs. *1* and *2*. In addition, Table VI of Ref. *1* presents average k_{ij} values that can be used, at least as a rough guide, for mixtures comprised of ketones, ethers, alcohols, and water.

Polar–Nonpolar Binaries. The B_{ij} for polar–nonpolar binaries is assumed to have no polar term (*1*), and therefore the calculation is carried out exactly as for nonpolar binaries; that is, with Equations 2–7.

Average values for the k_{ij} of polar–hydrocarbon binaries were given in 1974, in the first paper on the new correlation (*1*). They were as follows

i (j = *hydrocarbon)*	k_{ij}
ketones	0.13
ethers	0.10
alcohols	0.15
water	0.40

In addition, the data analysis in Ref. *2* suggested that $k_{ij} \cong 0.05$ for haloalkane–hydrocarbon binaries.

New B_{ij} data on binaries of ethanol (*21*), 1-butanol, and ethyl ether (*22*) have been reduced to k_{ij} values, as summarized in Table IV. The 1974 predictions are supported to a large extent by the new information.

Table IV. Optimum k_{ij} Values for Polar–Nonpolar Binaries[a]

j	k_{ij}	t *Range (°C) (No. Points)*
i = Ethanol		
H_2	0.16	25–75 (3)
Ar	0.17	25–75 (3)
Methane	0.15	25–75 (3)
Ethylene	0.15	25 (1)
Ethane	0.17	25–75 (3)
N_2O	0.12	25–75 (3)
i = 1-Butanol		
N_2	0.24	25 (1)
Ar	0.24	25 (1)
Methane	0.20	25 (1)
Ethane	0.17	25 (1)
i = Ethyl Ether		
N_2	0.22	25 (1)
Ar	0.19	25 (1)
Methane	0.12	25 (1)
Ethane	0.06	25 (1)

[a] Ethanol binaries: Ref. *21*. 1-Butanol and ethyl ether binaries: Ref. *22*.

However, the more extensive data on the alcohol binaries (from Table IV and from Ref. *1*) suggest certain trends in the k_{ij} values that warrant further examination.

The average k_{ij} value for the ten alcohol–hydrocarbon binaries is 0.16 ± 0.03, which is in good agreement with the 1974 value. Indeed, if the eight alcohol–inorganic binaries also are included, the average becomes 0.15 ± 0.04. However, it appears that k_{ij} increases in going from methanol to ethanol and then to 1-butanol.

The apparent trends in the k_{ij} values for the alcohol–nonpolar binaries are shown in Figure 5, where k_{ij} has been plotted vs. V_{c_j}, the critical volume of the nonpolar component. (Plotting k_{ij} vs. V_{c_i}/V_{c_j} did not bring the data close together or even suggest more definite trends.) At a given V_{c_j}, k_{ij} definitely increases with the carbon number of the alcohol. The k_{ij} also increases as V_{c_j} increases, at least for the methanol and ethanol binaries, although a limiting k_{ij} is approached at $V_{c_j} \gtrsim 300$ cm^3/gmol. This limiting value is about 0.22, close to the average for all four 1-butanol binaries (*see* Table IV and Figure 5).

Also plotted on Figure 5 are the k_{ij}'s for the water binaries. If a trend similar to that for methanol is assumed, the limiting k_{ij} value is about 0.40 (in agreement with the 1974 recommendation for water–hydrocarbon binaries).

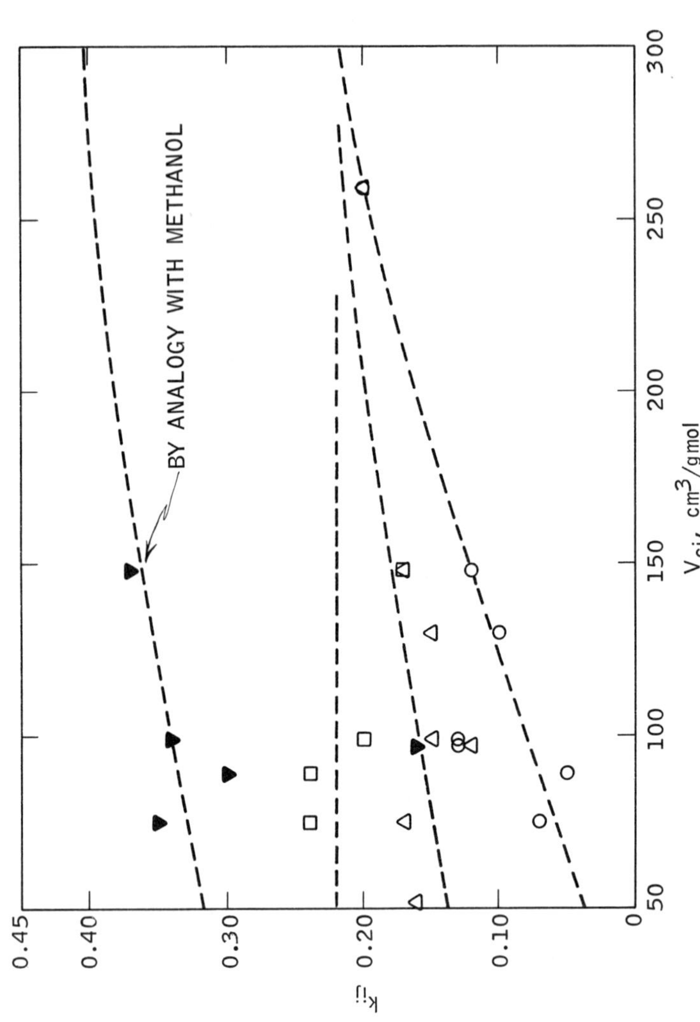

Figure 5. Optimum k_{ij} *values for nonpolar binaries of alcohols and water:* (○), *methanol;* (△), *ethanol;* (□), *1-butanol;* (▼), *water.*

The dashed lines in Figure 5 should not be given undue significance. They only present apparent trends and may prove useful in providing rough estimates for other alcohol–nonpolar or water–nonpolar binaries for which B_{ij} (and, hence, k_{ij}) data are not available. More information is needed to confirm and extend the usefulness of the trends indicated in Figure 5.

The more limited data for the hydrocarbon binaries of ethyl ether and ketones are relatively insensitive to V_{c_j}. In both cases, the present recommendations are essentially those made in 1974: 0.09 ± 0.02 for the ethyl ether–hydrocarbon and 0.13 ± 0.03 for the ketone–hydrocarbon binaries. However, it should be noted that the ethyl ether–inorganic binaries have a k_{ij} closer to 0.2 (*see* Table IV), while the k_{ij} for acetone–benzene is temperature-dependent because of complex formation (*see* Ref. *1*).

CO_2 Binaries. The final group of binaries that will be examined is that of CO_2 with nonpolar and polar compounds. The thermodynamic properties of CO_2 differ significantly from those of similar nonpolar compounds, and this is the result of its very strong quadrupole moment. This effect is seen more clearly in the behavior of CO_2 in mixtures. King and his students have reported B_{ij} data for several CO_2 binaries; the resulting k_{ij}'s have been in Table V.

The k_{ij} values in Table V belong in two groups. The first one is the hydrocarbon and CCl_4 binaries, for which the k_{ij} is higher than what would be expected for the propane binaries. (Propane is the nonpolar homomorph of CO_2; the k_{ij}'s of the propane binaries can be estimated with Equation 15.) However, the k_{ij}'s for the second group, the binaries with the polar compounds, are significantly lower than those for the corresponding propane binaries.

Table V. Optimum k_{ij} Values for CO_2 Binaries

j	k_{ij}	t *Range (°C) (No. Points)*	*References*
Naphthalene	0.20[a]	24– 73 (9)	*13*
Phenanthrene	0.23	39–141 (7)	*15*
Anthracene	0.24	65–176 (6)	*16*
CCl_4	0.20	25–75 (3)	*23*[b]
Methanol	0.01	15–60 (5)	*1, 24*
Ethanol	0.07	25–75 (3)	*21*
1-Butanol	0.09	25 (1)	*22*
Ethyl ether	−0.11	25 (1)	*22*
Water	0.14	25–100 (4)	*1, 25*

[a] B_{ij} data definitely suggest that k_{ij} is temperature-dependent (*see* text).
[b] Optimum k_{ij} values have also been determined for the following nonpolar binaries of CCl_4: 0.22 (H_2); 0.16 (N_2); 0.15 (Ar); 0.09 (methane); 0.08 (ethylene).

The apparent increase in the k_{ij} values for the aromatic hydrocarbon and CCl_4 binaries is most probably caused by the effect of the quadrupole moment on the properties of CO_2. That is, the strong quadrupole moment of CO_2 raises the values of all the k_{ij}'s for its nonpolar binaries. (The same is also true of the acetylene binaries.) However, the k_{ij}'s have a more complex behavior in the case of naphthalene–CO_2, where a definite dependence of k_{ij} on temperature was found (*13*):

t (°C)	k_{ij}
25	0.16
50	0.20
75	0.24

This is consistent with the observation of Najour and King (*13*) that "a relatively stable gas-phase, charge-transfer complex exists between naphthalene and CO_2, with naphthalene being the donor and CO_2 the acceptor." The effect of the complex weakens as the temperature is increased. Apparently, the complex reaction must be weaker in binaries of CO_2 with the other hydrocarbons and CCl_4 because their optimum k_{ij}'s are temperature-independent.

The behavior of the CO_2–polar binaries becomes much clearer when compared with the corresponding propane binaries. The k_{ij}'s for the two groups are as follows:

	CO_2	*Propane* (see *Figure 5*)	*Difference*
Methanol	0.01	0.16	−0.15
Ethanol	0.07	0.20	−0.13
1-Butanol	0.09	0.22	−0.13
Ethyl ether	−0.11	0.09 (average)	−0.20
Water	0.14	0.38	−0.24

The lowering of k_{ij} is essentially the same for all three alcohols, much larger for water, and surprisingly large for ethyl ether. As King and his students have pointed out, these effects are caused by specific chemical interaction between CO_2 and the polar molecules. However, the negative k_{ij} for CO_2–ethyl ether is not easily explainable. Negative k_{ij}'s normally indicate the existence of a strong complex, probably formed by hydrogen bonding; an ether is incapable of forming a hydrogen bond with CO_2.

Conclusions on Polar–Nonpolar Mixtures. The recommendations given in Ref. *1* for polar–hydrocarbon binaries are generally still valid. With the new k_{ij}'s reported here for alcohol–nonpolar binaries, however, it is possible to develop a correlation for the nonpolar binaries of water as well as for alcohols. This tentative correlation, which relates k_{ij} to V_{c_j} (j is the nonpolar component), is presented in Figure 5.

Finally, for the special case of the polar–CO_2 binaries, the k_{ij}'s are given by:

$$k_{ij}\ (\text{polar–CO}_2) = k_{ij}\ (\text{polar–propane}) - \Delta k_{ij}$$

For alcohols, $\Delta k_{ij} = 0.14 \pm 0.01$, while for ethers Δk_{ij} may be as high as 0.20.

Glossary of Symbols

Notation

a = parameter in the VDW equation of state
B = second virial coefficient
$f^{(0)}, f^{(1)}$ = dimensionless terms of Equation 2
I = first ionization potential
k_{ij} = characteristic binary constant; *see* Equation 5
n_c = carbon number
P = pressure
R = gas constant
$t(T)$ = temperature (absolute)
V = molar volume
y_i = vapor mole fraction of Component *i*
z = compressibility factor

Greek Letters

ϵ = energy parameter
σ = collision diameter
ϕ_i = fugacity coefficient of Component *i*
ω = acentric factor

Subscripts

c = critical property
i, *j* = property of Component *i*, *j*
ij = characteristic property used in the calculation of the second virial cross-coefficient
M = mixture property
R = reduced property

Acknowledgment

The author is grateful to Exxon Research & Engineering Company for the permission to publish this work.

Literature Cited

1. Tsonopoulos, C. *AIChE J.* **1974,** *20,* 263.
2. *Ibid.* **1975,** *21,* 827.
3. *Ibid.* **1978,** *24,* 1112.
4. Chueh, P. L.; Prausnitz, J. M. *Ind. Eng. Chem. Fundam.* **1967,** *6,* 492.
5. Prausnitz, J. M.; Chueh, P. L. "Computer Calculations for High-Pressure Vapor–Liquid Equilibria," Prentice–Hall: Englewood-Cliffs, NJ, 1968.
6. Hiza, M. J.; Duncan, A. G. *AIChE J.* **1970,** *16,* 733.
7. Chueh, P. L.; Prausnitz, J. M. *AIChE J.* **1967,** *13,* 1099.
8. Teja, A. S. *Chem. Eng. Sci.* **1978,** *33,* 609.
9. Robinson, R. L., Jr.; Hiza, M. J. *Adv. Cryog. Eng.* **1975,** *20,* 218.
10. Fender, B. E. F.; Halsey, G. D., Jr. *J. Chem. Phys.* **1962,** *36,* 1881.
11. Coan, C. R.; King, A. D., Jr. *J. Chromatogr.* **1969,** *44,* 429.
12. D'Avila, S. G.; Kaul, B. K.; Prausnitz, J. M. *J. Chem. Eng. Data* **1976,** *21,* 488.
13. Najour, G. C.; King, A. D., Jr. *J. Chem. Phys.* **1966,** *45,* 1915.
14. Kaul, B. K.; Prausnitz, J. M. *AIChE J.* **1978,** *24,* 223.
15. Bradley, H., Jr.; King, A. D., Jr. *J. Chem. Phys.* **1970,** *52,* 2851.
16. Najour, G. C.; King, A. D., Jr. *J. Chem. Phys.* **1970,** *52,* 5206.
17. King, A. D., J r.; Robertson, W. W. *J. Chem. Phys.* **1962,** *37,* 1453.
18. King, A. D., Jr. *J. Chem. Phys.* **1968,** *49,* 4083.
19. Bradley, H., Jr.; King, A. D., Jr. *J. Chem. Phys.* **1967,** *47,* 1189.
20. Landolt-Börnstein, "Zahlenwerte und Funktionen"; 6th ed.; Springer–Verlag: Berlin, 1951; Vol. 1, Part 3.
21. Gupta, S. K.; Leslie, R. D.; King, A. D., Jr. *J. Phys. Chem.* **1973,** *77,* 2011.
22. Massoudi, R.; King, A. D., Jr. *J. Phys. Chem.* **1973,** *77,* 2016.
23. Gupta, S. K.; King, A. D., Jr. *Can. J. Chem.* **1972,** *50,* 660.
24. Hemmaplardh, B.; King, A. D., Jr. *J. Phys. Chem.* **1972,** *76,* 2170.
25. Coan, C. R.; King, A. D., Jr. *J. Am. Chem. Soc.* **1971,** *93,* 1857.

RECEIVED October 5, 1978.

9

A New Equation of State

HIDEZUMI SUGIE

Nagoya Institute of Technology, Gokiso, Showa, Nagoya, Japan

BENJAMIN C.-Y. LU

University of Ottawa, Ottawa, Ontario, Canada

A new pressure-explicit equation of state suitable for calculating gas and liquid properties of nonpolar compounds was proposed. In its development, the conditions at the critical point and the Maxwell relationship at saturation were met, and PVT *data of carbon dioxide and Pitzer's table were used as guides for evaluating the values of the parameters. Furthermore, the parameters were generalized. Therefore, for pure compounds, only* T_c, P_c, *and* ω *were required for the calculation. The proposed equation successfully predicted the compressibility factors, the liquid fugacity coefficients, and the enthalpy departures for several arbitrarily chosen pure compounds.*

The purpose of this investigation is to develop a new pressure-explicit equation of state which: (1) yields acceptable values of Z_c, and satisfies the usual two initial pressure–volume derivatives at the critical point and the Maxwell relationship at saturation (equal fugacity for coexisting liquid and vapor phases); (2) is suitable for representing *PVT* behavior of liquid and gas phases over a wide range of temperature and pressure; and (3) can be integrated and differentiated easily for obtaining derived thermodynamic properties. It is anticipated that the parameters of the resulting equation can be generalized in terms of the critical properties and the acentric factor ω. The application is limited, however, to pure nonpolar compounds.

0-8412-0500-0/79/33-182-163$05.50/1

Development of the Proposed Equation of State

A suitable equation of state must satisfy certain limiting conditions and follow some general trends. One of the more important conditions is that the equation of state must reduce at low pressures and at all temperatures to the ideal gas equation. Hence

$$Z = \frac{PV}{RT} = 1 \tag{1}$$

as $P \to 0$ and $V \to \infty$. Expressing P in terms of a power series of $1/V$, the following equation was obtained:

$$P = \frac{h_1}{V} + \frac{h_2}{V^2} + \ldots + \frac{h_n}{V^n} \tag{2}$$

in which h_1 must be equal to RT.

Another condition is that the volume of all gases at high pressures approaches a limiting value which is "practically independent of the temperature and close to 0.26 V_c," as suggested by Redlich and Kwong (*1*). Keeping this in mind, the development of the new equation of state began using the following expression:

$$P = \frac{RT}{V - \mathrm{b}} + \mathrm{f}(T, V) \tag{3}$$

where

$$b = 0.26\, V_c \tag{4}$$

If $\mathrm{f}(T, V)$ could be represented by a group of terms such as that expressed in Equation 5,

$$\mathrm{f}(T, V) = \frac{h(T)}{V^{n_0} (V + \mathrm{k}_1)^{n_1} (V + \mathrm{k}_2)^{n_2} \ldots (V + \mathrm{k}_m)^{n_m}} + \ldots \tag{5}$$

the differentiation of Equation 3 and the integration of the thermodynamic expressions for evaluating fugacities and enthalpy departures, as represented by Equations 6 and 7, would be simplified.

$$\ln \phi = Z - 1 - \ln Z - \int_{\infty}^{V} (Z - 1) \frac{dV}{V} \tag{6}$$

$$(H^p - H^\circ)_T = PV - RT - \int_{\infty}^{V} \left[P - T \left(\frac{\partial P}{\partial T} \right)_v \right]_T dV \tag{7}$$

In Equation 5, $k_1 \ldots\ldots k_m$ are constants, while $n_o \ldots\ldots n_m$ are either zero or positive integers.

At the critical point, the critical isotherm shows a point of inflection. Hence

$$\left(\frac{\partial P}{\partial V}\right)_{T_c} = 0 \tag{8}$$

and

$$\left(\frac{\partial^2 P}{\partial V^2}\right)_{T_c} = 0 \tag{9}$$

The relationship between Z_c and ω may be expressed as follows:

$$Z_c = 0.291 - 0.080\omega \tag{10}$$

Expressing Equation 3 for the critical isotherm yields

$$P = \frac{RT_c}{V - b} + f(T_c, V) \tag{11}$$

which also must satisfy Equations 8 through 10 at the critical point. Consequently, it would be more convenient for mathematical manipulation if $f(T_c, V)$ was expressed in terms of three constants which could be determined by Equations 8 through 10.

It was proposed that $f(T_c, V)$ be represented by three truncated expressions of Equation 5, with each containing three terms in the denominator. In addition, the k's were assumed to be either $-b$, zero, or $+b$. Consequently,

$$f(T_c, V) = \frac{a(T_c)}{(V-b)^{n_1}V^{n_2}(V+b)^{n_3}} + \frac{C(T_c)}{(V-b)^{n_4}V^{n_5}(V+b)^{n_6}} + \frac{d(T_c)}{(V-b)^{n_7}V^{n_8}(V+b)^{n_9}} \tag{12}$$

The quantities $a(T_c)$, $c(T_c)$, and $d(T_c)$ were determined from Equations 8 through 10. The n's were taken to be either zero or positive integers. In addition, the following relationship between the n's must be satisfied:

$$2 \leqslant n_1 + n_2 + n_3 < n_4 + n_5 + n_6 < n_7 + n_8 + n_9 \tag{13}$$

However, the second virial coefficient derived from Equation 11 would not be temperature dependent if $n_1 + n_2 + n_3 \geqslant 3$. Consequently, in Equation 13,

$$n_1 + n_2 + n_3 = 2 \tag{14}$$

and

$$3 \leqslant n_4 + n_5 + n_6 < n_7 + n_8 + n_9 \tag{15}$$

Equation 11 then was used to fit the critical isotherm (i.e., the curve relating P and V at T_c) of carbon dioxide, as reported by Michels (2), using various values of n's under the conditions of Equations 14 and 15 and the values of $a(T_c)$, $c(T_c)$, and $d(T_c)$ determined from Equations 8 through 10. The best fit was obtained using the following values: $n_1 = 0$; $n_2 = 1$; $n_3 = 1$; $n_4 = 1$; $n_5 = 2$; $n_6 = 1$; $n_7 = 0$; $n_8 = 0$; and $n_9 = 7$. Hence, Equation 11 took the following form:

$$P = \frac{RT_c}{V - b} - \frac{a(T_c)}{V(V + b)} + \frac{c(T_c)}{(V - b)\, V^2\, (V + b)} - \frac{d(T_c)}{(V + b)^7} \tag{16}$$

The first two terms of the right-hand side of Equation 16 are in the same form as the well-known Redlich–Kwong (RK) equation of state (*1*).

The acentric factor of carbon dioxide is 0.225. In order to extend the applicability of Equation 16 to a wider range of ω $(0 < \omega < 0.5)$, the $(V + b)$ terms of Equation 16 were modified as expressed in Equation 17.

$$P = \frac{RT_c}{V - b} - \frac{a(T_c)}{V(V + b_1)} + \frac{c(T_c)}{(V - b)\, V^2\, (V + b_2)} - \frac{d(T_c)}{(V + b_3)^7} \tag{17}$$

where,

$$b_1 = (0.1181 + 0.4730\omega)\, V_c \tag{18}$$

$$b_2 = (0.2117 + 0.1611\omega)\, V_c \tag{19}$$

$$b_3 = (0.2515 + 0.0283\omega)\, V_c \tag{20}$$

The ω values used in this study are identical to those previously reported (*3*). The calculated values of critical isotherm pressures using Equation 17 are compared with Pitzer's table in Table I (*4*). The average absolute

Table I. Comparison of Calculated Crtical Isotherm Pressures with Pitzer's Table

Acentric Factor (ω)	*Region*[a]	*Average Absolute Deviation (%)* RK (1)	*This Work*
0.0	*v*	0.2	0.2
	l	16.0	6.7
0.1	*v*	0.3	0.2
	l	34.0	5.4
0.2	*v*	0.6	0.1
	l	58.0	4.0
0.3	*v*	0.8	0.0
	l	90.0	4.9
0.4	*v*	1.2	0.1
	l	130.0	5.2
0.5	*v*	—	0.1
	l	—	7.0

[a] The regions v: $P_r \leqslant 1$ (5 points) and l: $P_r > 1$ (15 points).

deviations obtained are much smaller than those obtained by the RK equation, especially at higher ω values. A comparison of the calculated and experimental compressibility factors along the critical isotherm for sulfur dioxide (*5*) and carbon dioxide (*2*) is shown in Table II. The critical constants of the substances investigated were obtained from Kudchadker et al. (*6*) and Mathews (*7*).

In order to expand the applicability of Equation 17 to isotherms other than the critical, it was necessary to determine the temperature dependence of $a(T)$, $c(T)$, and $d(T)$. Let

$$a(T) = a\, f_a(T) \tag{21}$$

$$c(T) = c\, f_c(T) \tag{22}$$

$$d(T) = d\, f_d(T) \tag{23}$$

and at $T = T_c$, $f_a(T_c) = f_c(T_c) = f_d(T_c) = 1$. In other words, $a = a(T_c)$, $c = c(T_c)$, and $d = d(T_c)$. Hence Equation 17 became

$$P = \frac{RT}{V-b} - \frac{af_a(T)}{V(V+b_1)} + \frac{cf_c(T)}{V^2\,(V-b)\,(V+b_2)} - \frac{df_d(T)}{(V+b_3)^7} \tag{24}$$

Table II. Comparison of Experimental and Calculated Compressibility Factors along the Critical Isotherm

Component	*Region*	*Number of Data Points*	*Average Absolute Deviation (%)*	
			RK (1)	*This Work*
Sulfur dioxide (*5*)	$P_r \leqslant 1$	16	1.46	0.30
	$P_r > 1$	49	12.91	0.89
Carbon dioxide (*2*)	$P_r \leqslant 1$	16	3.01	0.89
	$P_r > 1$	8	10.80	1.83

In order to obtain a suitable expression for $f_a(T)$, Equation 24 was used to obtain the following expression for the second virial coefficient:

$$B = \lim_{\rho \to 0} \left(\frac{\partial Z}{\partial \rho} \right)_T = b - af_a\,(T)/RT \tag{25}$$

which then was used to fit the Pitzer and Curl correlation of the second virial coefficient (*8*):

$$\begin{aligned} \frac{BP_c}{RT_c} &= (0.1445 + 0.073\omega) - (0.330 - 0.46\omega)/T_r \\ &- (0.1385 + 0.50\omega)/T_r^2 - (0.0121 + 0.097\omega)/T_r^3 \\ &- 0.0073\omega/T_r^8 \end{aligned} \tag{26}$$

The expression obtained for $f_a(T)$ was as follows:

$$\begin{aligned} f_a(T) = &- (0.1711 + 0.2147\omega)\,T_r + (0.8340 - 1.2211\omega) \\ &+ (0.2630 + 1.1065\omega)/T_r + (0.0741 + 0.3120\omega)/T_r^2 \\ &+ 0.0173\omega/T_r^7 \end{aligned} \tag{27}$$

when $T = T_c$, $T_r = 1$ and Equation 27 reduces to $f_a(T_c) = 1$.

Next, Equation 24 was rearranged by splitting the third term of the right-hand side of the equation into two terms. Hence,

$$P = \frac{RT}{V - b} - \frac{af_a(T)}{V(V + b_1)} - \frac{ef_e(T)}{V^2(V - b)} + \frac{ef_g(T)}{V^2(V + b_2)} - \frac{df_d(T)}{(V + b_3)^7} \tag{28}$$

where

$$e = \frac{c}{b + b_2} \tag{29}$$

In order to represent *PVT* data over a wide temperature range by means of Equation 28, it was necessary to make both the quantity b, in the third term of the right-hand side of the equation, and the quantity b_3 dependent on temperature. Thus,

$$P = \frac{RT}{V-b} - \frac{a f_a(T)}{V(V+b_1)} - \frac{e f_e(T)}{V^2(V-b')} + \frac{e f_g(T)}{V^2(V+b_2)} - \frac{d f_d(T)}{(V+b_3')^7} \quad (30)$$

which is the final expression of the proposed equation.

The evaluation of the temperature-dependent quantities $f_e(T)$, $f_g(T)$, $f_d(T)$, b', and b_3' was based on: (1) satisfying the Maxwell relationship at saturation (the fugacity of the liquid calculated from Equation 30 should be equal to the fugacity of the vapor calculated from the same equation); (2) satisfying the generalized correlation of the saturated liquid volume proposed earlier by Lu et al. (*9*); and (3) fitting *PVT* data as correlated in Pitzer's tables (*4*) over the complete range of T_r and P_r. The following set of temperature functions finally was obtained after numerous fitting trials were made:

$$b' = 0.26\, T_r^{0.2}\, V_c \quad (31)$$

$$b_3' = (0.2515 + 0.0283\omega)\, T_r V_c \quad (32)$$

$$f_e(T) = T_r^{-4} \quad (33)$$

$$f_d(T) = T_r^{-1.8} \quad (34)$$

$$f_g(T) = x_1 + x_2/T_r^2 + x_3/T_r^4 + x_4/T_r^6 + x_5/T_r^7 \quad (35)$$

$$\begin{aligned} x_1 &= -0.3588 + 0.4982\omega + 0.8208\omega^2 \\ x_2 &= 0.2993 + 0.3038\omega - 0.6829\omega^2 \\ x_3 &= 0.9826 - 0.7758\omega - 0.0343\omega^2 \\ x_4 &= 0.0883 - 0.0106\omega - 0.1054\omega^2 \\ x_5 &= -0.0114 - 0.0156\omega + 0.0018\omega^2 \end{aligned} \quad (36)$$

For the purpose of satisfying the Maxwell relationship more precisely, an adjustment was made on the temperature-dependent function $f_a(T)$. The final expression obtained for $f_a(T)$ is as follows:

$$\begin{aligned} f_a(T) = &-(0.1664 + 0.0043\omega)T_r + (0.8137 - 1.2204\omega) \\ &+ (0.2861 + 1.1297\omega)/T_r + (0.0666 + 0.2977\omega)/T_r \\ &+ 0.0173\omega/T_r^7 \end{aligned} \quad (37)$$

It should be mentioned that the difference between Equations 37 and 27 is very small. The second virial coefficient calculated from Equation 25 using the new expression of $f_a(T)$ still agrees very well with the correlation of Pitzer and Curl (Equation 26).

Numerical Values of the Parameters of the Proposed Equations

As mentioned above, the final expression of the proposed equation is represented by Equation 30:

$$P = \frac{RT}{V-b} - \frac{af_a(T)}{V(V+b_1)} - \frac{ef_e(T)}{V^2(V-b')} + \frac{ef_g(T)}{V^2(V+b_2)} - \frac{df_d(T)}{(V+b_3')^7} \quad (30)$$

in which

$$\begin{aligned}
b &= 0.26V_c \\
b_1 &= (0.1181 + 0.4730\omega)V_c \\
b_2 &= (0.2117 + 0.1611\omega)V_c \\
b' &= 0.26T_r^{0.2}V_c \\
b_3' &= (0.2515 + 0.0283\omega)T_rV_c \\
af_a(T) &= a_1T + a_2 + a_3/T + a_4/T^2 + a_5/T^7 \\
a_1 &= a^*y_1RV_c \\
a_2 &= a^*y_2RT_cV_c \\
a_3 &= a^*y_3RT_c^2V_c \\
a_4 &= a^*y_4RT_c^3V_c \\
a_5 &= a^*y_5RT_c^8V_c \\
y_1 &= -0.1664 - 0.2243\omega \\
y_2 &= 0.8137 - 1.2204\omega \\
y_3 &= 0.2861 + 1.1297\omega \\
y_4 &= 0.0666 + 0.2977\omega \\
y_5 &= 0.0173\omega \\
ef_e(T) &= eT^{-4} \\
e &= e^*RT_c^5V_c^2 \\
e^* &= c^*/(0.4717 + 0.1611\omega) \\
ef_g(T) &= g_1 + g_2/T^2 + g_3/T^4 + g_4/T^6 + g_5/T^7 \\
g_1 &= e^*x_1RT_cV_c^2 \\
g_2 &= e^*x_2RT_c^3V_c^2 \\
g_3 &= e^*x_3RT_c^5V_c^2 \\
g_4 &= e^*x_4RT_c^7V_c^2 \\
g_5 &= e^*x_5RT_c^8V_c^2 \\
x_1 &= -0.3588 + 0.4982\omega + 0.8208\omega^2 \\
x_2 &= 0.2993 + 0.3038\omega - 0.6829\omega^2 \\
x_3 &= 0.9826 - 0.7758\omega - 0.0343\omega^2 \\
x_4 &= 0.0883 - 0.0106\omega - 0.1054\omega^2 \\
x_5 &= -0.0114 - 0.0156\omega + 0.0018\omega^2
\end{aligned}$$

and

$$\begin{aligned}
df_d(T) &= dT^{-1.8} \\
d &= d^*RT_c^{2.8}V_c^6
\end{aligned}$$

In the above expressions,

$$V_c = Z_c R T_c / P_c$$
$$Z_c = 0.291 - 0.080\omega$$

The values of a^*, c^*, and d^* were determined from Equations 8 through 10. The values obtained for several ω values at regular intervals are listed below to serve as examples:

ω	a*	c*	d*
0.0	1.3827	0.2973	0.7464
0.1	1.4342	0.2859	0.7443
0.2	1.4872	0.2761	0.7443
0.3	1.5419	0.2680	0.7467
0.4	1.5982	0.2615	0.7517
0.5	1.6561	0.2564	0.7594

Testing of the Proposed Equation

The applicability of the proposed equation was tested in terms of its predicted values of the compressibility factors, liquid fugacity coefficients, and isothermal enthalpy departures of pure compounds.

Compressibility Factors. A total of 2772 Z values of Pitzer's table was used to test the capability of the proposed equation for calculating compressibility factors of pure nonpolar compounds.

The calculated values obtained from Equation 30, together with those obtained from the equations of Redlich et al. (*10*), Edmister et al. (*11*), Redlich and Kwong (*1*), and Sugie et al. (*12*) are compared with the Z values of Pitzer's work in Table III. The proposed equation provides the smallest standard deviation.

Table III. Comparison of Calculated Compressibility Factors with Pitzer's Table by Various Methods for a Total of 2772 Data Points at 22 T_r and 21 P_r Conditions[a]

	Standard Deviation
Ref. *10*	0.013
Ref. *11*	0.025
Ref. *1*	0.054
Ref. *12*	0.024
This work	0.0108

[a] $0.8 < T_r < 4$, $0.2 < P_r < 9$; $\omega = 0.0$–0.5.

A comparison of the calculated and experimental Z values of propane (*13*) and sulfur dioxide (*5*), in the gas region is shown in Table IV. In addition, the calculated results obtained from the equations of Redlich and Kwong (*1*), Redlich and Dunlop (*14*), Gray et al. (*15*), and Sugie et al. (*12*) also are included in Table IV. The proposed equation yields the best results.

Table IV. Comparison of Experimental and Calculated Z Values for Propane and Sulfur Dioxide

		Average Absolute Deviation (%)				
T	P*(psia)*	*Ref.* 1	*Ref.* 14	*Ref.* 15	*Ref.* 12	*This Work*
		Propane (13)				
220°F	20–6000	6.3	5.2	1.9	1.0	0.80
340°F	20–6000	1.7	1.5	1.9	1.2	1.47
460°F	20–6000	1.0	0.4	4.2	1.8	1.67
Overall		3.0	2.4	2.7	1.3	1.31
		Sulfur dioxide (5)				
157.5°C	10–300	13.5	6.4	3.3	0.8	0.77
200°C	10–300	4.9	3.3	0.9	2.0	1.02
250°C	10–300	2.2	1.0	3.3	1.2	1.32
Overall		6.9	3.6	2.5	1.3	1.04

In addition, the calculated and the experimental Z values of hydrogen sulfide (*16*), in the regions including gas, vapor, and liquid, are compared in Table V. The calculated results obtained from the equations of Redlich and Kwong (*1*) and Sugie et al. (*12*) also are included in this table for comparison. Again, the proposed equation yields the best results.

Table V. Comparison of Experimental and Calculated Z Values for Hydrogen Sulfide (16)

			Average Absolute Deviation (%)		
T_r	P_r	*Number of Points*	*Ref.* 1	*Ref.* 12	*This Work*
0.744	0.01–7.7	34	3.64	1.53	1.37
0.834	0.01–7.7	34	3.94	1.52	1.01
0.893	0.01–7.7	34	4.00	1.40	0.84
1.012	0.01–7.7	34	3.84	0.55	0.53
1.102	0.01–7.7	34	2.54	0.88	0.66
1.191	0.01–7.7	34	1.69	0.66	0.85
Overall			3.27	1.09	0.88

Fugacities. The fugacity coefficient, ϕ, which is equal to the ratio of fugacity to pressure, can be calculated from Equation 38.

$$\begin{aligned}\ln \phi &= Z - 1 - \ln Z - \int_{\infty}^{V} (Z-1)\frac{dV}{V} \\ &= Z - 1 - \ln Z - \ln \frac{V-b}{V} + \frac{af_a(T)}{RTb_1} \ln \frac{V}{V+b_1} \\ &\quad - \frac{ef_e(T)}{RTb'}\left[\frac{1}{b'}\ln\frac{V-b'}{V} + \frac{1}{V}\right] \\ &\quad - \frac{ef_g(T)}{RTb_2}\left[\frac{1}{b_2}\ln\frac{V}{V+b_2} + \frac{1}{V}\right] - \frac{df_d(T)}{6RT(V+b_3')^6}\end{aligned} \tag{38}$$

The above equation was used to obtain ϕ values, at regular intervals of T_r from $T_r = 0.5$ to $T_r = 1$, for liquid methane, ethane, propane, n-butane, and n-pentane. The calculated values are compared with some of the available tabulations in the literature (*17, 18, 20*) in Figures 1 through 5. Excellent agreement was obtained.

In addition, ϕ values were calculated using Equation 38 for the same five compounds but at lower temperatures ($T_r = 0.4$ and 0.3). A similar comparison is shown in Figures 6 and 7. Good agreement generally is obtained with the exception of methane.

Isothermal Enthalpy Departures. The isothermal enthalpy departures from the ideal-gas state were calculated for six pure, saturated liquids (methane, ethane, propane, n-butane, i-butane, and n-pentane) using Equation 39. The proposed equation was, of course, used in its derivation

$$\begin{aligned}(H^P - H^\circ)_T &= PV - RT - \int_{\infty}^{V}\left[P - T\left(\frac{\partial P}{\partial T}\right)_V\right] dV \\ &= \frac{bRT}{V-b} - \frac{af_a(T)}{V+b_1} + \frac{ef_e(T)}{V(V-b')} \\ &\quad - \frac{ef_g(T)}{V(V+b_2)} - \frac{df_d(T)V}{(V+b_3')^7} + \frac{F_1}{b_1}\ln\frac{V}{V+b_1} \\ &\quad - \frac{F_2}{b'}\left[\frac{1}{b'}\ln\frac{V-b'}{V} + \frac{1}{V}\right] \\ &\quad - \frac{0.2cf_e(T)}{b'}\left[\frac{2}{b'}\ln\frac{V-b'}{V} + \frac{1}{V} + \frac{1}{V-b'}\right] \\ &\quad - \frac{F_3}{b_2}\left[\frac{1}{b_2}\ln\frac{V}{V+b_2} + \frac{1}{V}\right] - \frac{F_4}{6(V+b_3')^6} \\ &\quad - \frac{df_d(T)b_3'}{(V+b_3')^7}\end{aligned} \tag{39}$$

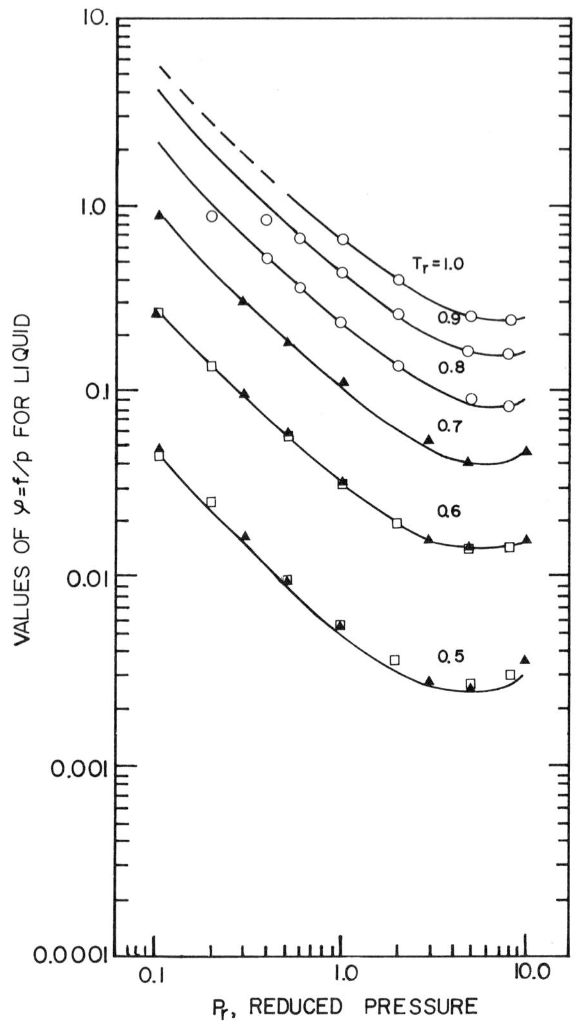

Figure 1. Fugacity coefficients of pure liquid methane (ω = 0.013) in the T_r *range of 0.5 to 1.0: (——), this work; (○), Pitzer tabulation; (▲), Chao tabulation; and (□), Kobayashi tabulation.*

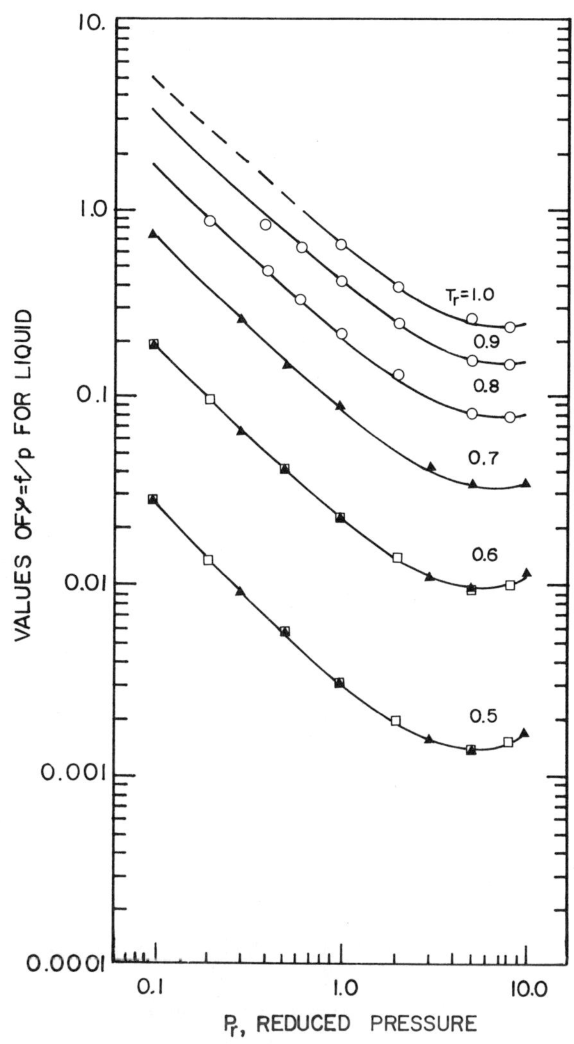

Figure 2. Fugacity coefficients of pure liquid ethane (ω = 0.105) in the T_r *range of 0.5 to 1.0: (——), this work; (○), Pitzer tabulation; (▲), Chao tabulation; and (□), Kobayashi tabulation.*

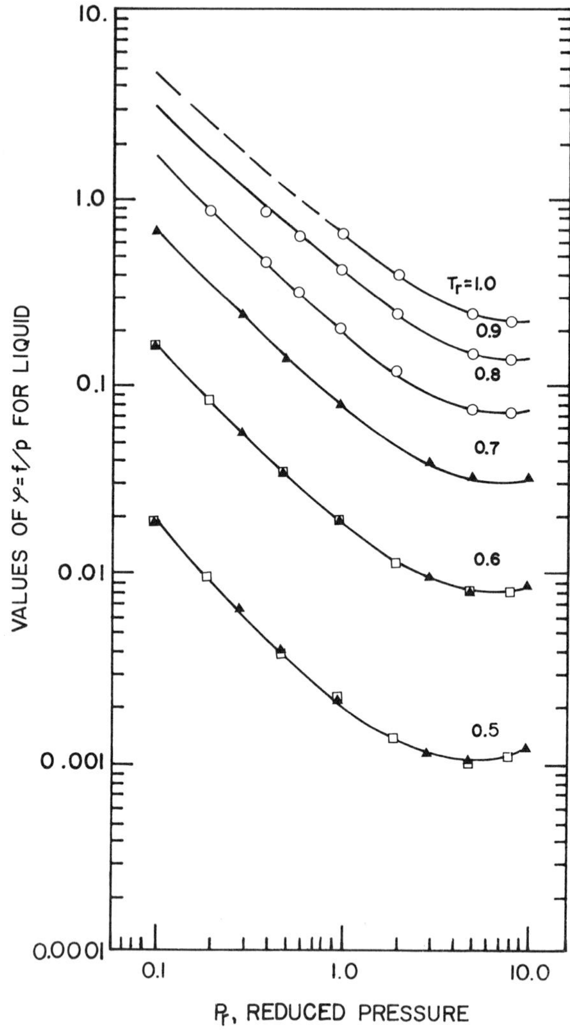

Figure 3. Fugacity coefficients of pure liquid propane (ω = 0.153) in the T_r range of 0.5 to 1.0: (——), this work; (○), Pitzer tabulation; (▲), Chao tabulation; and (□), Kobayashi tabulation.

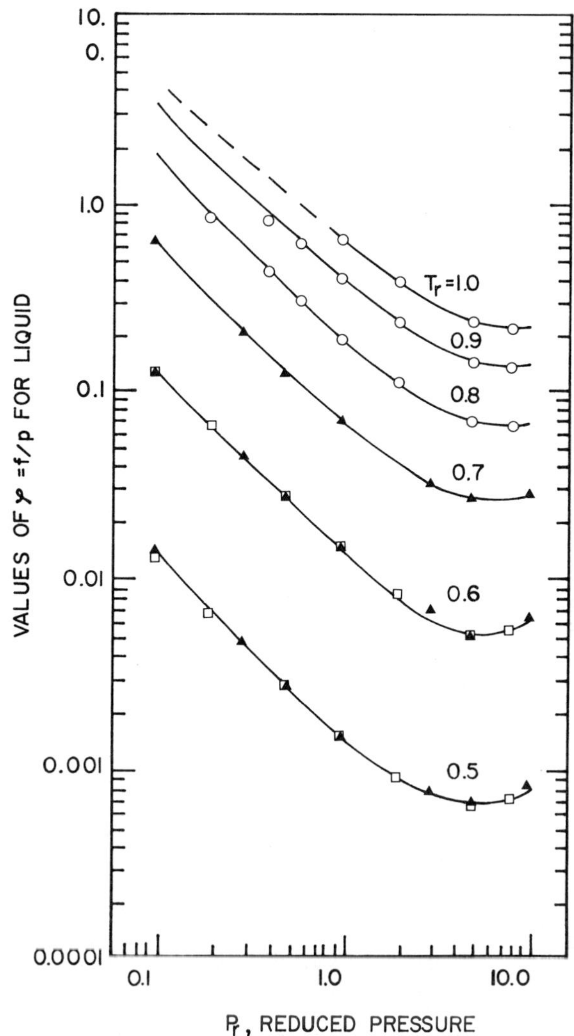

Figure 4. Fugacity coefficients of pure liquid n-*butane* ($\omega = 0.200$) *in the* T_r *range of 0.5 to 1.0: (——), this work; (○), Pitzer tabulation; (▲), Chao tabulation; and (□), Kobayashi tabulation.*

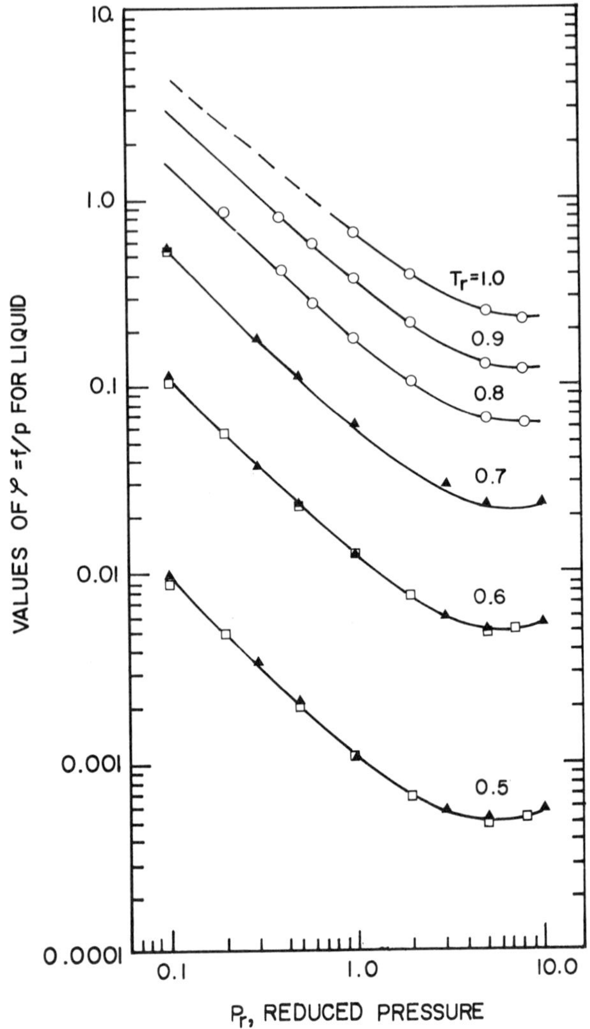

Figure 5. Fugacity coefficients of pure liquid n-pentane *(ω = 0.252) in the* T_r *range of 0.5 to 1.0: (——), this work; (○), Pitzer tabulation; (▲), Chao tabulation; and (□), Kobayashi tabulation.*

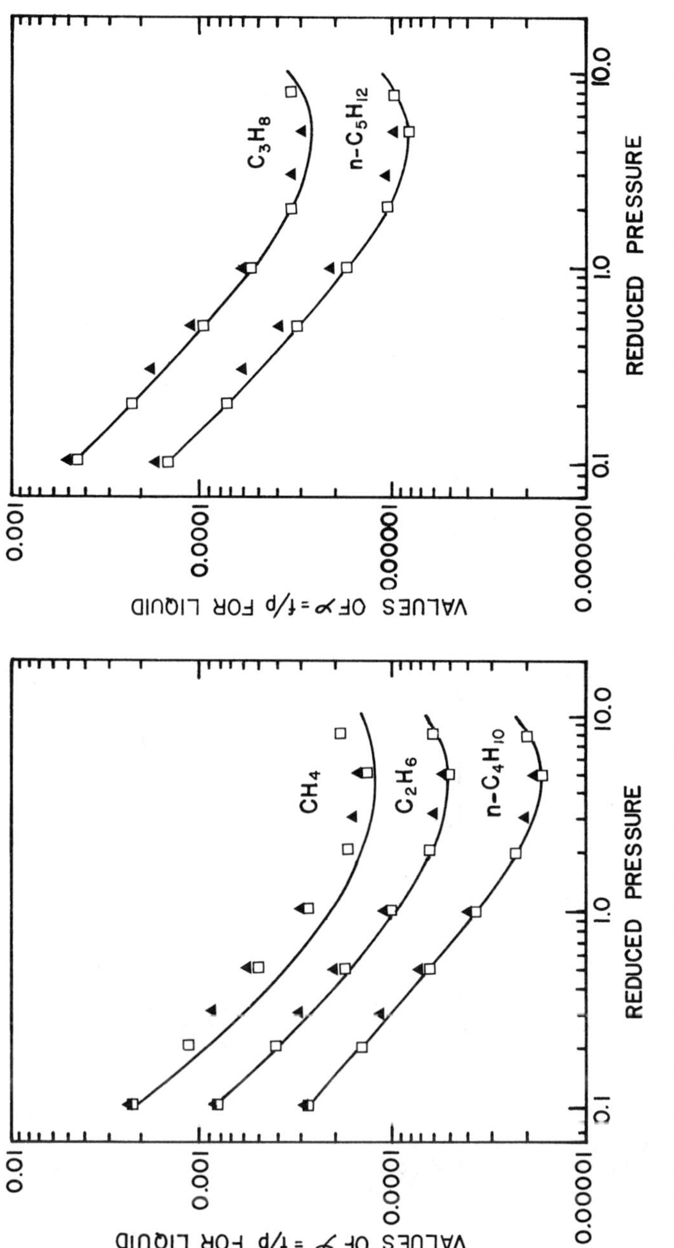

Figure 5. Fugacity coefficients of pure liquid methane, ethane, propane, n-*butane, and* n-*pentane at* $T_r = 0.4$: (——), *this work;* (▲), *Chao tabulation; and* (□), *Kobayashi tabulation.*

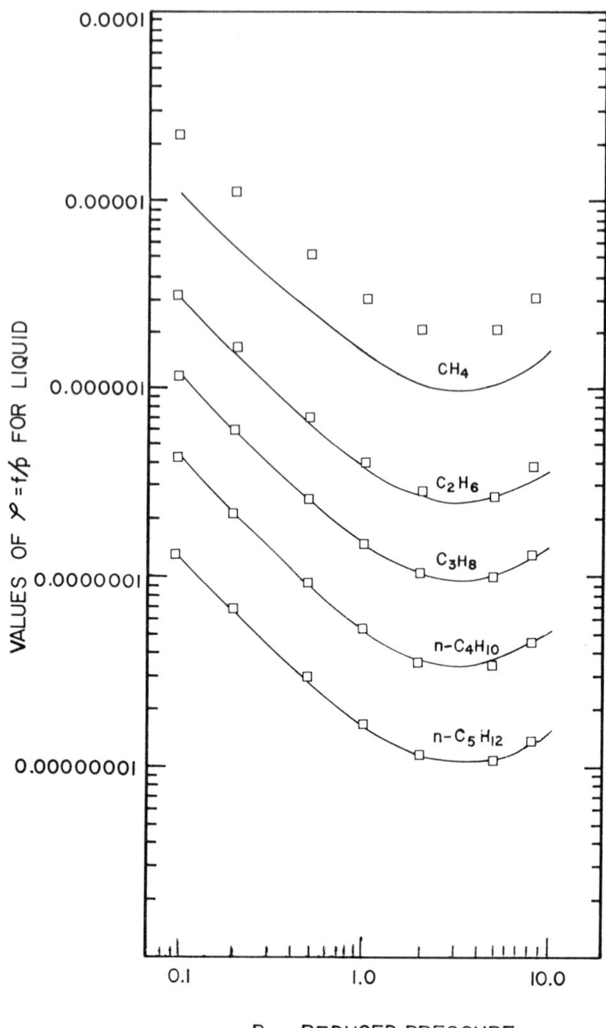

Figure 7. Fugacity coefficients of pure liquid methane, ethane, propane, n-*butane, and* n-*pentane at* $T_r = 0.3$: *(——), this work and (□), Kobayashi tabulation.*

in which

$$F_1 = af_a(T) - T\left(\frac{\partial af_a(T)}{\partial T}\right) = a_2 + 2a_3/T + 3a_4/T^2 + 8a_5/T^7$$

$$F_2 = ef_e(T) - T\left(\frac{\partial ef_e(T)}{\partial T}\right) = 5e/T^4$$

$$F_3 = ef_g(T) - T\left(\frac{\partial ef_g(T)}{\partial T}\right) = g_1 + 3g_2/T^2 + 5g_3/T^4 + 7g_4/T^6 + 8g_5/T^7$$

$$F_4 = df_d(T) - T\left(\frac{\partial df_d(T)}{\partial T}\right) = 2.8\, d/T^{1.8}$$

The values obtained from Equation 39 and several other methods (*11, 12, 19, 21, 22, 23, 24, 25*) are compared with the available experimental values in the literature (*26, 27, 28, 29*). The comparison, expressed in terms of average absolute deviations, is presented in Table VI. However, the results obtained by the proposed equation are only fair.

Discussion and Conclusion

This investigation follows our efforts previously made on the modification of the RK equation of state (*3, 30*). The repulsive term of the RK equation was retained with the anticipation that the original terms would be preserved as part of the new equation. This practice may be subject to modifications in future endeavors. The repulsive term may be replaced by a more suitable term such as that proposed by Carnahan and Starling (*31*).

The proposed equation was not compared with any of the more recent cubic equations, such as the Soave modification of the RK equation (*32*), the Peng and Robinson equation (*33*), and the Fuller equation (*34*), because all of these equations do not yield acceptable values of Z_c.

Some of the ω values used in this study differ slightly from those suggested by Passut and Danner (*35*). However, these small differences hardly affected the calculated results.

In conclusion, a new pressure-explicit equation of state has been successfully developed as intended. It is suitable for representing *PVT* behavior of liquid and gas phases over a wide range of temperature and pressure for pure, nonpolar compounds. Furthermore, the parameters of the proposed equation are generalized in terms of the critical properties and the acentric factor.

Table VI. Comparison of Experimental and Calculated

Component	T_r	P_r	*No. of Data Pts.*
Methane (*26*)	0.757–0.990	0.172–0.945	5
Ethane (*27*)	0.654–0.982	0.045–0.894	7
Propane (*26, 27, 28*)	0.660–0.983	0.041–0.891	13
N-Butane (*27*)	0.653–0.967	0.032–0.794	7
iso-Butane (*27*)	0.757–0.990	0.034–0.872	7
N-Pentane (*29*)	0.662–0.970	0.032–0.802	14
Overall			53

Glossary of Symbols

a, c, d = quantities represented by Equations 21, 22, and 23, respectively
a^*, c^*, d^* = quantities determined from Equations 8, 9, and 10
B = second virial coefficient
b = parameter of Equation 3
b_1, b_2, b_3 = parameters of Equation 17
b', b'_3 = parameters of Equation 30
e = quantity defined by Equation 29
e^* = $c^*/(0.4717 + 0.1611\omega)$
$F_1, \ldots F_4$ = functions of Equation 39
H° = enthalpy at ideal gas state
H^P = enthalpy at pressure P
$h_1 \ldots h_n$ = parameters of Equation 1
$k_o \ldots k_m$ = constants of Equation 5
$n_1 \ldots n_m$ = zero or positive integer
P = pressure
P_c = critical pressure
R = gas constant
T = temperature
T_c = critical temperature
T_r = reduced temperature
V = volume
V_c = critical volume
$x_1 \ldots x_5$ = functions represented by Equation 36
Z = compressibility factor
Z_c = compressibility factor at the critical point

$(H^0 - H^P)_T$ Values for Pure Saturated Liquids

Average Absolute Deviation (Btu/lb)

Ref. 19	*Ref.* 21	*Ref.* 22	*Ref.* 23	*Refs.* 24, 25	*Ref.* 11	*Ref.* 12	*This Work*
3.7	9.8	10.0	2.6	7.0	32.1	3.0	4.7
1.7	2.2	8.1	5.0	5.1	9.3	1.9	4.3
1.2	3.5	3.4	4.6	2.7	5.1	1.5	4.6
2.2	3.4	9.4	4.0	4.4	4.6	2.1	4.7
2.3	6.1	13.6	4.4	—	—	2.5	5.5
1.9	4.6	11.7	4.8	3.1	3.4	1.5	5.0
2.2	4.9	9.4	4.2	4.5	11.0	2.1	4.8

Greek Letters

ρ = density
ϕ = fugacity coefficient
ω = acentric factor

Literature Cited

1. Redlich, O.; Kwong, J. N. S. *Chem. Rev.* **1949,** *44,* 233.
2. Michels, C. "Some Physical Properties of Compressed Carbon Dioxide"; University of Amsterdam: Amsterdam, 1937.
3. Sugie, H.; Lu, B. C.-Y. *Ind. Eng. Chem., Fundam.* **1970**, 9, 428.
4. Pitzer, K. S.; Lippman, D. Z.; Curl, R. F., Jr.; Huggins, C. M.; Petersen, D. E. *J. Am. Chem. Soc.* **1955,** *77,* 3433.
5. Kang, T. L.; Hirth, L. T.; Kobe, K. A.; McKetta, J. J. *J. Chem. Eng. Data* **1961,** *6,* 220.
6. Kudchadker, A. P.; Alani, G. H.; Zwolinski. *Chem. Rev.* **1968,** *68,* 659.
7. Mathews, J. F. *Chem. Rev.* **1972,** *72,* 71.
8. Pitzer, K. S.; Curl, R. F., Jr. *J. Am. Chem. Soc.* **1957,** *79,* 2369.
9. Lu, B. C.-Y.; Ruether, J. A.; Hsi, C.; Chiu, C. H. *J. Chem. Eng. Data* **1973,** *18,* 241.
10. Redlich, O.; Ackerman, F. J.; Gunn, R. D.; Jacobson, M.; Lau, S. *Ind. Eng. Chem., Fundam.* **1965,** *4,* 369.
11. Edmister, W. C.; Vairogs, J.; Klekers, A. J. *AIChE J.* **1968,** *14,* 479.
12. Sugie, H.; Ono, K.; Hiraide, M.; Lu, B. C.-Y., presented at the 7th Annual Meeting of Society of Chemical Engineers, Japan, 1973.
13. Johnston, H. L.; White, D. *Trans. ASME* **1950,** *72,* 785.
14. Redlich, O.; Dunlop, A. K. *Chem. Eng. Prog., Symp. Ser.* **1963,** *59*(44), 95.
15. Gray, R. S., Jr.; Rent, N. H.; Zudkevitch, D. *AIChE J.* **1970,** *16,* 991.
16. Reamer, H. H.; Sage, B. H.; Lacey, W. N. *Ind. Eng. Chem.* **1950,** *42,* 140.
17. Chao, K. C.; Greenkorn, R. A.; Olabisi, O.; Hensel, B. H. *AIChE J.* **1971,** *17,* 353.
18. Carruth, G. F.; Kobayashi, R. *Ind. Eng. Chem., Fundam.* **1972,** *11,* 509.
19. Lee, B.-I.; Edmister, W. C. *Ind. Eng. Chem., Fundam.* **1971,** *10,* 229.

20. Lewis, G. N.; Randall, M. "Thermodynamics," 2nd ed.; Revised by Pitzer, K. S., Brewer, L.; McGraw-Hill: New York, 1961.
21. Stevens, W. F.; Thodos, G. *AIChE J.* **1963,** *9,* 293.
22. Yen, L. C.; Alexander, R. E. *AIChE J.* **1965,** *11,* 334.
23. Erbar, J. H.; Persyn, C. L.; Edmister, W. C., Proceedings of the 43rd Annual Convention Natural Gas Processors Association, March 1964.
24. Benedict, M.; Webb, G. B.; Rubin, L. C. *J. Chem. Phys.* **1940,** *8,* 334.
25. Benedict, M.; Webb, G. B.; Rubin, L. C. *J. Chem. Phys.* **1942,** *10,* 747.
26. Jones, M. L., Jr.; Mage, D. T.; Faulkner, R. C., Jr.; Katz, D. L. *Chem. Eng. Prog., Symp. Ser.* **1963,** *59*(44), 52.
27. Canjar, L. N.; Manning, F. S. "Thermochemical Properties and Reduced Correlations for Gases"; Gulf Publishing Co.: Houston, 1967.
28. Yesavage, V. R., Ph.D. Thesis, University of Michigan, Ann Arbor, 1969.
29. Brydon, J. W.; Walen, N.; Canjar, L. N. *Chem. Eng. Prog., Symp. Ser.* **1953,** *49*(7), 151.
30. Sugie, H.; Lu, B. C.-Y. *AIChE J.* **1971,** *17,* 1068.
31. Carnahan, N. F.; Starling, K. E. *J. Chem. Phys.* **1969,** *51,* 1184.
32. Soave, G. *Chem. Eng. Sci.* **1972,** *27,* 1197.
33. Peng, D.-Y.; Robinson, D. B. *Ind. Eng. Chem., Fundam.* **1976,** *15,* 59.
34. Fuller, G. G. *Ind. Eng. Chem., Fundam.* **1976,** *15,* 254.
35. Passut, C. A.; Daner, R. P. *Ind. Eng. Chem., Process Des. Dev.* **1973,** *12,* 365.

RECEIVED October 8, 1978.

10

Calculation of Three-Phase Solid–Liquid–Vapor Equilibrium Using an Equation of State

DING–YU PENG and DONALD B. ROBINSON

Department of Chemical Engineering, University of Alberta,
Edmonton, Alberta, Canada T6G 2G6

An efficient computation algorithm is proposed for predicting the initial formation of pure solids in hydrocarbon and related systems. In order to use this method, only the density and vapor pressure of the pure solid are required together with the equation-of-state parameters for the solid-forming material. The use of the algorithm is illustrated with the Peng–Robinson equation of state for systems containing carbon dioxide as the solid-forming component at cryogenic temperatures. Good agreement was obtained between the predicted results and the experimental literature values. The predicted triple point for pure carbon dioxide was 72.1 psia at −71.2°F as compared with the literature value of 76.9 psia at −69.9°F.

The widespread use of low temperature for the processing of natural gas has been stimulated by economic incentives and it has been made possible by advances in materials and process technology. The need to be able to accurately describe the behavior of these hydrocarbon mixtures at cryogenic conditions is evident. In addition to the usual vapor–liquid equilibrium relationships, a knowledge of the solubility of heavier hydrocarbons and other potential solid-forming gases such as carbon dioxide or hydrogen sulfide is necessary for the safe and effective operation of processing plants. The presence of solid carbon dioxide, for example, may foul heat exchangers or interfere with the operation of distillation equipment to the point that costly plant shutdowns and maintenance may be required.

In order to avoid these potentially damaging conditions, it is useful to be able to delineate the solid-forming regions for any given processing mixture. A number of studies dealing with the correlation of the solubility

0-8412-0500-0/79/33-182-185$05.00/1

of solids in liquid hydrocarbons at cryogenic conditions have been reported (*1, 2, 3, 4, 5*). Some of these are based on the Scatchard–Hildebrand regular solution theory and others on the Flory–Huggins equation for athermal mixtures. Regardless of the basis, however, the equilibrium criterion in all cases is that the free energy of the pure solid must equal its partial molal free energy in the solution. In the work described in this chapter, a computation algorithm based on equating the fugacities of the pure solid and the component in the solution is presented for use in predicting incipient solid formation of a component from an equilibrium vapor or liquid solution.

Thermodynamic Relationships

The equilibrium condition for a solid–liquid–vapor (SLV) system at a specified temperature and pressure may be written in terms of the fugacities of the solid-forming component as

$$f_k^S = f_k^L = f_k^V \tag{1}$$

where f_k^S is the fugacity of component k in the solid phase, f_k^L is the fugacity of component k in the liquid phase, and f_k^V is that of the same component in the vapor phase. It is understood that the fugacities of all the other components must follow the equation

$$f_j^L = f_j^V \tag{2}$$

All of these fugacities are evaluated at the temperature and pressure of the system. The fugacities of component k for the fluid phases can be calculated using the rigorous thermodynamic equation

$$\ln \frac{f_k}{x_k P} = \int_V^\infty \left[\frac{1}{RT} \left(\frac{\partial P}{\partial n_k} \right) - \frac{1}{v} \right] dv - \ln Z \tag{3}$$

Normally one can assume that the solid phase is not a solid solution; consequently, the solid phase is essentially pure component k whose fugacity follows the equation

$$\ln \frac{f_k^S}{P} = \int_0^P \left(\frac{v_k}{RT} - \frac{1}{P} \right) dP \tag{4}$$

This can be written as

$$\ln \frac{f_k^S}{P} = \int_0^{P_k^\circ} \left(\frac{v_k^V}{RT} - \frac{1}{P} \right) dP + \int_{P_k^\circ}^P \left(\frac{v_k^S}{RT} - \frac{1}{P} \right) dP \tag{5}$$

where P_k^* is the vapor pressure of component k at the system temperature, v_k^V is the molar volume of component k in its vapor state, and v_k^S is the molar volume of solid component k. In Equation 5, the first integral represents the fugacity coefficient of the pure component k at its vapor pressure while the second integral accounts for the compression effect upon the solid at the system pressure. This is greater than the solid–vapor equilibrium pressure for this component at the system temperature. Neglecting the pressure dependency of the molar volume of solid k, we readily obtain

$$f_k^S\ (P) = f_k^V\ (P_k^*) \exp\left[\frac{f_k^S\ (P - P_k^*)}{RT}\right] \qquad (6)$$

The above equation simply states that the fugacity of solid component k at pressure P equals the fugacity of pure k at its vapor pressure multiplied by a Poynting correction factor.

Evaluation of Fugacities Using an Equation of State. The fugacities of the components in the fluid phases are related to the volumetric and phase behavior of the mixture while the fugacity of the solid component depends only on the *PVT* relationship of the pure component. Theoretically it is possible to evaluate the fugacities using experimental volumetric and/or phase equilibrium data in conjunction with Equations 3 and 6. However, these data are normally either unavailable or insufficient and an equation-of-state model has to be used to compute the fugacities.

Many equations of state are available for calculating the quantities in Equations 3 and 6. When selecting a suitable equation of state, consideration should be given to the fact that the capability of an equation of state to accurately represent the SLV three-phase system hinges on the ability of the same equation to describe the simpler two-phase liquid–vapor system. In this study we have chosen the Peng–Robinson equation of state (*6*) to model the phase and volumetric behavior of the fluid phases of mixture. The volumetric behavior of the pure component k in its vapor-phase region at the solid-formation temperature also was represented in terms of this closed-form equation. The Peng–Robinson equation has been applied successfully to the prediction of liquid–liquid–vapor (LLV) equilibrium (*7*), hydrate formation (*8, 9*), and mixture critical-point determination (*10*) in addition to the usual vapor–liquid equilibrium calculations. This equation has the following form:

$$P = \frac{RT}{v - b} - \frac{a(T)}{v(v + b) + b(v - b)} \qquad (7)$$

and when the classical derivatives at the critical point are imposed it yields

$$a(T_c) = 0.45724 \frac{R^2 T_c^{\,2}}{P_c} \tag{8}$$

$$b = 0.07780 \frac{RT_c}{P_c} \tag{9}$$

At temperatures other than the critical point

$$a(T) = a(T_c)\ \alpha(T_r, \omega) \tag{10}$$

$$\alpha(T_r, \omega) = [1 + \kappa(1 - \mathrm{T_r}^{1/2})]^2 \tag{11}$$

$$\kappa = 0.37464 + 1.54226\omega - 0.26992\omega^2 \tag{12}$$

The mixing rules for use with these equations are

$$b_{\mathrm{m}} = \sum_i x_i b_i \tag{13}$$

$$a_{\mathrm{m}} = \sum_i \sum_j x_i x_j\ (a_i a_j)^{1/2}\ (1 - \delta_{ij}) \tag{14}$$

where δ_{ij}'s are fitted binary interaction parameters based on binary vapor–liquid equilibrium data.

Substituting Equations 7 through 14 into Equation 3 we have

$$\ln \frac{\mathrm{f}_k}{x_k P} = \frac{b_k}{b_{\mathrm{m}}}(Z_{\mathrm{m}} - 1) - ln(Z_{\mathrm{m}} - B_{\mathrm{m}}) - \frac{A_{\mathrm{m}}}{2\zeta 2\ B_{\mathrm{m}}}\left[\frac{2\sum_i x_i a_{ik}}{a_{\mathrm{m}}} - \frac{\mathrm{b}_k}{\mathrm{b_m}}\right] \ln\left(\frac{Z_{\mathrm{m}} + 2.414 B_{\mathrm{m}}}{Z_{\mathrm{m}} - 0.414 B_{\mathrm{m}}}\right) \tag{15}$$

where

$$A_{\mathrm{m}} = \frac{a_{\mathrm{m}} P}{R^2 T^2} \tag{16}$$

and

$$B_{\mathrm{m}} = \frac{b_{\mathrm{m}} P}{RT} \tag{17}$$

Similarly, applying Equations 7 through 12 to Equation 6, we have

$$f_k^S = P_k^* \exp\left[\Phi + \frac{v_k^S (P - P_k^*)}{RT}\right] \tag{18}$$

where Φ is the lagarithm of the fugacity coefficient of pure component k at its vapor pressure corresponding to the solid-formation temperature.

Computation Procedure

From the outset the relationships between the fugacity and the state variables are highly nonlinear. To determine the composition of each phase for a SLV system such that Equations 1 and 2 are satisfied requires that an iterative method be used. Because of the constraints imposed on the system by the phase rule somewhat different procedures were used in this study to compute the SLV equilibrium condition for multicomponent systems and for binary systems, respectively. Both procedures calculate the fluid-phase compositions of a given mixture at the incipient solid-formation condition.

Multicomponent System. According to the phase rule, there are N-1 degrees of freedom for an N-component system which is in SLV equilibrium. Consequently, the equilibrium condition for a system containing three or more components can be determined when both the temperature and the pressure are independently specified. The computation proceeds as follows: the fugacity of the compressed pure solid at the specified condition is calculated according to Equation 18. A vapor–liquid equilibrium flash calculation is then performed at the specified system temperature and pressure for a mixture with known composition. The fugacities of each component in both phases are calculated according to Equation 15. Comparison of the fugacities of component k is made at this stage to check if this component could indeed form a solid phase at the specified condition. If the fugacity of pure solid k as obtained from Equation 18 is less than that of the same component in the fluid phases as calculated from Equation 15 indicating the precipitation of this component, the composition of the given mixture is modified according to the following equations to reduce the concentration of this solid-forming component:

$$z_j^{(q+1)} = z_j^{(q)} \cdot \sigma^{(q)} \tag{19}$$

and

$$z_k^{(q+1)} = z_k^{(q)} - \epsilon^{(q)} \tag{20}$$

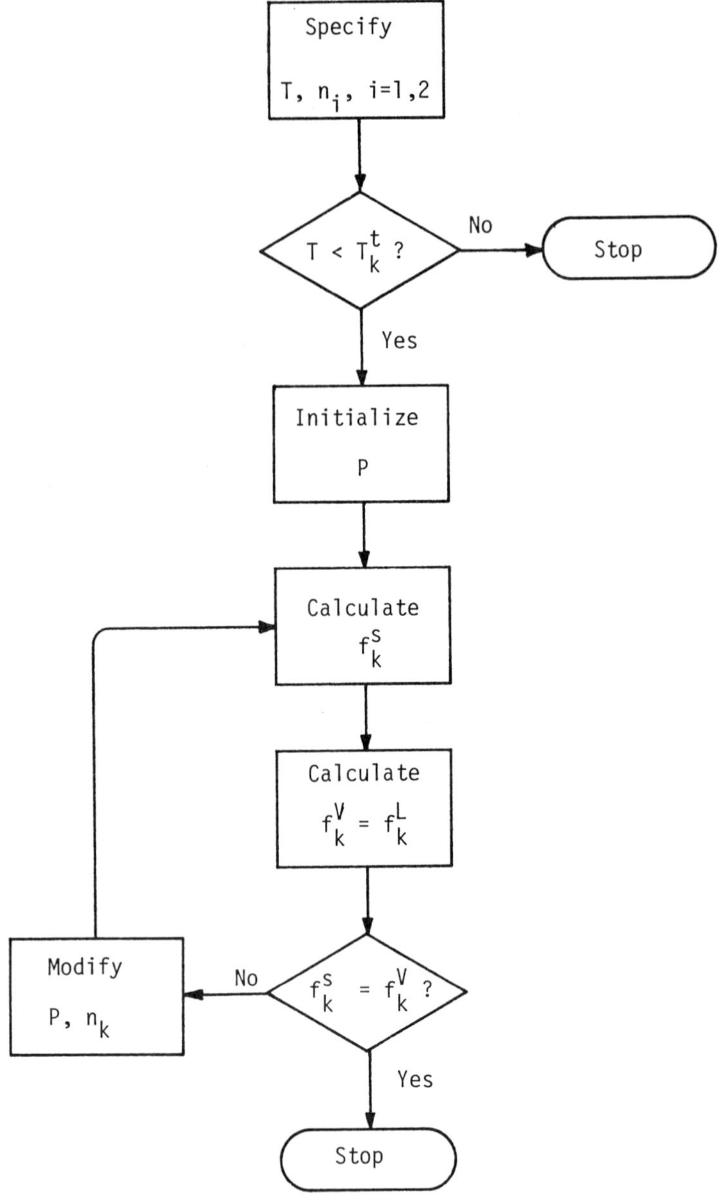

Figure 1. Flow diagram showing computation procedure used for determining initial solid forming condition

where

$$\epsilon = \frac{f_k^V - f_k^S}{\left(\frac{\partial f_k^V}{\partial z_k}\right)} \qquad (21)$$

$$\sigma = 1 + \frac{\epsilon}{1 - z_k} \qquad (22)$$

A vapor–liquid equilibrium flash calculation is again carried out to obtain new fugacity values using Equation 15. The process of comparing fugacities and adjusting mixture composition is repeated until Equation 1 is satisfied and the SLV equilibrium condition is considered to have been found.

Binary System. Owing to the fact that a binary system which is in SLV equilibrium is univariant, one can not specify the system pressure when the temperature has been fixed already, and vice versa. The pressure (or temperature) has to be determined along with the composition of each phase in an iterative manner. The scheme used to search for the phase compositions in a binary system is essentially the same as that used in the multicomponent system except that the pressure value is modified at each iteration until Equation 1 is satisfied. The pressure is corrected according to the following equation which is based on the one-dimensional Newton's method using the ratio of the fugacity values as the objective function:

$$P^{(q+1)} = P^{(q)}\left[2 - \frac{f_k^{V(q)}}{f_k^{S(q)}}\right] \qquad (23)$$

A flow chart illustrating the algorithm used to compute the SLV equilibrium for binary systems is shown in Figure 1.

Results

Using the proposed procedure in conjunction with literature values for the density (*11*) and vapor pressure (*12*) of solid carbon dioxide, the solid-formation conditions have been determined for a number of mixtures containing carbon dioxide as the solid-forming component. The binary interaction parameters used in Equation 14 were the same as those used previously for two-phase vapor–liquid equilibrium systems (*6*). The value for methane–carbon dioxide was 0.110 and that for ethane–carbon dioxide was 0.130. Excellent agreement has been obtained between the calculated results and the experimental data found in the literature. As shown in Figure 2, the predicted SLV locus for the methane–carbon

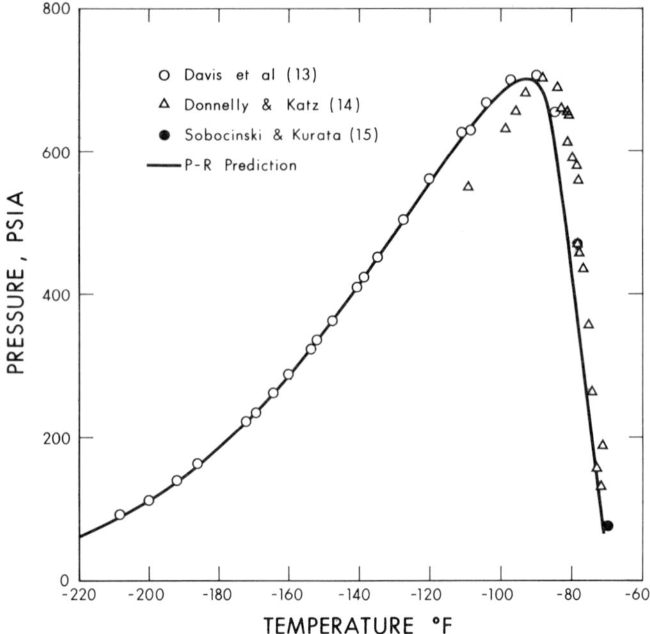

Figure 2. Experimental and predicted solid–liquid–vapor locus for the methane–carbon dioxide system: (○), Davis et al. (13); (△), Donnelly & Katz (14); (●), Sobocinski & Kurata (15); (——), P–R prediction.

dioxide binary system agrees very well with that obtained by Davis et al. (*13*) at temperatures up to −90°F. At temperatures higher than this value and up to the triple point of carbon dioxide, slight discrepancies exist between the predicted solid-formation temperatures and those determined by Davis et al. (*13*) and by Donnelly and Katz (*14*) at specified pressures. However, the maximum difference between the predictions and the data of Davis et al. in this region was only about 2.5°F. The predicted triple point for pure carbon dioxide was 72.1 psia at −71.2°F as compared with the reported value of 76.9 psia −69.9° ± 0.05°F (*15*). The liquid-phase composition of the methane–ethane–carbon dioxide ternary system at the SLV equilibrium condition is presented in Figure 3. The agreement between the predictions and the data is quite good except for the highest temperature isotherm where the predicted carbon dioxide solubilities are higher than the experimental values. This discrepancy may be caused by the fact that at high temperatures there is substantial enhancement of the solubility of carbon dioxide in hydrocarbon solvents as compared with low temperatures. As an example, the liquid composition of the methane–carbon dioxide binary system at the SLV equilibrium condition is shown in Figure 4 where the carbon dioxide content of the liquid phase changes about 3% per degree change in temperature.

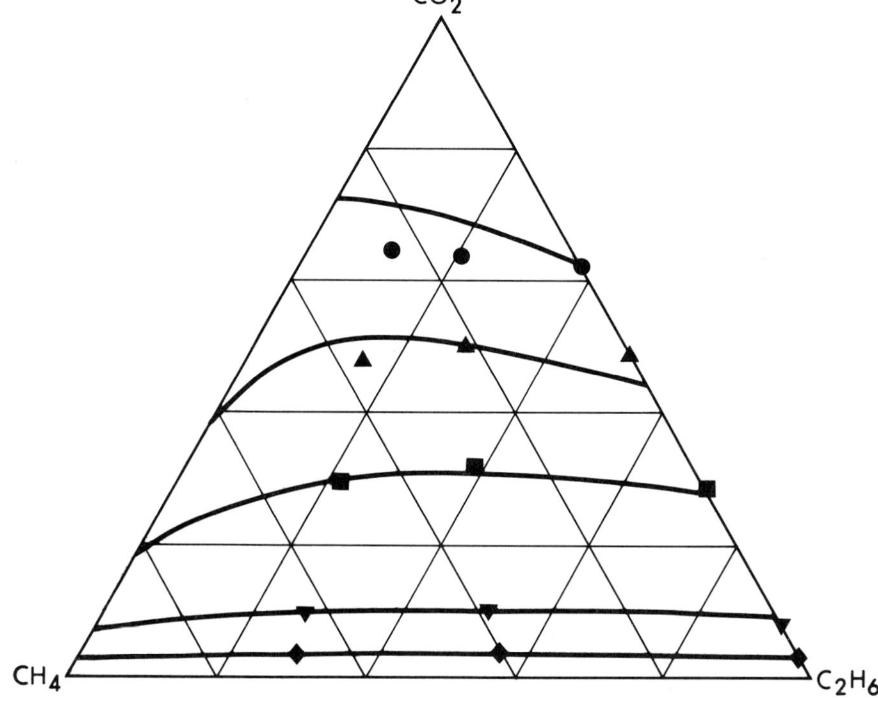

Figure 3. Liquid composition at the solid–liquid–vapor condition in the methane–ethane–carbon dioxide ternary system: (●), −84.9°F; (▲), −90°F; (■), −100°F; (▼), −129.9°F; (◆), −160°F.

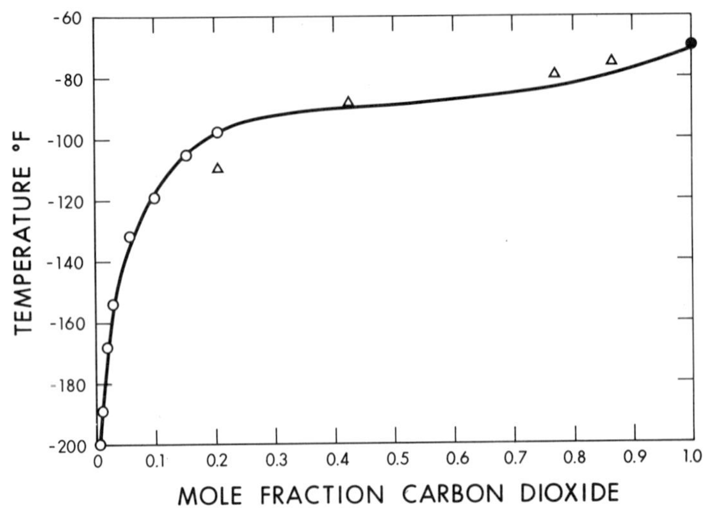

Figure 4. Carbon dioxide content of the liquid phase at the solid–liquid–vapor condition in the methane–carbon dioxide system: (○), Davis et al. (13); (△), Donnelly & Katz (14); (●), Sobocinski & Kurata (15); and (——), P–R prediction.

Conclusion

An efficient procedure for determining the solid-formation condition in a three-phase SLV system has been presented. The Peng–Robinson equation of state generates reliable results for hydrocarbon mixtures containing carbon dioxide as the solid-forming component. Although only the results for carbon dioxide solid are presented as examples, the proposed procedure is applicable to other solid-forming components such as hydrogen sulfide and heavy hydrocarbons. The extension of the procedure to predict the incipient solid forming condition in solid–liquid and solid–vapor equilibrium or for systems involving more than one solid phase is straightforward.

Acknowledgment

The financial assistance received for this work from the National Research Council of Canada is appreciated.

Glossary of Symbols

a = attraction parameter
b = van der Waals co-volume
f = fugacity
n = mole number

N = number of components
P = pressure
R = gas constant
T = temperature
v = molar volume
x = mole fraction
Z = compressibility factor

Greek Letters

α = scaling factor defined by Equation 11
δ = interaction parameter
κ = characterization constant defined by Equation 12
Φ = logarithm of the fugacity coefficient of pure component
ω = acentric factor

Superscripts

L = liquid
S = solid
t = triple point
V = vapor
* = saturation

Subscripts

c = critical property
i, j, k = component identification
m = mixture property
r = reduced property

Literature Cited

1. Null, H. R. *Chem. Eng. Prog. Symp. Ser.* **1967,** *63*(81), 52.
2. Cheung, H.; Zander, E. H. *Chem. Eng. Prog. Symp. Ser.* **1968,** *64*(88), 34.
3. Puri, S.; Kohn, J. P. *J. Chem. Eng. Data* **1970,** *15*, 372.
4. Jensen, R. H.; Kurata, F. *A.I.Ch.E. J.* **1971,** *17*, 357.
5. deMateo, A.; Kurata, F. *Ind. Eng. Chem. Process Des. Dev.* **1975,** *14*, 137.
6. Peng, D.; Robinson, D. B. *Ind. Eng. Chem., Fundam.* **1976,** *15*, 59.
7. Peng, D.; Robinson, D. B. *Can. J. Chem. Eng.* **1976,** *54*, 595.
8. Ng, H.-J.; Robinson, D. B. *Ind. Eng. Chem., Fundam.* **1976,** *15*, 293.
9. Ng, H.-J.; Robinson, D. B. *A.I.Ch.E. J.* **1977,** *23*, 477.
10. Peng, D.; Robinson, D. B. *A.I.Ch.E. J* **1977,** *23*, 137.
11. Perry, J. H. "Chemical Engineers' Handbook," 3rd ed.; McGraw-Hill: New York, 1950; p. 254.
12. Weast, R. C. "Handbook of Chemistry and Physics," 48th ed.; The Chemical Rubber Co.: Cleveland, OH, 1968; p. D-141.
13. Davis, J. A.; Rodewald, N.; Kurata, F. *A.I.Ch.E. J.* **1962,** *8*, 537.
14. Donnelly, H. G.; Katz, D. L. *Ind. Eng. Chem.* **1954,** *46*, 511.
15. Sobocinski, D. P.; Kurata, F. *A.I.Ch.E. J.* **1959,** *5*, 545.

RECEIVED August 10, 1978.

11

Applications of the Augmented van der Waals Theory of Fluids: Tests of Some Combining Rules for Mixtures

ALEKSANDER KREGLEWSKI and STEPHEN S. CHEN

Thermodynamics Research Center, Texas A&M University,
College Station, TX 77843

An accurate equation of state of fluids is used to test the combining rules for the interaction energy $\bar{u}_{ij}$ *of mixtures derived from the theories of London energy between small or large (chain) molecules. The tests are based mostly on comparisons with the thermodynamic excess functions of binary systems at* $P = 0$. *For long chains the parameter* η/k, *determining the dependence of* $\bar{u}/k$ *on the temperature, depends on the reduced density* $\rho^* = V^\circ/V$ *of the system (where* V° *is the close-packed volume) and* $\eta/k \to 0$ *when* $\rho^* > 0.75$ *(solid state).*

The thermodynamic properties of mixtures are extremely sensitive to the values of mixed-pair interaction energies $u_{ij}(r)$ relative to those acting between the molecules of the pure components, $u_{ii}(r)$ and $u_{jj}(r)$, and to the average collision diameter σ_x of the mixture obtained from σ_{ii}, σ_{jj}, and σ_{ij}. A relatively small error in u_{ij} greatly affects the phase diagram obtained. If the predicted value of u_{ij} is too low by 2%, the phase diagram may, for example, exhibit an azeotropic mixture or even a separation into three phases, liquid | liquid | gas, while the real diagram shows only two phases and no azeotrope.

The purpose of this chapter is to test the combining rules that result from three approximations for random mixtures of fluids. The square-well (SW) approximation yields very good results for systems of small molecules. The complexity of the problem for large molecules, aggravated by the dependence of $u(r)$ on the reduced density of the system, is outlined.

0-8412-0500-0/79/33-182-197$05.00/1

Systems of Small Molecules

We shall limit our consideration to systems with London (dispersion) energies only. For such systems London (*1, 2*) derived the famous relation

$$u^{\circ}_{ij}(r) = -\frac{3}{2}\frac{p_i p_j}{r^6}\left[\frac{1}{h\nu_i} + \frac{1}{h\nu_j}\right]^{-1} \tag{1}$$

in which p_i and p_j are the polarizabilities of the two molecules, r is the distance between their centers, and $h\nu$ are certain energies which probably are related to the ionization potentials I. Except for He and H_2, $h\nu$ is not equal or proportional to I. The superscript "o" in u_{ij}° indicates that the relation was derived for small spherical molecules.

Taking into account that a hard repulsion exists at close distances r, we may write

$$u_{ij}(r) = \infty \qquad \text{at } r \leqslant \sigma \tag{2}$$

and

$$u_{ij}(r) = -\bar{u}_{ij}(\sigma_{ij}/r)^6 \qquad \text{at } r > \sigma \tag{3}$$

where $\bar{u}_{ij}$ is the minimum value of $u_{ij}(r)$. For small spherical molecules $\bar{u}_{ij} = \bar{u}_{ij}^{\circ}$. By comparing Equations 1 and 3 for the pure components, one obtains

$$h\nu_i = \frac{4}{3}\,\bar{u}^{\circ}_{ii}\,\sigma^6_{ii}/p^2_i \text{ and } h\nu_j = \frac{4}{3}\,\bar{u}^{\circ}_{jj}\,\sigma^6_{jj}/p^2_j$$

Upon substitution into Equation 1 the unknown quantities $h\nu$ are eliminated.

$$\bar{u}^{\circ}_{ij} = 2\left[\frac{1}{\bar{u}^{\circ}_{ii}}\frac{p_i}{p_j}\left(\frac{\sigma_{ij}}{\sigma_{ii}}\right)^6 + \frac{1}{\bar{u}^{\circ}_{jj}}\frac{p_j}{p_i}\left(\frac{\sigma_{ij}}{\sigma_{jj}}\right)^6\right]^{-1} \tag{4}$$

This combining rule was derived in a different manner by Kohler (*3*).

It is now well known (*4*) that the so-called random mixing approximation, which is based on the Lennard–Jones potential, greatly exaggerates the effect of size differences of the molecules on u_{ij}. Since the mixtures are random in all the models considered in this chapter, it is more appropriate to call that in which $u(r)$ varies with $(\sigma/r)^6$ the Lennard–Jones (LJ) approximation.

In the SW approximation and at a constant width (σ/r) of the well for all substances we have

$$u_{ij}(r) = \text{constant } \overline{u}_{ij} \qquad (r > \sigma); \tag{5}$$

and instead of Equation 4 we obtain

$$\overline{u}^{\circ}_{ij} = 2\left[\frac{1}{\overline{u}^{\circ}_{ii}}\frac{p_i}{p_j} + \frac{1}{\overline{u}^{\circ}_{jj}}\frac{p_j}{p_i}\right]^{-1} \text{(small molecules)} \tag{6}$$

Also, the average $\bar{u}_m$ of the mixture becomes independent of σ

$$\overline{u}_m = \sum_i \sum_j x_i x_j \overline{u}_{ij} \tag{7}$$

where $\bar{u}_{ij} = \bar{u}_{ij}^{\circ}$ for small molecules.

In the theory of van der Waals (VDW) the parameter a is proportional to $\bar{u}\sigma^3$, where $\bar{u}$ is the minimum value of an unknown potential $u(r)$. Leland et al. (*3*) have shown that the VDW approximation is much better than the LJ approximation. In this approximation the combining rules always are applied to a_{ij}, usually $a_{ij} = k_{ij}(a_{ii}a_{jj})^{1/2}$, where k_{ij} is a constant because nothing better may be derived when $u(r)$ is not known.

Kac et al. (*5*) have shown that the attraction term in the VDW equation of state, a/RTV, is obtained when the intermolecular forces are infinitely weak, $du(r)/dr = 0$. This is also the characteristic feature of the SW potential which differs from the Kac potential in that the attraction is cut off at a certain (r/σ), usually at 3/2. Hence, Equation 6 is the proper combining rule for $\bar{u}_{ij}$ in a_{ij}.

On the contrary, if we insist on coupling $\bar{u}_{ij}$ with σ_{ij} and using the VDW approximation for a_m, the effect on the calculated properties of mixtures is the same as if $\bar{u}_{ij}(r)$ were expressed by a hypothetical potential $\bar{u}_{ij}(\sigma_{ij}/r)^3$. This leads to a combining rule given by Equation 4 in which all of the $\sigma^{6\text{'s}}$ are replaced by $\sigma^{3\text{'s}}$. This is called the rule for the VDW approximation.

The rules for the VDW and SW approximations are compared with the experimental data for mixtures of noble gases in Table I. The experimental values of $\bar{u}_{ij}$ are selected by Smith et al. (*6*) from molecular beam scattering, second virial coefficient, and viscosity data. The values of $\bar{u}/k$ of the pure components are: He, 10.5 K; Ar, 140 K; Kr, 196 K; and Xe, 265 K. The values of σ given by Chen et al. (*6*) are: He, 2.65 Å; Ne, 2.75 Å; Ar, 3.34 Å; Kr, 3.64 Å; and Xe, 3.81 Å. The polarizabilities were taken from the Landolt–Börnstein Tabellen (1960). By comparing the

Table I. Values of $\bar{u}^{\circ}_{ij}$ for Mixed Interactions between Noble Gases (K)

	Experimental	*Equation 6*	*VDW*[a]	*Equation 8*
He + Ar	30.2	29.0	33.0	38.6
He + Kr	30.2	28.8	35.2	45.6
He + Xe	28.0	25.3	33.8	54.3

[a] Equation 4 with all σ^6's replaced by σ^3. The $\sigma^3{}_{ij}$ is calculated from Equation 10.

values of $\bar{u}_{ij}$ given by Smith et al. with those of Chen et al. (which are lower) we estimate their uncertainty to be about ±10%. These data show that Equation 6 is the best approximation. The last column in Table I gives the geometric mean values

$$\bar{u}_{ij} = (\bar{u}_{ii}\bar{u}_{jj})^{1/2} \tag{8}$$

The comparisons with other existing combining rules were made by Smith et al. and are not repeated here.

The two most commonly used rules for σ_{ij}, namely that owing to Lorentz,

$$\sigma_{ij} = (\sigma_{ii} + \sigma_{jj})/2, \tag{9}$$

and to van der Waals

$$\sigma^3{}_{ij} = (\sigma^3{}_{ii} + \sigma^3{}_{jj})/2 \tag{10}$$

are compared with the experimental data given by Chen et al. (7) in Table II. Equation 10 appears to be much better than Equation 9 although it is oversimplified in that it neglects the effect of $\bar{u}_{ij}$ on σ_{ij}. The average σ_m of the mixture, derived from Equation 10, is

$$\sigma^3{}_m = \sum_i x_i\, \sigma^3{}_{ii} \tag{11}$$

Table II. Collision Diameters σ_{ij} for Mixed Interactions between Noble Gases (in Å)

	Experimental	*Equation 9*	*Equation 10*
He + Ne	2.73	2.70	2.70
He + Ar	3.09	3.00	3.03
He + Kr	3.27	3.14	3.22
He + Xe	3.61	3.23	3.33

In addition to the above direct tests, the rule given by Equation 6 was tested by calculating the thermodynamic excess functions of mixtures. The relations are given by Kreglewski and Chen (*8*). The basic one for the residual Helmholtz energy of a mixture $A_m{}^r$ is

$$\frac{A_m{}^r}{RT} = (\alpha_m{}^2 - 1)\ln(1 - \xi_m) + \frac{(\alpha_m{}^2 + 3\alpha_m)\xi_m - 3\alpha_m\xi_m{}^2}{(1-\xi_m)^2}$$

$$+ \sum_n \sum_m D_{nm}\left(\frac{\bar{u}_m}{kT}\right)^n \left(\frac{V^\circ{}_m}{V_m}\right)^m \qquad (12)$$

The first two terms were derived from Boublik's (*9*) relation for the compressibility factor of hard convex bodies. Here α is a constant depending on the shape of the molecules ($\alpha = 1$ for spherical molecules), for mixtures assumed to be $\alpha_m = \sum_i x_i\alpha_i$.

The term $\xi_m = 0.74048\ V_m{}^\circ/V_m = 1/6\ \pi N_o\sigma_m{}^3/V_m$, where $V_m{}^\circ$ is the close-packed volume, N_o is the Avogadro number, and V_m is the molar volume of the system. V° is a simple function of the temperature (T) (*10*) with a characteristic value $V^{\circ\circ}$ at $T = 0$ K. The last term in Equation 12 was introduced by Alder et al. (*11*). D_{nm} are 24 universal constants common for all substances whose radial and higher distribution functions are the same functions of $\bar{u}/kT$ and the reduced density $\rho^* = V^\circ/V$. As shown by Chen and Kreglewski (*10*) and Simnick, Lin, and Chao (*12*), Equation 12 is the most accurate known equation with four characteristic constants: α, $V^{\circ\circ}$ (V° at $T = 0$ K), $\bar{u}^\circ/k$, and η/k (*see* Equations 13 and 14). They also have shown (*10*) that in order to obtain agreement with second virial coefficient data of the gas and the internal energy or the enthalpy of the liquid, it is necessary to assume that $u(r)$ is a function of T as required by the theory of noncentral forces between nonspherical molecules (*13*)

$$\bar{u}^L{}_{ij} - \bar{u}^\circ{}_{ij}(1 + \eta^L{}_{ij}/kT) \qquad (13)$$

where L indicates London interactions. For a pure fluid (*10*)

$$\eta^L/k \approx 0.60\ \omega T^c \qquad (14)$$

where ω is the acentric factor and T^c is the critical temperature in K. They also have concluded that

$$\eta^L{}_{ij} \approx (\eta^L{}_{ii} + \eta^L{}_{jj})/2 \qquad (15)$$

The values of the characteristic constants for Ar, Kr, N_2, and CH_4 are given in Ref. *10*. For Xe, O_2, and all the substances considered in Table IV the constants were estimated as follows. The effect of the value of α on the excess functions is negligible; therefore, α was set equal to unity. The η/k was estimated from Equation 14 and $V^{\circ\circ}$ from the liquid molar volume at the reduced temperature $T/T^c = 0.6$.

The energy $\bar{u}^\circ/k \approx T^c$ only if $\eta/k = 0$. In other cases it is not related so directly to a macroscopic property but it can be estimated by relying on the accuracy of Equation 12. When all of the characteristic constants are known, Equation 12 yields accurate molar volumes of the saturated liquid and vapor phases obtained by an iteration at constant T and P from the Gibbs equilibrium condition. Hence, if α, $V^{\circ\circ}$, and η/k have been estimated already, $\bar{u}^\circ/k$ is found by varying its value until the liquid volume calculated from Equation 12 or from the compressibility factor Z agrees with the observed value at a certain T, say, $T/T^c = 0.6$.

The values of η/k for the components given in Table III are 1 K for CH_4, 2 K for O_2, 3 K for N_2, and zero for Ar, Kr, and Xe. The values of $\bar{u}_{ij}^L$ in Equation 13 were calculated from Equations 6 and 15. Considering that a small error in the combining rules leads to large errors in G^E and H^E, the agreement with the observed values is very satisfactory. Exceptionally, for the $N_2 + CH_4$ system Equation 15 yields $\eta_{12}^L = 2$ K and values of the excess functions (*8*) that are too small. Nitrogen is a quadrupolar molecule and this effect varies with $(kT)^{-1}$ analogously to the London energy. If the value of η_{11}/k is ascribed to quadrupole interactions, η_{11}^Q/k, we have N_2: $\eta_{11}/k = \eta_{11}^Q/k = 3$, $\eta_{11}^L/k = 0$; CH_4: $\eta_{22}/k = \eta_{22}^L/k = 1$, $\eta_{22}^Q/k = 0$. Since $\eta_{12}^Q \sim (\eta_{11}^Q\,\eta_{22}^Q)^{1/2}$, we obtain $\eta_{12}/k = (\eta_{12}^L + \eta_{12}^Q)/k = 0.5$ K instead of $\eta_{12}/k = \eta_{12}^L = 2$ K for the total.

Table IV. Excess Functions of

Components (1) (2)	T/K	η_{22}/k	ι_{12}
n-C_6 + n-C_{12}	293	218[e]	[f]
	293	0	1.33
n-C_6 + n-C_{16}	293	305[e]	[f]
	293	0	0.91
cy-C_5 + cy-C_8	298	101[e]	[f]
	298	0	1.275
cy-C_5 + OMCTS	298	194[e]	1.11
	298	28	0.834

[a] Binary systems at $x = 0.5$.
[b] The values kept constant are $\eta/k = 88$ of n-C_6, $\eta/k = 59$ K of cyclo-C_5 and u°/k and V° of all the components.
[c] Weighing of interactions: mole fractions.

Table III. Excess Functions of Liquid Systems of Small Molecules[a]

System	*Observed Values*[c]				*Predicted Values, Equation 6*		
	T/K	G^E	H^E	V^E	G^E	H^E	V^E
Ar + Kr	103.94	82.5	—	—	87	47	−0.19
	115.77	83.9	—	−0.53	93	26	−0.51
	115.77	82.4	—	−0.464	93	26	−0.51
Kr + Xe	161.36	103	—	−0.459	135	64	−0.51
	161.36	114	—	−0.695	135	64	−0.51
Ar + O_2	83.82	37	60	+0.14	35	50	+0.12
Ar + CH_4	91	72	103	+0.18	79	100	+0.15
	115.77	76	—	—	79	69	−0.12
Kr + CH_4	115.77	29	55	−0.025	22	22	−0.02
N_2 + CH_4	90.7	169.6	—	—	166[d]	190[d]	−0.71[d]
	91.5	—	138	—	166[d]	186[d]	−0.75[d]
	105.0	—	101	−1.16	169[d]	86[d]	−1.95[d]

[a] Binary systems at $x = 0.5$.
[b] G_E and H_E are in joules mole^{-1} and V_E is in cubic centimeters mole^{-1}.
[c] The references to the observed values are given in Refs. *8* and *17*. Also, some of the calculated values were taken from Ref. *8*.
[d] Values calculated with the assumption that η/k of nitrogen is caused by the quadrupole moment (*see* the text).

Systems of Large Molecules

When one or more components are large molecules, Equation 6 begins to fail. Departures already are noted for ethane ($\eta/k = 19$ K). The reason may be that only a part of the large molecule is involved in the mixed interaction $\bar{u}_{ij}$. Accordingly, the concept of interactions between segments of molecules is popular. Since for molecules with London

Liquid Systems of Large Molecules[a]

Calculated[b, c]			*Observed*[d]		
G^E	H^E	V^E	G^E	H^E	V^E
—	—	—	−25	+46	−0.31
−20	+31	−0.31			
—	—	—	−65	+129	−0.48
−50	+95	−0.46			
—	—	—	+2	−40	−0.28
+1	−34	−0.20			
−209	+516	+2.36	−208	+212	+0.05
−215	+334	+0.06			

[d] The references to the experimental data are given in Ref. *17*.
[e] From Equation 14.
[f] Wrong results for any value of ι_{12}.

energies (only) p_i is an approximately linear function of σ_{ii}^3 (expressed by the critical volume V_i^c or the close-packed volume V_i°), the combining rule for interactions ϵ_{ij}° between segments of equal size and, implicitly, equal polarizabilities is

$$\epsilon^\circ_{ij} = 2\,[(1/\epsilon^\circ_{ii}) + (1/\epsilon^\circ_{jj})]^{-1} \qquad (16)$$

Simultaneously, the weighing of interactions by means of the mole fractions in Equation 7 must be replaced by a more proper measure of the probabilities of ϵ_{ii}, ϵ_{jj}, and ϵ_{ij} interactions. The site fractions, as defined by Flory (*14*), are used for systems of chain molecules but they contain the parameter s_i/s_j that has to be fitted to a given system. Kreglewski and Kay (*15*) obtained very good results for the critical constants of mixtures by using Equation 16 and crudely defined surface fractions. The agreement with observed data is so good that the concept should not be abandoned only because the surface fractions have a poor theoretical foundation. There must exist a bridge between the mole fractions (appropriate for small spherical molecules) and the site fractions for long chains.

The concept of segment interactions has not been used in our present work. Our considerations are limited to the exact treatment based on total interactions u_{ij} and mole fractions. According to Salem (*16*) the London energy between two long straight or circular chains at close distances is proportional to $m\,u^\circ(r)$ where $u^\circ(r)$ is given by Equation 1 and m is the number of chain units. The same transformation that led to Equation 6 yields

$$\bar{u}^\circ_{ij} = 2\,[(1/\bar{u}^\circ_{ii})(1/\iota_{ij}) + (1/\bar{u}^\circ_{jj})\,\iota_{ij}]^{-1} \quad \text{(large molecules)} \qquad (17)$$

where $\iota_{ij} = p_j m_j/(p_i m_i)$. The values of ι_{ij} required to fit the observed values of G^E of the n-hexane + n-dodecane or n-hexadecane and the cyclopentane + cyclooctane or octamethyl-cyclotetrasiloxane (OMCTS) system are close to unity and so they differ very much from the theoretical values.

There is a more serious problem than the necessity of using empirical values of ι_{ij}. The values of $\bar{u}^\circ/k$, η/k, and V° of the components were estimated from the properties of the liquid state. The values of η/k, estimated by means of Equation 14, range from 59 K for cyclopentane to 305 K for hexadecane. The value for η_{ij} was either calculated from Equation 15 or deliberately varied within reasonable limits. Also ι_{ij} was varied. We have established that none of the combinations of η_{ij} and ι_{ij}

yield simultaneously good values of G^E, H^E, and V^E. The results are not improved when mole fractions in Equation 7 are replaced by surface or site fractions.

The results made it clear that the values of η/k of the pure components must be changed. This conclusion is strongly endorsed by the following results obtained recently by one of the authors. For both spherical and chain molecules up to *n*-pentane or cyclopentane the same values of η/k, obtained from Equation 14, yield accurate values of the second virial coefficients $\beta\ (T)$ (reduced density $\rho^* = 0$) and the residual internal energies U^r of the liquid (at $\rho^* > 0.52$). The acentric factors are determined at $\rho^* \approx 0.52$. For larger molecules, beginning with benzene or *n*-C_6, η/k values required to fit $\beta\ (T)$ data become larger than Equation 14 suggests whereas that fitting U^r data of the liquid becomes slightly smaller (at constant $\bar{u}^\circ/k$ and V°). That is, η/k begins to depend on ρ^*. The value of η/k varies with ρ^* more rapidly for *n*-alkanes (*n*-C_7, etc.) than for cycloalkanes (cy-C_7, etc.). The phenomenon is related clearly to restricted rotation at high densities. When the chain is sufficiently long and $\rho^* > 0.75$ (solid state) we can expect $\eta/k \rightarrow 0$ (completely restricted rotations).

Accordingly, we have retained the values of η/k of *n*-C_6 and cy-C_5 calculated from Equation 14, but the values for *n*-C_{12}, *n*-C_{16}, cy-C_8, and OMCTS were decreased to an appropriate level. Simultaneously, $\eta_{ij} = (\eta_{ii} + \eta_{jj})/2$ and ι_{ij} were varied until the three properties G^E, H^E, and V^E agreed within $\pm$ 20 J mol^{-1} and ± 0.05 cm^3 mol^{-1} with the observed values.

The results are compared in Table IV. The first row for each system shows the errors in H^E and V^E when ι_{12} is fitted to G^E and η_{22} is that calculated from Equation 14. The second row shows the improvement of the results when η_{22} is properly diminished. For three of the systems $\eta_{22} = 0$ is required. In the fourth system, a decrease of η_{22}/k below 28 K would improve H^E; however, V^E would become negative. The value of ι_{12} appears to vary in an unpredictable manner. When the surface fractions are used (with the same values of η_{22}) then always $\iota_{12} > 1$ in qualitative agreement with the theory of Salem. However, ι_{12} cannot be predicted when the systems are treated as random mixtures. It is shown elsewhere (*18*) that the properties of mixtures of large molecules can be predicted with nearly the same accuracy as those of small molecules by introducing an approximate correction to $A_m{}^r$ owing to nonrandom mixing.

In all theories of polymer solutions $\bar{u}_{ij}$ *or* ϵ_{ij} always are assumed to be independent of the temperature, apparently contradicting the theory of noncentral forces. Our results show that there is no contradiction and that this assumption is allowed for long straight or circular chains at high densities (liquids below their normal boiling temperatures).

Glossary of Symbols

T = temperature in K
P = pressure in bar
V = volume of the system
V° = close-packed volume ($V^\circ = V^{\circ\circ}$ at $T = 0$)
A = Helmholtz free energy, $A(T, V)$
U = internal energy, $U(T, V)$
G = Gibbs free energy, $G(T, P)$
H = enthalpy, $H(T, P)$
I = ionization potential
k = Boltzmann constant ($\mathrm{R} = \mathrm{N_0 k}$ is the gas constant)
r = intermolecular distance
$u(r)$ = pair interaction energy
$\bar{u}, \bar{u}^\circ$ = minimum value of $u(r)$ ($\bar{u}/\mathrm{k}$ in K)
η/k = parameter of noncentral energy (in K)
σ = collision diameter of a molecule
p = mean polarizability of a molecule
ω = acentric factor
x = mole fraction
ρ^* = reduced density, V°/V

Superscripts

r = residual property (real fluid minus perfect gas at the same T and P or T and V)
E = excess property (real mixture minus ideal mixture at the same T, P, and x)
c = gas–liquid critical constants

Subscripts

i, j = interacting species ($\mathrm{i} = 1,2, \ldots m$; $j = i$ or $j \neq i$)
m = molar property of a fluid or per mole of a m-component mixture, e.g., $V_m/\mathrm{cm^3\ mol^{-1}}$; also an average value of a molecular property of a mixture, e.g., $\bar{u}_\mathrm{m}$ or σ_m.
Excess properties, e.g., G^E, are also molar but the subscript is deleted.

Acknowledgments

Dr. Chen participated in an early stage of this work. He died in September, 1977.

Thanks to B. J. Zwolinski, former Director of the Thermodynamics Research Center, for encouraging this research and to C. Chen for her help in computer programming. Partial support of the Texas Engineering Experiment Station is gratefully acknowledged.

Literature Cited

1. London, F. *Z. Phys.* **1930**, *63*, 245.
2. London, F. *Z. Phys. Chem., Abt. B* **1930**, *11*, 222.
3. Kohler, F. *Monatsh. Chem.* **1957**, *88*, 857.
4. Leland, T. W., Jr.; Rowlinson, J. S.; Sather, G. A. *Trans. Faraday Soc.* **1968**, *64*, 1447.
5. Kac, M.; Uhlenbeck, G. E.; Hemmer, P. C. *J. Math. Phys.* **1963**, *4*, 216, 229.
6. Smith, K. M.; Rulis, A. M.; Scoles, G.; Aziz, R. A. *J. Chem. Phys.* **1977**, *67*, 155.
7. Chen, C. H.; Siska, P. E.; Lee, Y. T. *J. Chem. Phys.* **1973**, *59*, 601.
8. Kreglewski, A.; Chen, S. S. *J. Chim. Phys.* **1978**, *75*, 347.
9. Boublik, T. *J. Chem. Phys.* **1975**, *63*, 4084.
10. Chen, S. S.; Kreglewski, A. *Ber. Bunsenges. Phys. Chem.* **1977**, *81*, 1048.
11. Alder, B. J.; Young, D. A.; Mark, M. A. *J. Chem. Phys.* **1972**, *56*, 3013.
12. Simnick, J. J.; Lin, H. L.; Chao, K. C.; Chapter 12 in this book.
13. Rowlinson, J. S. "Liquids and Liquid Mixtures"; Butterworths: London, 1969.
14. Flory, P. J. *J. Am. Chem. Soc.* **1965**, *87*, 1833.
15. Kreglewski, A.; Kay, W. B. *J. Phys. Chem.* **1969**, *73*, 3359.
16. Salem, L. *J. Chem. Phys.* **1962**, *37*, 2100.
17. Kreglewski, A.; Wilholt, R. C.; Zwolinski, B. J. *J. Phys. Chem.* **1973**, *77*, 2212.
18. Kreglewski, A. *J. Chim. Phys.* **1979**, *76* (in press).

RECEIVED August 8, 1978.

12

The BACK Equation of State and Phase Equilibria in Pure Fluids and Mixtures

J. J. SIMNICK, H. M. LIN, and K. CHU CHAO

School of Chemical Engineering, Purdue University, West Lafayette, IN 47907

The Boublik–Alder–Chen–Kreglewski augmented hard-core equation is applied to pure fluids and mixtures with special attention to the representation of phase equilibria. Equation constants are determined for twelve substances, and the three constants which are required of nonpolar, nonquantum fluids are correlated with the critical properties and acentric factor. The equation describes mixture-phase equilibria with the introduction of mixing rules and the use of up to two interaction constants for each binary system.

Many equations of state have been proposed for the representation of thermodynamic properties of pure fluids and mixtures. The success of several equations in the quantitative description of some fluid mixture systems has added incentive to the further development of equations of state in recent years. The new equation of Kreglewski and Chen (*1*) appears particularly attractive for several reasons. It is highly accurate in fitting the *PVT* behavior of a number of substances. Only a few equation constants are required for each substance, and these are properties of the molecules (or very closely related to them) about which much is already known. In this work we apply the equation to pure fluids and mixtures of some common substances. Special attention is paid to the representation of phase equilibria.

The BACK Equation

The Boublik–Alder–Chen–Kreglewski (BACK) Equation is an augmented hard-core equation of the form

$$\frac{P\tilde{V}}{RT} = z = z^{\mathrm{h}} + z^{\mathrm{a}} \tag{1}$$

0-8412-0500-0/79/33-182-209$06.25/1

Equation 1 expresses that the compressibility factor of a real fluid is the sum of a repulsive term and an attractive term. Chen and Kreglewski (*1*) suggested using Boublik's hard-core equation z^{h} for the repulsive term (*2*) and to use the polynomial of Alder et al. (*3*) for the attractive term z^{a}. Thus

$$z^{h} = \frac{1 + (3\alpha - 2)\xi + (3\alpha^2 - 3\alpha + 1)\xi^2 - \alpha^2 \xi^3}{(1 - \xi)^3} \tag{2}$$

$$z^{a} = \sum_{N}^{4} \sum_{M}^{9} M \, D_{NM} \left(\frac{u}{kT}\right)^{N} \left(\frac{\tilde{V}^{\circ}}{\tilde{V}}\right)^{M} \tag{3}$$

The density of the fluid enters in Equation 2 in the form of ξ defined by

$$\xi = 0.74048 \, \tilde{V}^{\circ}/\tilde{V} \tag{4}$$

where $\tilde{V}^{\circ}$ is the close-packed volume of the molecular hard cores. The shape of a hard core is expressed by α, which is defined to be the surface integral of the radius of curvature divided by three times the molecular volume. It is a constant for each molecule and is equal to unity for spheres but greater than one for other convex bodies. Equation 2 reduces to the Carnahan–Starling (*4*) hard-sphere equation for $\alpha = 1$.

The constants D_{NM} in Equation 3 originally were determined by Alder et al. (*3*) to fit their computer-generated data. Chen and Kreglewski (*1*) redetermined the constants based on data on argon. The latter set of constants is used in this chapter. Since their values have been reported in Ref. *1*, they will not be repeated here.

The best description of liquids and compressed gases requires the molecular hard-core volume to be a decreasing function of temperature. Thus Chen and Kreglewski express $\tilde{V}^{\circ}$ by means of

$$\tilde{V}^{\circ} = \tilde{V}^{\circ\circ} \left[1 - C \exp\left(-3u^{\circ}/kT\right)\right]^3 \tag{5}$$

The characteristic energy u in Equation 3 is independent of temperature for spherical molecules. However, for nonspherical molecules u depends on temperature and Chen and Kreglewski use

$$\frac{u}{k} = \frac{u^{\circ}}{k}\left(1 + \frac{\eta}{kT}\right) \tag{6}$$

in which $\eta = 0$ for spheres, and $\eta > 0$ for acentric molecules.

Five constants must be known for each substance: $\widetilde{V}^{\circ\circ}$, α, u°/k, η/k, and C. Of these five only three must be determined from fitting experimental data, and these are $\widetilde{V}^{\circ\circ}$, α, and u°/k. Chen and Kreglewski suggested assigning values to the other two constants: C was given the same value, 0.12, for all nonpolar substances and $\eta/k = 0.6\ \omega\ T_c$. Chen and Kreglewski reported values of the constants for 11 substances (*1*).

The accuracy of the BACK equation for the representation of *PVT* data is tested with argon. Figure 1 shows the quantitative agreement that is obtained. Comparison with experimental data such as in Figure 1, however accurate, is nevertheless fragmentary on account of the limited amount of data on any one substance. In order to reveal the behavior of the equation over a wide range of conditions, we compare the computed compressibility of argon with Pitzer's generalized correlation for simple

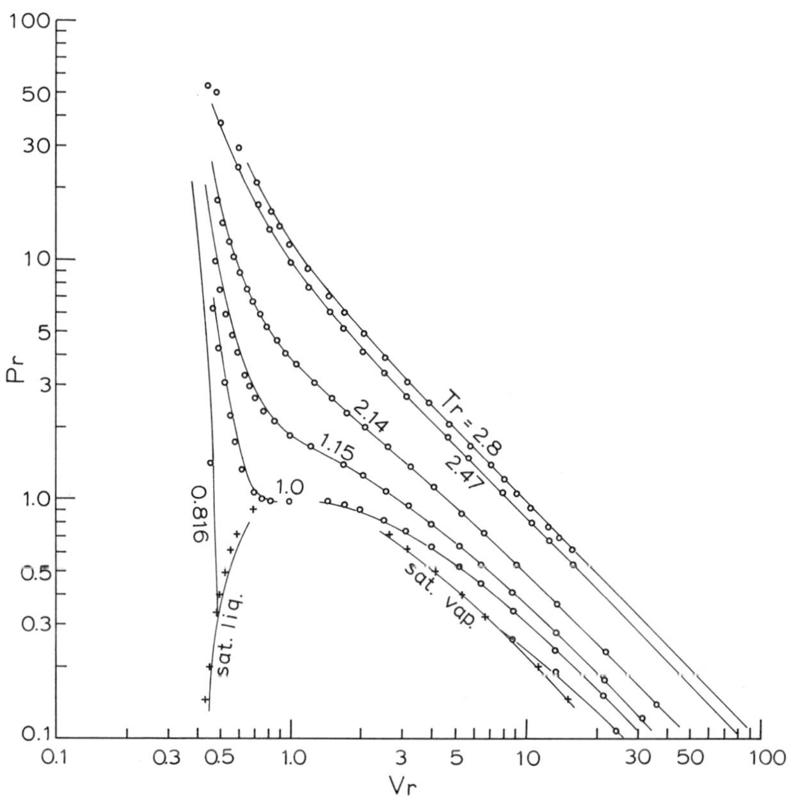

Figure 1. BACK equation and PVT data for argon: (○), Michels data (31); (+), Gibbons Correlation (32); (——), BACK Equation.

fluids of which argon is one. Figure 2 shows the comparison. In the small insert in the figure we show the limiting behavior of the gas as $P \rightarrow 0$. Clearly the second virial coefficient is represented by the equation.

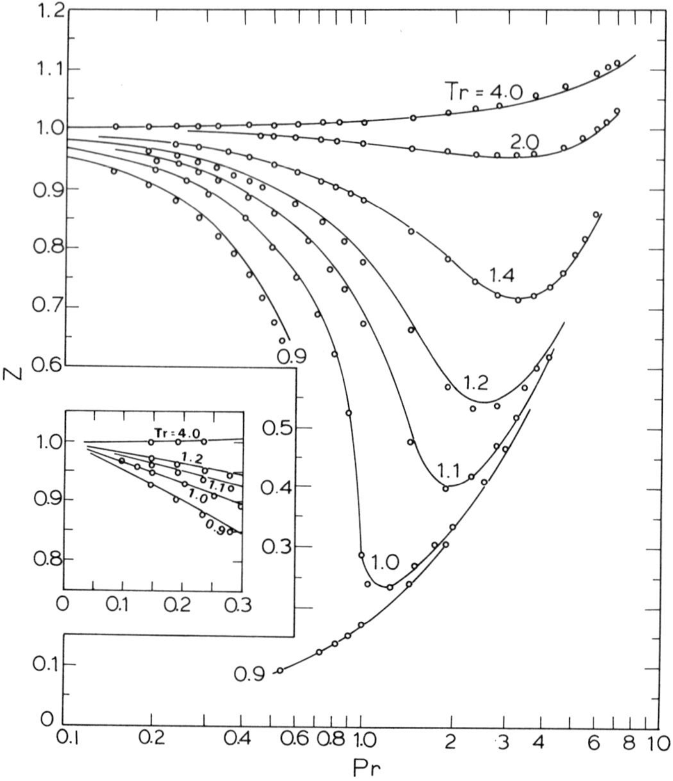

Figure 2. BACK equation for argon and generalized correlation: (○), BACK Equation; (——), Pitzer.

Determining Equation Constants

Equation constants are determined for 12 substances in this chapter and the results are shown in Table I.

To determine the three constants $\tilde{V}^{\circ\circ}$, α, and u°/k for a substance, we use the critical constants, vapor pressure, and liquid-density data. An objective function is defined as the sum of squares of the relative deviations of those calculated from experimental values. The three equation constants are found when minimizing the objective function. For use in

Table I. BACK Equation Constants

Compound	$\tilde{V}^{oo}$, *cm³/mol*	α	u^{o}/k, K	η/k, K[a]	*C*
Hydrogen	13.625	1.0004	39.171	0.499	0.241
n-Pentane	65.751	1.0566	435.83	70.72	0.12
i-Pentane	64.958	1.0565	432.20	62.71	0.12
neo-Pentane	65.518	1.0498	409.59	51.28	0.12
(Kreglewski)	67.740	1.075	418.74	50	0.12
n-Hexane	77.228	1.0720	468.33	90.11	0.12
n-Heptane	88.351	1.0799	491.00	113.77	0.12
n-Octane	96.556	1.0981	517.52	134.50	0.12
n-Decane	110.72	1.1349	558.07	181.57	0.12
Benzene	54.383	1.0587	532.12	71.50	0.12
Toluene	67.013	1.0621	552.43	91.24	0.12
Cyclohexane	64.772	1.0583	522.46	70.72	0.12
m-Xylene	79.037	1.0705	563.23	122.54	0.12

[a] $\eta/k = 0.6\ \omega\ T_c$.

the objective function, the critical constants are determined from the BACK equation by solving numerically the BACK equation itself and Equations 7 and 8,

$$\left(\frac{\partial P}{\partial V}\right)_T = 0 \tag{7}$$

$$\left(\frac{\partial^2 P}{\partial V^2}\right)_T = 0 \tag{8}$$

The vapor pressures are calculated from the BACK equation by numerically solving for the vapor and the liquid densities, ρ_G and ρ_L, simultaneously from the two following equations at a fixed temperature

$$\mu_G = \mu_L \tag{9}$$

$$P_G = P_L \tag{10}$$

where μ denotes chemical potential and P is the pressure.

The temperature range of the vapor-pressure and liquid-density data used in the calculations are shown in Table II. The temperatures are chosen to cover the reduced temperature range of approximately 0.60 to 1.0 at even intervals. Also shown in Table II are the relative deviations of the calculated vapor pressures, and saturated liquid and vapor volumes.

The calculated critical properties are generally in good agreement with accepted experimental values. The average absolute deviation amounts to 1.2% for T_c, 3.2% for P_c, and 4.5% for V_c. The smallest

Table II. Temperature Range

Component	*Temperature Range (K)*	*Relative Deviations*[a] *(%)* *Vapor Pressure* *rms*[b]	*AAD*[c]	*BIAS*[d]
n-Pentane	309–455	2.9	2.6	−1.6
i-Pentane	301–420	0.9	0.8	−0.2
neo-Pentane	282–420	1.5	1.3	−0.4
n-Hexane	303–493	3.2	2.8	−2.1
n-Heptane	323–523	3.1	2.8	−1.7
n-Octane	333–553	3.3	3.1	−1.6
n-Decane	373–603	4.9	4.5	−2.3
Benzene	343–543	1.0	0.9	−0.3
Toluene	353–573	1.8	1.6	−0.5
m-Xylene	373–603	2.7	2.4	−0.5
Cyclohexane	333–533	1.2	1.1	−0.1

[a] Relative deviations = dev = (experimental value − calculated value)/experimental value; NOB = number of observations.
[b] The abbreviation rms = $(\Sigma \text{dev}^2/\text{NOB})^{1/2}$.

deviations are observed for cyclohexane, 0.5% for T_c, 0.6% for P_c, and 1.3% for V_c. The largest deviations are observed for *n*-decane, 2.6% for T_c, 9.2% for P_c, and 13.6% for V_c. There is a tendency for the normal paraffins to show greater deviations as the chain length is increased.

As a check of our procedure for determining equation constants we include neo-pentane in this work for which equation constants have been reported by Chen and Kreglewski. The set of constants from this work as well as the set by Chen and Kreglewski are both presented in Table I, and they are in close agreement. Slight differences appear to be caused by the different data used. Kreglewski and Chen used API Research Project 44 tables while we used the recent data of Das (*6*).

We are interested in using the BACK equation for hydrogen mixtures. Therefore we have determined equation constants for hydrogen, and these are included in Table I. *PVT* data (*7*) at temperatures of 111–2778 K and pressures up to 1020 atm are used in this determination. Neither vapor-pressure nor critical-point data are used to avoid complications owing to quantum effects. It is found necessary to adopt an unusual value of the constant C of 0.241. With this C value the calculated pressure shows a relative root-mean-squared deviation of 0.5% and a relative bias of less than 0.1%.

A sensitivity analysis has been made of the calculated results to the values of the equation constants using benzene data. The most sensitive constant is (u°/k). A variation of 1% from its optimal value increases the error of calculated vapor pressures by 6%, and of calculated liquid

and Fitting of Pure Fluid Data

Relative Deviations[a] *(%)*

Liquid Volume			Vapor Volume			
rms	*AAD*	*BIAS*	*rms*	*AAD*	*BIAS*	*Refs.*
0.3	0.2	0.05	3.7	3.0	−0.5	*8*
0.8	0.8	0.8	2.1	1.5	−1.1	*9*
1.3	1.3	1.3	1.6	1.3	−0.7	*6*
0.7	0.6	0.6	4.6	3.5	−1.3	*10*
2.2	2.0	2.0	5.4	4.0	−1.1	*10*
6.1	6.0	6.0	7.9	5.7	−1.4	*10*
13.7	13.7	13.7	—	—	—	*11*
0.3	0.2	−0.1	2.4	1.9	−0.2	*10*
1.1	1.0	−0.3	—	—	—	*11–16*
1.3	1.1	−0.8	—	—	—	*11, 12, 16, 17*
1.2	1.1	1.1	—	—	—	*11*

[c] AAD = Σ |dev|/NOB.
[d] BIAS = Σ dev/NOB.

volumes by 1.0%. Subsequently α is next in importance. A 1% variation in its value produces a response of about 3% in calculated vapor pressures and 1.4% in calculated liquid volumes. A 1% variation in $V^{\circ\circ}$ produces an approximately equal percent response in liquid volumes, but only one tenth as much relative change in vapor pressure. A 1% change in (η/k) makes a difference of about 1.5% in vapor pressures, but only .3% in liquid volumes. Both vapor pressure and liquid volume are insensitive to C.

Correlation of the Equation Constants

Values of the BACK equation constants are now available for 22 substances. Suitable correlations of the constants can add greatly to the usefulness of the equation.

We have found $\tilde{V}^{\circ\circ}$ to be correlated with $\tilde{V}_c$. Figure 3 shows the result. A simple proportionality exists for most of the substances with $\tilde{V}^{\circ\circ} = 0.21\ \tilde{V}_c$. The higher normal paraffins starting with C_8 show a tendency to deviate from the linear relationship.

The interaction energy u°/k was found by Kreglewski and Chen to be equal to the critical temperature for small molecules. In Figure 4 we show u°/k as a function of T_c. The simple equality holds up to propane, above which the hydrocarbons tend to show a curve downward.

Figure 5 shows that the shape parameter α correlates with the acentric factor ω.

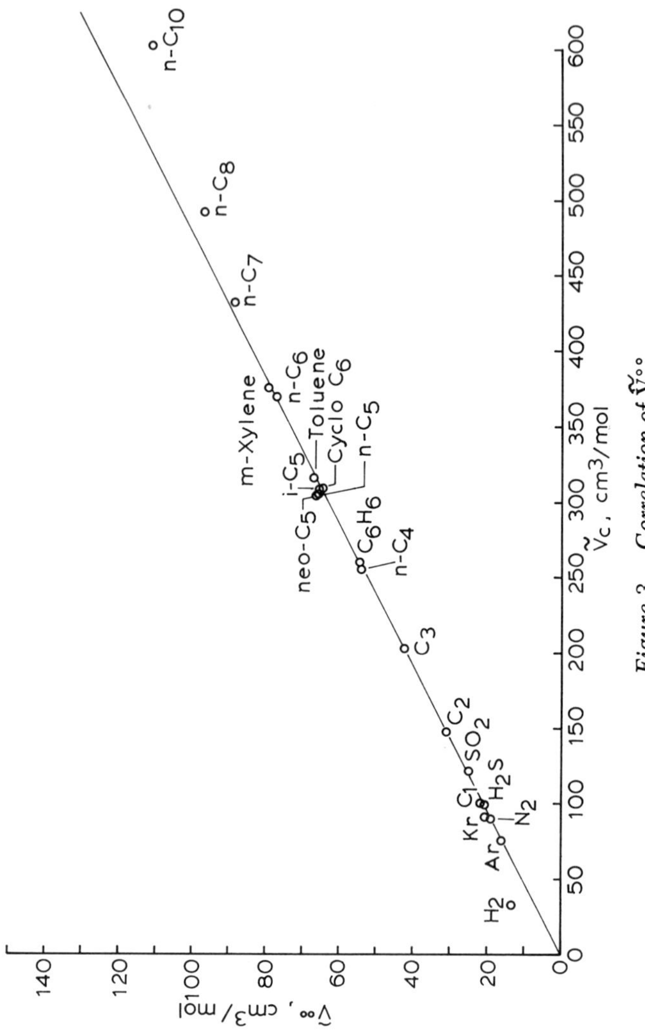

Figure 3. Correlation of $\tilde{V}_\infty$

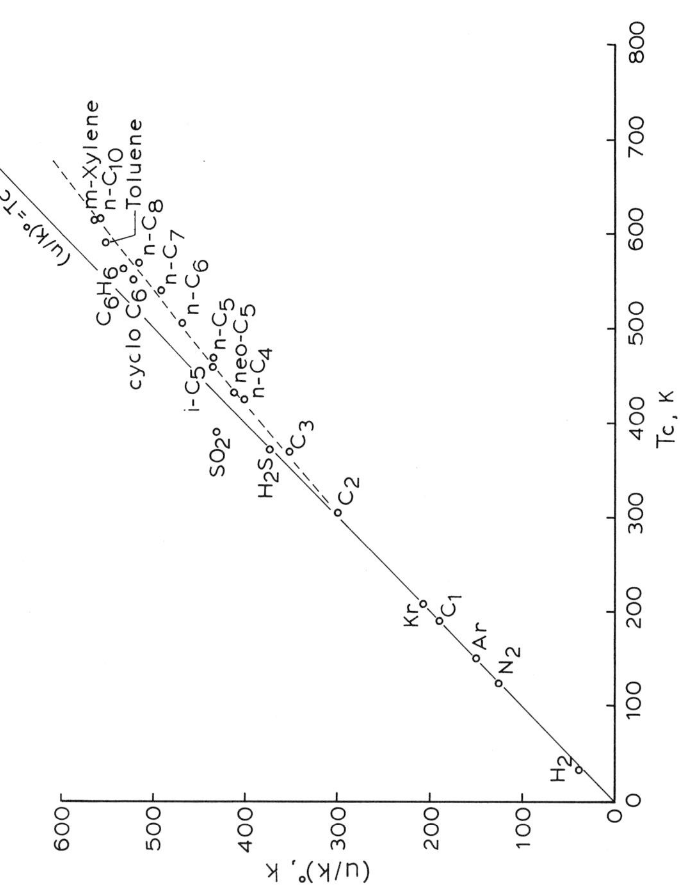

Figure 4. Correlation of u°/*k*

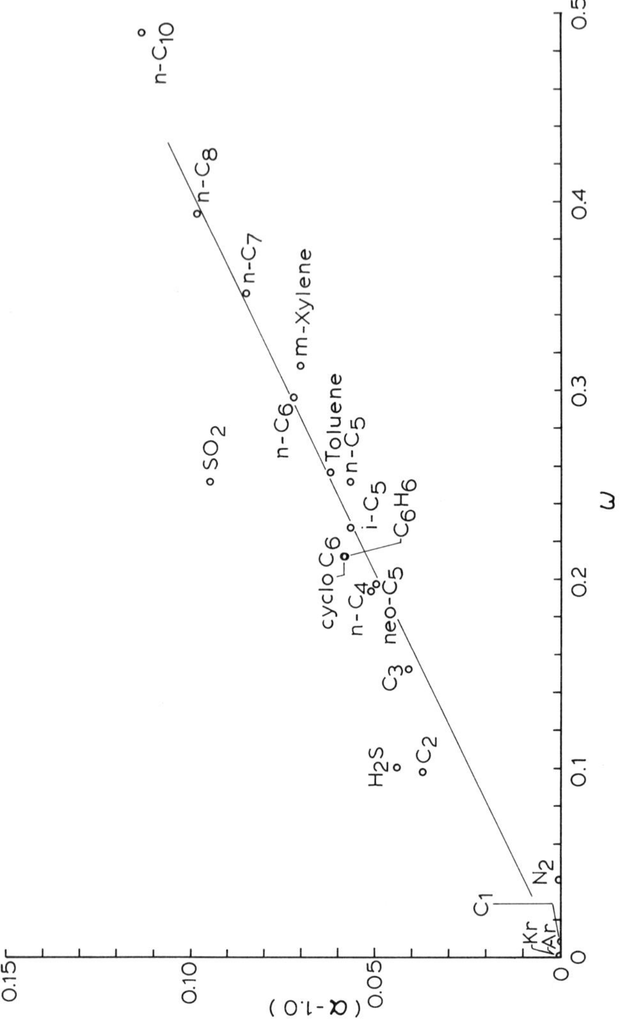

Figure 5. Correlation of α

Mixing Rules

The BACK equation is extended to mixtures with the introduction of mixing rules for the equation constants. The following mixing rules are used in this work:

$$\alpha_m = \sum_i N_i \alpha_i \tag{11}$$

$$\tilde{V}^\circ{}_m = \sum_i \sum_j N_i N_j \tilde{V}^\circ{}_{ij} \tag{12}$$

$$\left(\frac{u}{k}\right)_m = \sum_i \sum_j N_i N_j \left(\frac{u}{k}\right)_{ij} \tag{13}$$

where N stands for mole fraction and subscript m denotes mixtures.

The linear combination of α by Equation 11 was used also by Kreglewski and Chen (5). The mixing of $\tilde{V}^\circ$ and (u/k) by Equations 12 and 13 eliminates the separate mixing of $\tilde{V}^{\circ\circ}$, u°/k, C, and η.

The cross interaction terms $\tilde{V}_{ij}^\circ$ and $(u/k)_{ij}$ with $i \neq j$ that appear in the mixing rules are related to the pure fluid quantities by

$$\tilde{V}^\circ{}_{ij} = (1 + \nu_{ij}) \left(\frac{\sqrt[3]{\tilde{V}^\circ{}_{ii}} + \sqrt[3]{\tilde{V}^\circ{}_{jj}}}{2} \right)^3 \tag{14}$$

and

$$\left(\frac{u}{k}\right)_{ij} = (1 - \kappa_{ij}) \left[\left(\frac{u}{k}\right)_{ii} \left(\frac{u}{k}\right)_{jj} \right]^{1/2} \tag{15}$$

Equations 14 and 15 define ν_{ij} and κ_{ij} as the dimensionless cross-interaction constants. At the present stage they have to be determined from fitting mixturc data.

Fugacity and K-Value

The fugacity f_i of Component i in a fluid mixture is expressed conveniently in terms of a fugacity coefficient ϕ_i

$$f_i^V = \phi_i^V y_i P \tag{16}$$

$$f_i^L = \phi_i^L x_i P \tag{17}$$

Table III. Range of Conditions

System	*Temperature Range (K)*	*Pressure Range (atm)*	K_1 *rms*[a]
$N_2 + Ar$	140–240	1–26	2.9%
$N_2 + C_1$	138–183	6–48	2.6 2.3
$C_1 + C_2$	158–283	1.76–68	5.4 4.0
$C_1 + nC_4$	244–394	3–122	9.8 9.5
$C_1 + nC_6$	310–410	3–17	4.7
$H_2 + nC_4$	327–394	27–166	5.6 6.4
$H_2 + C_6H_6$	433–533	19–175	5.6

[a] The abbreviation rms (root-mean-square deviation) $= (\sum^{NOB} \text{dev}^2/\text{NOB})^{1/2}$, where dev $\equiv \left(\frac{K^{\text{calc}} - K^{\text{exp}}}{K^{\text{exp}}}\right)$; NOB = number of observations.

When equilibrium exists between a gas mixture and a liquid solution.

$$f_i^{V} = f_i^{L} \tag{18}$$

for all i components in the system. Combining Equations 16–18 gives an expression of the K-value of i,

$$K_i \equiv y_i/x_i = \phi_i^{L}/\phi_i^{V} \tag{19}$$

The fugacity coefficient is derived from the BACK equation with the use of the mixing rules of Equations 11–15 by following standard procedures of classical thermodynamics. The result is given below. For brevity we have left out the subscript m from quantities that apply to the fluid mixture as a whole; thus, e.g., $z = z_m$.

and Fitting of Mixture Data

K_2 *rms*	κ_{ij}	ν_{ij}	*Refs.*
4.8%	-7.8×10^{-4}	-6.1×10^{-3}	*18*
1.4	0.02150	0	*28*
0.8	0.02949	0.0555	
8.2	−0.03114	0	*19, 20*
8.4	−0.04119	−0.1198	
6.8	−0.09262	0	*21–26*
7.3	−0.09452	−0.02367	
4.1	−0.17019	0	*27*
9.4	−0.9886	0	*29*
5.7	−0.9027	0.5116	
6.4	−1.0907	0.7475	*30*

$$
\begin{aligned}
RT \ln \phi_i &= RT \ln \phi - RTz + RT \\
&+ RT \sum_N^4 \sum_M^9 \mathrm{D}_{NM} \left(\frac{u}{\mathrm{k}T}\right)^N \left(\frac{\widetilde{V}^\circ}{\widetilde{V}}\right)^M \psi_{MN} \\
&+ RT\,\Theta \left\{\frac{6\alpha^2}{3(1-\xi)^3} + \frac{-6\alpha^2+6\alpha}{2(1-\xi)^2} + \frac{-3\alpha+1}{(1-\xi)} + \alpha^2 - 1\right\} \\
&+ RT\left(\frac{\partial \alpha}{\partial N_i}\right)\left\{2\alpha \ln(1-\xi) + \frac{4\alpha}{2(1-\xi)^2} + \frac{3-2\alpha}{(1-\xi)} - 3\right\}
\end{aligned}
\tag{20}
$$

where

$$
\psi_{\mathrm{MN}} \equiv \left\{ N \left(\frac{u}{\mathrm{k}}\right)^{-1} \left(\frac{\partial \frac{u}{\mathrm{k}}}{\partial N_i}\right) + M\left(1 + \frac{1}{\widetilde{V}^\circ}\frac{\partial \widetilde{V}^\circ}{\partial N_i}\right)\right\} \tag{21}
$$

$$
\frac{\partial \left(\frac{u}{\mathrm{k}}\right)}{\partial N_i} = 2\left\{\sum_j N_j \left(\frac{u}{\mathrm{k}}\right)_{ji} - \sum_j \sum_p N_j N_p \left(\frac{u}{\mathrm{k}}\right)_{jp}\right\} \tag{22}
$$

$$\frac{\partial V^{\circ}}{\partial N_i} = 2\left\{\sum_j N_j V^{\circ}_{ji} - \sum_j \sum_p N_j N_p V^{\circ}_{jp}\right\} \tag{23}$$

$$\frac{\partial \alpha}{\partial N_i} = \alpha_i - \sum_j N_j \alpha_j \tag{24}$$

$$\Theta = \left\{1 + \frac{1}{V^{\circ}} \frac{\partial V^{\circ}}{\partial N_i}\right\} \tag{25}$$

$$\ln \phi = (\alpha^2 - 1) \ln (1 - \xi) + \frac{(\alpha^2 + 3\alpha)\xi - 3\alpha\xi^2}{(1 - \xi)^2} + \sum_N^4 \sum_M^9 \mathrm{D}_{NM} \left(\frac{u}{\mathrm{k}T}\right)^N \left(\frac{V^{\circ}}{V}\right)^M + z - 1 - \ln z \tag{26}$$

Some Binary Mixtures

Using the BACK equation, we have studied the phase equilibrium of several binary mixtures for which experimental data are available over an extended range of conditions. Table III presents the mixture systems, the temperature and pressure ranges of the data, the overall fitting of K-values by the BACK equation, and the interaction constants ν and κ. The BACK equation constants for the pure components have been reported either in Table I or in Ref. *1*.

The interaction constants ν and κ in Table III are determined for each binary system by fitting the experimental K-values of both components in the least square sense for the relative deviations.

The simplest mixtures in Table III are the two bniary systems of nitrogen with argon and with methane. Comparison of calculated results with literature data are shown in Figures 6 and 7. The molecules are all quite small and similar in interaction energies. The interaction parameters have small values. For nitrogen + argon, both parameters are practically zero. For nitrogen + methane, κ has a small but significant value. But it makes little difference to the high accuracy if ν is set equal to zero. Figure 7 shows that quantitative agreement is obtained even in the retrograde region.

Four binary systems of methane are included in Table III. The use of a zero value of ν is tested on three of them and found to give about the same results as the best nonzero value. It appears that $\nu = 0$ for all these systems, and only one interaction constant, κ, needs to be determined for each of these pairs. In Figures 8 and 9 we compare calculated *K*-values with experimental data for methane + ethane and methane + *n*-butane, respectively.

Two binary mixtures of hydrogen have been studied. A large positive value of ν and a substantial negative value of κ are obtained for both systems. Having $\nu = 0$ gives definitely inferior results. The BACK equation gives an excellent representation of the two systems with the use of both interaction constants, as shown in Figures 10 and 11.

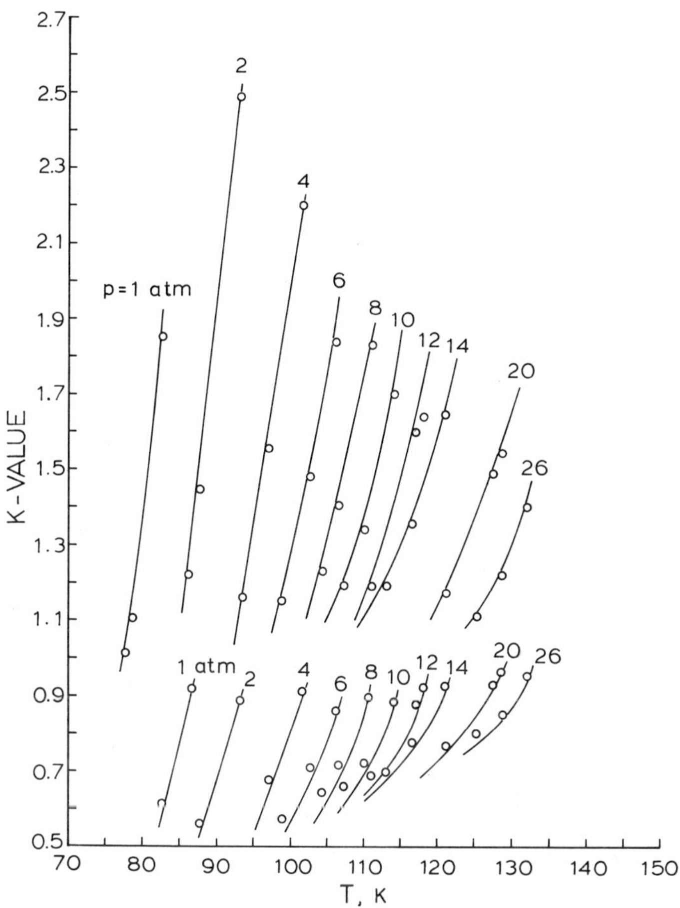

Figure 6. Experimental and BACK-predicted K*-values for N_2 + Ar: (○), Wilson et al. (18); (——), BACK Equation; $\kappa = -7.8 \times 10^{-4}$; $\nu = -6.1 \times 10^{-3}$.*

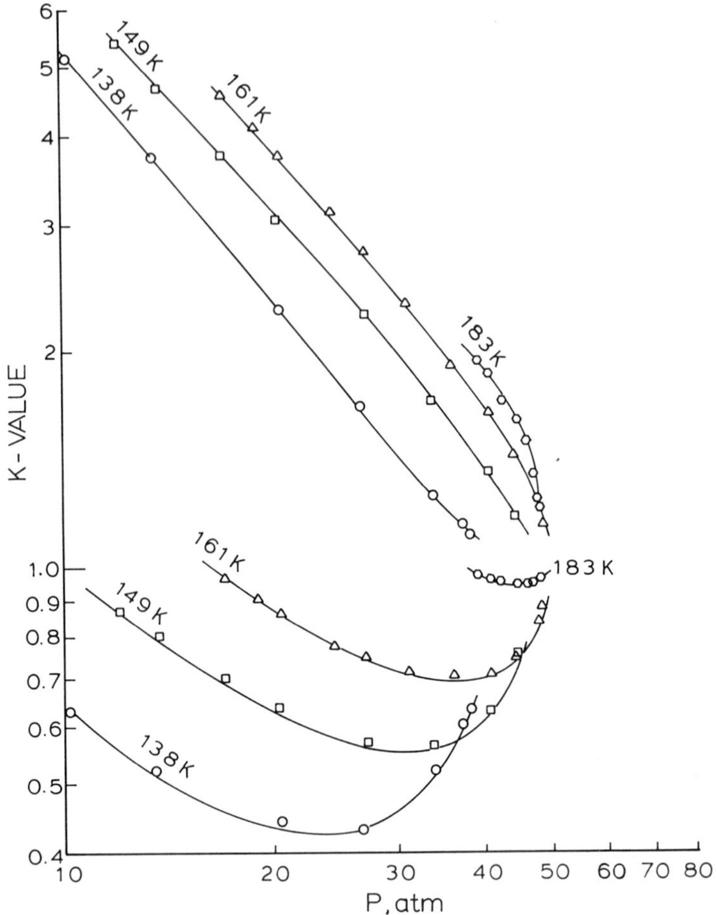

Figure 7. Experimental and BACK-predicted K-values for $N_2 + CH_4$: (○, □, △, ⬡), experimental; (——), BACK Equation; $\kappa = 0.0215$; $\nu = 0$.

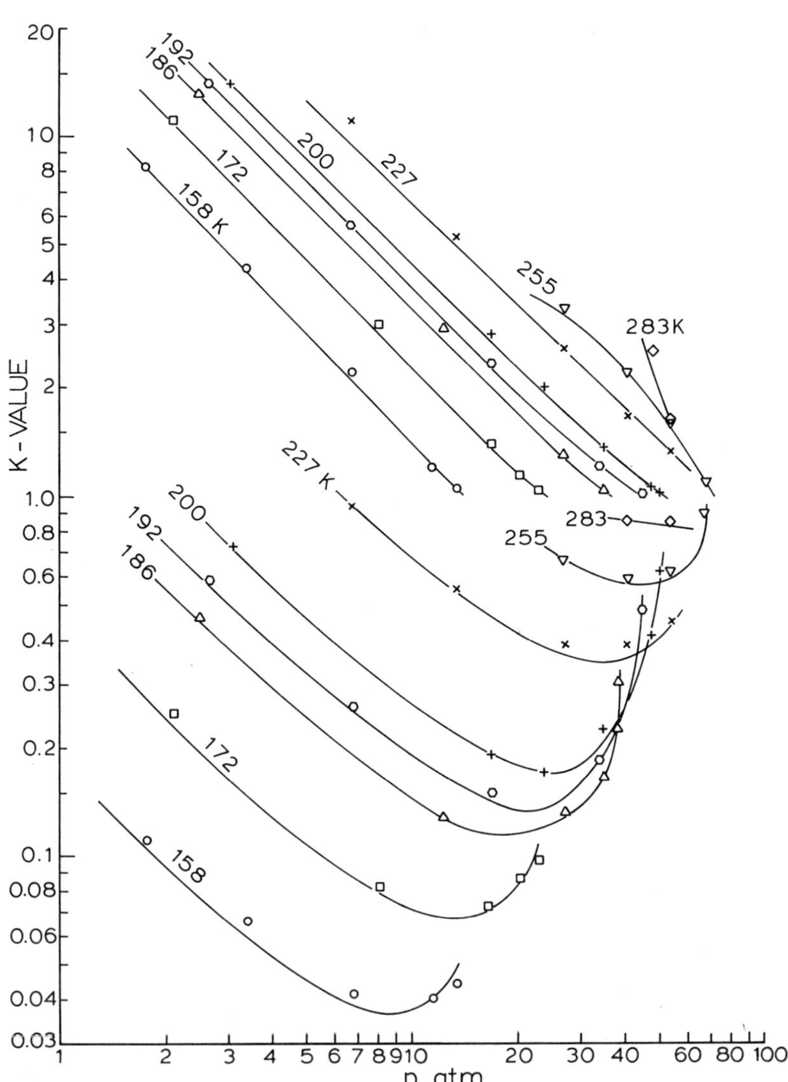

Figure 8. Experimental and BACK-predicted K-values for methane + ethane: (○, □, △, ⬡, +, ×), experimental; (——), BACK Equation; κ = −0.0311; ν = 0.

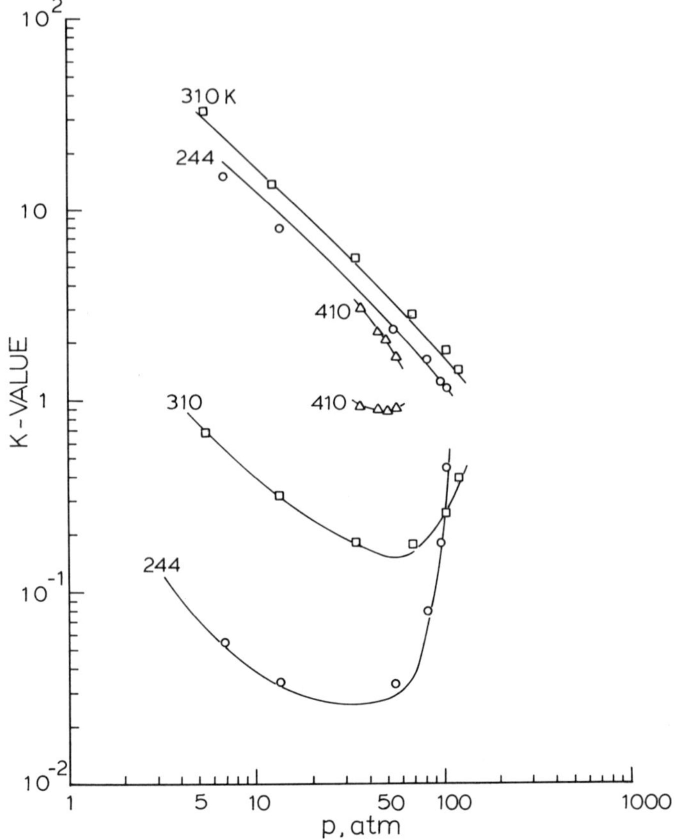

Figure 9. Experimental and BACK-predicted K*-values for methane +* n*-butane: (○, □, △), experimental; (——), BACK Equation; $\kappa = -0.093$; $\nu = 0$.*

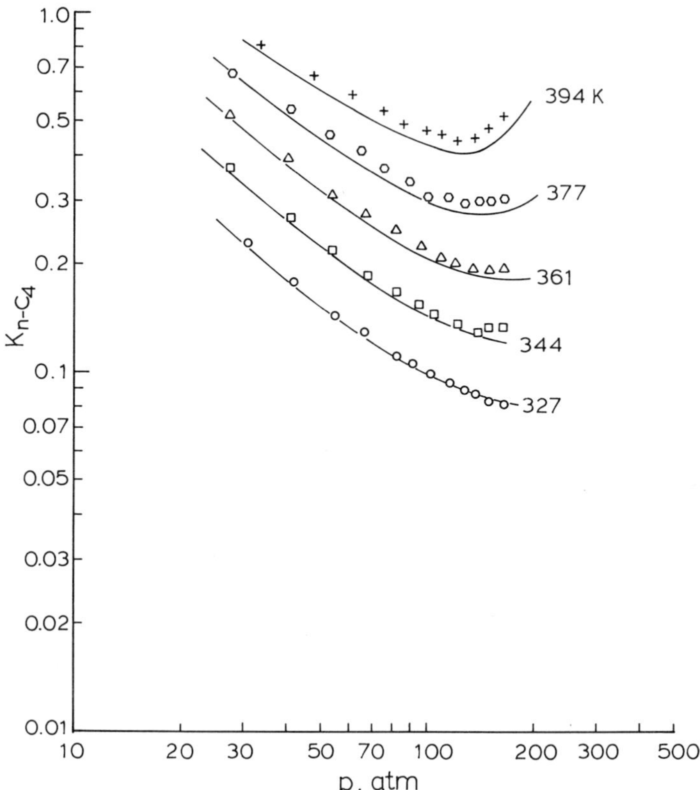

Figure 10a. Comparison of K-values of hydrogen in H_2 | n-butane with BACK equation: (○, □, △, ⬡, +), experimental; (——), BACK Equation; κ = −0.9027; ν = 0.5116.

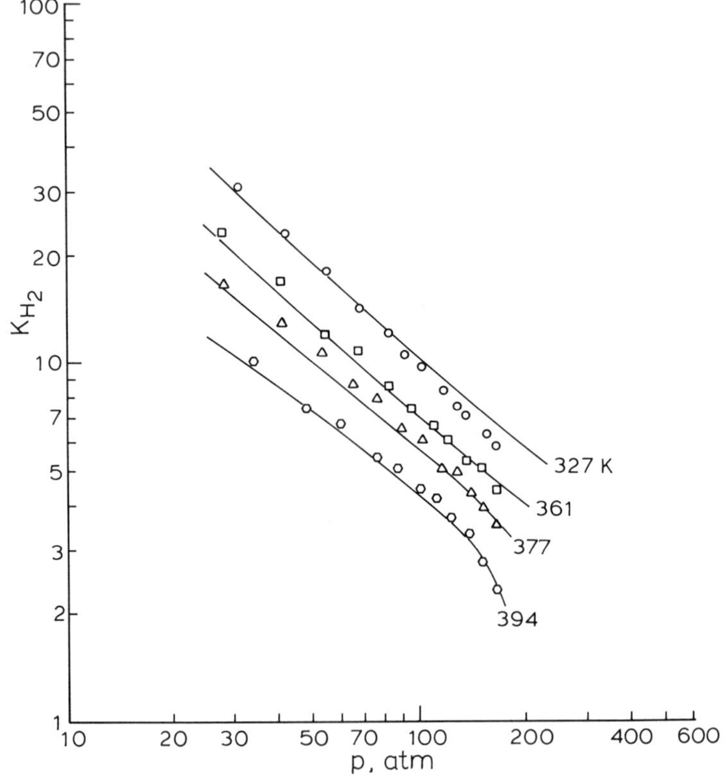

Figure 10b. Comparison of K-values of n-*butane in* H_2 + n-*butane with BACK equation: (○, □, △, ⬡), experimental; (——), BACK Equation;* $\kappa = -0.9027$; $\nu = 0.5116$.

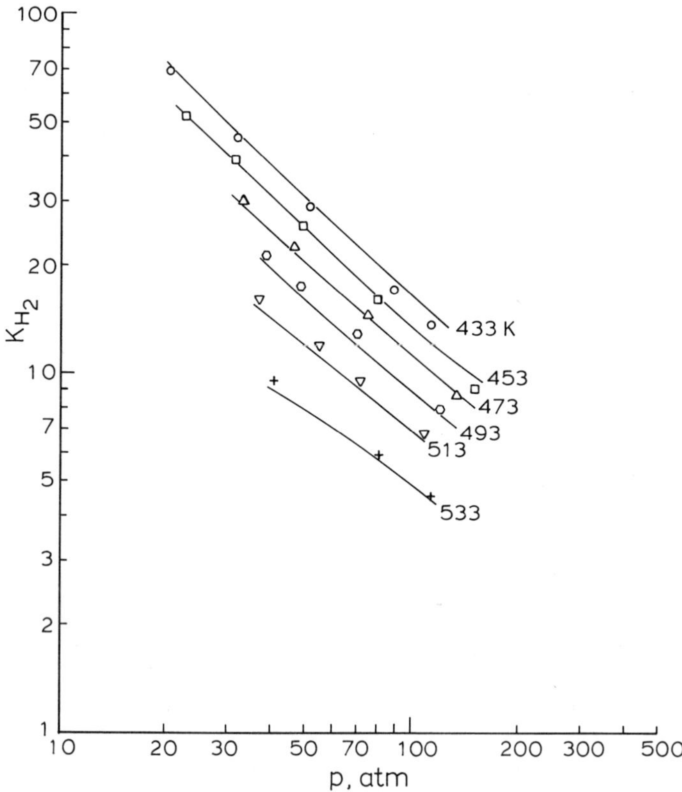

Figure 11a. Comparison of K-values of hydrogen in $H_2 + C_6H_6$ with BACK equation: (○, □, △, ⬡, ▽, +), experimental; (——), BACK Equation; $\kappa = -1.091$; $v = 0.748$.

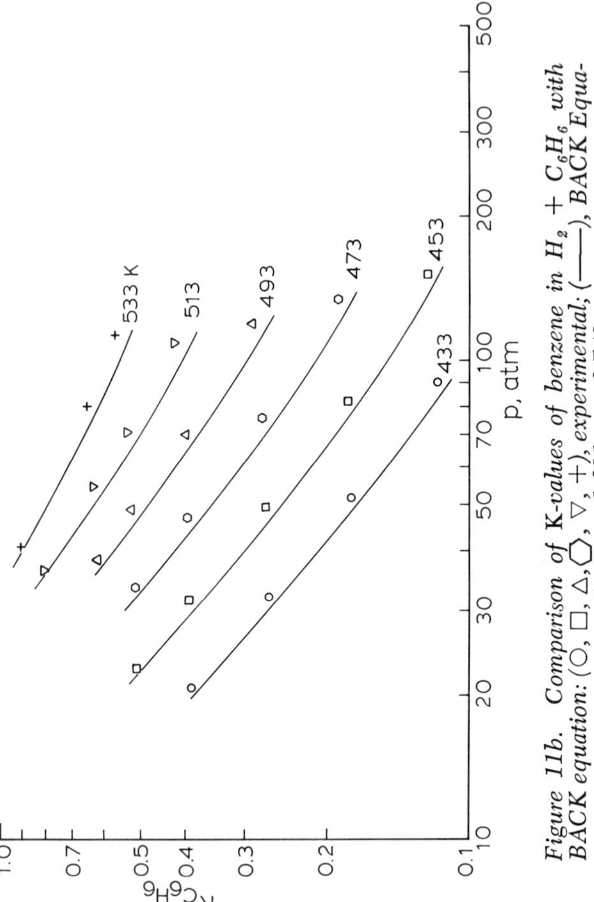

Figure 11b. Comparison of K-values of benzene in H_2 + C_6H_6 with BACK equation: (○, □, △, ⬡, ▽, +), experimental; (——), BACK Equation; $\kappa = 1.091$; $\nu = 0.748$.

Discussion and Conclusion

Even though the BACK equation constants are molecular properties and are known for some molecules, we prefer to treat them as empirical constants determined for the best fitting of thermodynamic data. We do so because molecular properties generally are not known with accuracy. The constants thus obtained correlate with properties of the fluid that are related closely to molecular properties.

The BACK equation is capable of accurately describing the phase equilibria of some pure fluids and mixtures. The accuracy appears the best for fluids of globular molecules, but not quite as good for long-chain molecules. Normal octane seems to be significantly less well fitted than the shorter chains and *n*-decane is even worse. However, even the relatively poor accuracy here appears to be superior to that attained by any other equations of state when applied to mixtures.

Glossary of Symbols

C = constant in Equation 5
D_{NM} = universal constants in Equation 3
K = equilibrium ratio = y/x
M = index for Equation 3
N = index for Equation 3
N = mole fraction
R = universal gas constant
T = absolute temperature
V = volume
f = fugacity
k = Boltzmann's constant
P = pressure
u = interaction energy
x = liquid-phase mole fraction
y = vapor-phase mole fraction
z = compressibility

Greek Letters

α = spherocylinder constant in Equation 2
η = constant for interaction energy temperature dependence, Equation 6
κ = interaction energy mixing parameter
μ = chemical potential
ν = interaction volume-mixing parameter
ϕ = fugacity coefficient

ξ = reduced volume, Equation 4
ρ = density
ω = acentric factor

Subscripts

c = critical property
$\left.\begin{matrix} i \\ j \\ k \end{matrix}\right\}$ = component indices
G = gas- or vapor-phase property
L = liquid-phase property
m = mixture property
r = reduced property

Superscripts

V = vapor- or gas-phase property
L = liquid-phase property
h = hard sphere
a = attractive
° = molecular property, as in $V°$, $u°/k$

Acknowledgment

Funds for this research were supplied by the Electric Power Research Institute through research project RP-367.

Literature Cited

1. Chen, S. S.; Kreglewski, A. *Ber. Bunsenges.* **1977,** *81* (10), 1048.
2. Boublik, T. *J. Chem. Phys.* **1975,** *63* (9), 4048.
3. Alder, B. J.; Young, D. A.; Mark, M. A. *J. Chem. Phys.* **1972,** *56* (6), 3013.
4. Carnahan, N. F.; Starling, K. E. *J. Chem. Phys.* **1969,** *51* (6), 1184.
5. Kreglewski, A.; Chen, S. S. *J. Chim. Phys.* **1978,** *75* (4), 347.
6. Das, T. R.; Reed, C. O.; Eubank, P. T. *J. Chem. Eng. Data* **1977,** *22* (1), 16.
7. McCarty, R. D. "Hydrogen Technological Survey—Thermophysical Properties"; *NASA Spec. Publ.* **1975,** 3089.
8. Das, T. R.; Reed, C. O., Jr.; Eubank, P. T. *J. Chem. Eng. Data,* **1977,** *22* (1), 3.
9. Ibid, p. 9.
10. Young, S., "Pysico-Chemical Constants of Pure Organic Compounds," 2nd ed.; Timmermanns, J., Ed.; Elsevier: Amsterdam, 1950.
11. Zwolinski, B. J., Ed. "Selected Values of Physical and Thermodynamic Properties of Hydrocarbons"; 1953–1977 API Research Project 44, Texas A&M University.
12. Francis, A. *Ind. Eng. Chem.* **1957,** *49* (10), 1779.

13. Krase, N. W.; Goodman, J. B. *Ind. Eng. Chem.* **1930,** *22*(13), 13.
14. Zmaczynski, M. A. "Physico-Chemical Constants of Pure Organic Compounds," 2nd ed.; Timmermanns, J., Ed.; Elsevier: Amsterdam, 1950; p. 151.
15. Griswold, J.; Andrea, D.; Klein, V. A. "Physico-Chemical Constants of Pure Organic Compounds," 2nd ed.; Timmermanns, J., Ed.; Elsevier: Amsterdam, 1965; Vol. 2, p. 99.
16. School of Chemical Engineering, Purdue University, IN, unpublished data.
17. Glaser, F.; Ruland, H. *Chem. Ing. Tech.* **1957,** *29,* 772.
18. Wilson, G. M.; Silverberg, P. M.; Zellner, M. G. *Adv. Cryog. Eng.* **1965,** *10,* 192.
19. Wichterle, I.; Kobayashi, R. *J. Chem. Eng. Data* **1972,** *17*(1), 9.
20. Price, R. A.; Kobayashi, R. *J. Chem. Eng. Data* **1959,** *4*(1), 40.
21. Elliot, D. G.; Chen, R. J. J.; Chappelear, P. S.; Kobayashi, R. *J. Chem. Eng. Data* **1974,** *19*(1), 71.
22. Karhre, L. C. *J. Chem. Eng. Data* **1974,** *19*(1), 67.
23. Sage, B. H.; Hicks, B. L.; Lacey, W. N. *Ind. Eng. Chem.* **1940,** *32*(3), 1085.
24. Weise, H. C.; Jacobs, J.; Sage, B. H. *J. Chem. Eng. Data* **1970,** *15*(1), 1970.
25. Roberts, L. R.; Wang, R. H.; Azarnoosh, A.; McKetta, J. J. *J. Chem. Eng. Data* **1962,** *7*(4), 484.
26. Sage, B. H.; Budenholzer, R. A.; Lacey, W. N. *Ind. Eng. Chem.* **1940,** *32*(9), 1262.
27. Gunn, D. D.; McKetta, J. J.; Ata, N. *AIChE J.* **1974,** *20*(2), 347.
28. GPA Tech. Publ. TP-4, "Low Temperature Data from Rice University for Vapor-Liquid and P-V-T Behavior," April, 1974.
29. Klink, A. E.; Cheh, H. Y.; Amick, E. H. *AIChE J.* **1975,** *21*(6), 1142.
30. Connolly, J. F. *J. Chem. Phys.* **1962,** *36*(11), 2897.
31. Michels, A.; Levelt, J. M.; DeGraff, W. *Physica* **1958,** *XXIV,* 659.
32. Gibbons, R. M.; Kuebler, G. P. "Research on Materials Essential to Cryocooler Technology—Thermophysical and Transport Properties of Argon, Neon, Nitrogen, and Helium-4"; **1968,** AFML-TR-68-370.

RECEIVED September 5, 1978.

13

An Equation of State for Polar Mixtures: Calculation of High-Pressure Vapor–Liquid Equilibria of Trace Polar Solutes in Hydrocarbon Mixtures

K. W. WON and C. K. WALKER

Fluor Engineers and Constructors, Inc., Irvine, CA 92730

A modification of the Soave equation of state

$$P = \frac{RT}{V - b} - \frac{A_n(T) + A_p(T)}{V(V + b)}$$

has been used to calculate vapor–liquid phase equilibria for water, methanol, and mercaptans when these compounds are present in small amounts in hydrocarbon streams. The results of these calculations are compared with experimental data with encouraging results. The equation-of-state method for solving such problems offers advantages over the normal activity coefficient approach.

In energy and environmental-related technology it is often necessary to remove trace amounts of nonhydrocarbon polar solutes such as water and mercaptans from hydrocarbon streams. While equations of state have played a very important role in the correlation of phase equilibrium (K-ratios) of nonpolar and slightly polar mixtures, much less attention has been given to an equation of state for the more difficult problem of polar mixtures.

It has been customary to apply an activity coefficient method to aid in the prediction of vapor–liquid equilibria of polar mixtures. At high pressures approaching the critical state of the fluid mixture, the activity coefficient method requires such thermodynamic properties as partial molar volumes or partial molar heats of solution that are very difficult,

0-8412-0500-0/79/33-182-235$05.00/1

if not impossible, to obtain. When the fluid mixture consists mainly of hydrocarbons with trace amounts of polar chemicals, the single-equation-of-state approach appears to be advantageous.

This chapter illustrates the use of an equation of state previously suggested and shows its application to several important industrial problems.

Vapor–Liquid Equilibrium K-Ratio

For any component i in a mixture, the equilibrium vapor concentration y_i and liquid concentration x_i are governed by the relations:

$$f_i^V = f_i^L \tag{1}$$

$$f_i^V = y_i \phi_i^V P \tag{2}$$

$$f_i^L = x_i \phi_i^L P \tag{3}$$

where f_i and ϕ_i are the fugacity and fugacity coefficient of component i, P is the total pressure, and the superscripts L and V indicate the liquid or vapor phase.

Using these relations, the vapor–liquid equilibrium (VLE) K-ratio is determined by

$$K_i = \frac{y_i}{x_i} = \frac{\phi_i^L}{\phi_i^V} \tag{4}$$

The vapor–liquid equilibrium K-ratio can be calculated via the fugacity coefficients from an equation of state as a function of temperature, pressure, and compositions of liquid and vapor mixtures.

Polar Equation of State

The equation of state used here is a further modification of the equation given by Soave (*1*). This modification was proposed by Won (*2*) for gases in a previous paper, but it is applied to both gases and liquids in this chapter. It can be written as:

$$P = \frac{RT}{V - b} - \frac{A_n(T) + A_p(T)}{V(V + b)} \tag{5}$$

where the nonpolar and polar contributions are separated.

For a polar component,

$$A_n(T) = [A(T_c) - A_p(T_c)][1 + \alpha(1 - T_r^{0.5})]^2 \quad (6)$$

$$A_p(T) = A_p(T_c)/T_r^3 \quad (7)$$

subject to

$$A(T_c) = A_n(T_c) + A_p(T_c) = 0.4278 \frac{(RT_c)^2}{P_c} \quad (8)$$

where

$$b = 0.0867\ RT_c/P_c \quad (9)$$

and α is treated as an independent parameter.

For a nonpolar component

$$A_p(T_c) = 0 \quad (10)$$

$$\alpha = 0.48 + 1.57\omega - 0.176\omega^2 \quad (11)$$

For mixtures,

$$A(T) = \sum_i \sum_j y_i y_j A_{ij}(T) \quad (12)$$

and

$$b = \sum_i y_i b_i \quad (13)$$

where

$$A_{ij}(T) = \sqrt{A_{ni}(T_c)A_{nj}(T_c)}\ [1 + \alpha_i(1 - T_{ri}^{0.5})] [1 + \alpha_j(1 - T_{rj}^{0.5})] + \frac{\sqrt{A_{pi}(T_c)A_{pj}(T_c)}}{(T_{ri}T_{rj})^{1.5}} \quad (14)$$

Substitution of Equations 5, 12, and 13 into the thermodynamic definition of ϕ (3) gives

$$\ln \phi_i = \ln\left(\frac{V}{V-b}\right) + \frac{b_j}{V-b} - \frac{2\Sigma y_j A_{ij} \ln\left(\frac{V+b}{V}\right)}{bRT} + \frac{Ab_i}{RTb^2}\left[\ln\left(\frac{V+b}{V}\right) - \frac{b}{V+b}\right] - \ln\left(\frac{PV}{RT}\right) \quad (15)$$

The molar volume V of each equilibrium phase is calculated by solving Equation 5. At a given temperature, pressure, and composition of the mixture, three values for V will be obtained, the largest of which is the vapor molar volume and the smallest the liquid molar volume. Although the calculated liquid molar volume, V^L, does not accurately reproduce experimental data on molar volume, the fugacity coefficients, ϕ_i^L, calculated from the inaccurate V^L, are in good agreement with those derived from experimental K-data.

Determination of the Polar Energy Parameter

The polar energy parameter, $A_p(T_c)$, can be determined in one of several different ways depending on the availability of reliable data.

1. Second virial coefficients, B, at several temperatures are the most commonly available experimental data for polar solutes. Substitution of Equation 5 into the definition of second virial coefficients (3) gives:

$$B = b - \frac{A_n(T) + A_p(T)}{RT} \tag{16}$$

The right-hand side of Equation 16 contains two temperature-dependent functions, $A_n(T)$ and $A_p(T)$, defined by Equations 6 and 7. The temperature dependence of experimental second virial coefficients allows the determination of the polar energy parameter, $A_p(T_c)$.

2. Second virial cross efficients between the polar chemical and the light gases appear in the literature less often than second virial coefficients for pure polar substances. However, these coefficients (B_{ij}) can be used to determine $A_p(T_c)$ of the polar solute. By substituting Equations 5, 12, and 13 in Equation 16 and by using the definition of the second virial cross coefficient (3):

$$B_{ij} = \frac{b_i + b_j}{2} - \frac{A_{ij}(T)}{RT} \tag{17}$$

Equations 17 and 14 can be fit to the experimental second virial cross coefficients to find $A_p(T_c)$.

3. Phase-equilibrium (K-ratio) data also can be used to determine the polar energy parameter. Equations 4 and 15 are used to calculate the phase equilibrium K-ratio and the parameter $A_p(T_c)$ that fits the experimental data is chosen.

The first and second methods allow the prediction of K-values from the gas phase, while the third offers a technique of K-value extrapolation to other temperatures and also to other hydrocarbon solvents.

Calculation Procedure for Vapor–Liquid Equilibrium K-Ratios

At a given temperature, pressure, and assumed composition, Equation 5 is solved for *V* and the fugacity coefficients of each component in the mixture in both vapor and liquid are calculated by Equation 15. The K-ratio for each component is calculated by Equation 4. The Rachford–Rice (*4*) form of the flash equation is used to calculate the amount of vapor per mole of the overall mixture v/F;

$$f\left(\frac{v}{F}\right) = \sum \frac{z_i(K_i - 1)}{(K_i - 1)\left(\frac{v}{F}\right) + 1} = 0 \qquad (18)$$

A mass balance determines the equilibrium composition of vapor and liquid which are used to calculate a new set of equilibrium K-ratios. These steps are repeated until the K-ratios and the vaporized fraction do not vary. The tolerance of the function f(V/F), set to less than 10^{-4}, was sufficient for most cases.

Results

Water Solute in Hydrocarbon-Rich Vapor and Liquid. The pure component parameters of water solute $A_p(T_c)$ and α were determined by using Equations 5 and 16 to fit the gas-phase volumetric properties of steam (*5*) and the second virial cross coefficients of steam and light gases such as methane, ethane, and nitrogen (*6*). The least-squares minimization technique was used to find the parameters that gave the minimum deviations between calculated and experimental pressures and second virial cross coefficients. (Table I lists the parameters for pure steam and of other compounds used in this study.)

Extensive thermodynamic properties of steam are available in the steam tables and these data offer a good comparison for the validity of the equation of state in the calculation of other thermodynamic properties. Figure 1 shows the fugacity coefficients of steam calculated from Equation 15. For comparison, Figure 1 also shows the fugacity coefficients of steam calculated from the steam tables.

Figure 2 shows the pressure and temperature domain of accuracies in calculating the enthalpy of steam. When the pressure is less than 50 atm, the equation of state calculates the enthalpy departure of steam to within 1.1 cal/g.

Table I. Equation-of-State Parameters

	$An(T_c)$ *Atm.* $(L/mol)^2$	$Ap(T_c)$ *Atm.* $(L/mol)^2$	b *(cc/mol)*	α *Dimensionless*
Water	2.93	2.59	21.1	−0.47
Methanol	5.09	4.57	46.5	1.76
Ethanol	5.73	6.47	58.3	1.52
Methyl Mercaptan	7.75	1.16	46.8	0.68
Ethyl Mercaptan	10.6	2.6	65.5	0.80
Methane	2.29	0.0	29.6	0.50
Ethane	5.57	0.0	45	0.64
Propane	9.4	0.0	63	0.72
n-Butane	13.9	0.0	81	0.78
n-Hexane	24.8	0.0	121	0.93
1-Butene	12.73	0.06	75.3	0.78
Benzene	18.6	0.13	82.2	0.81

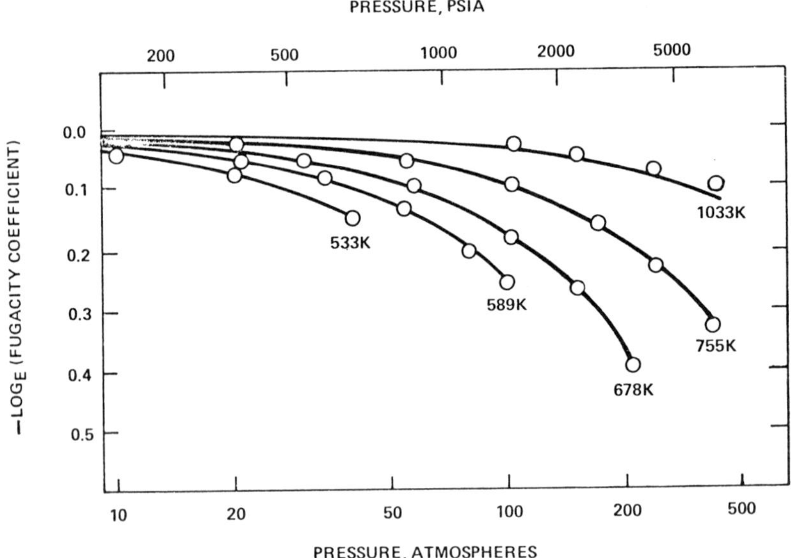

Figure 1. Fugacity coefficients for gaseous water: (—), predicted; (○), calculated from steam tables.

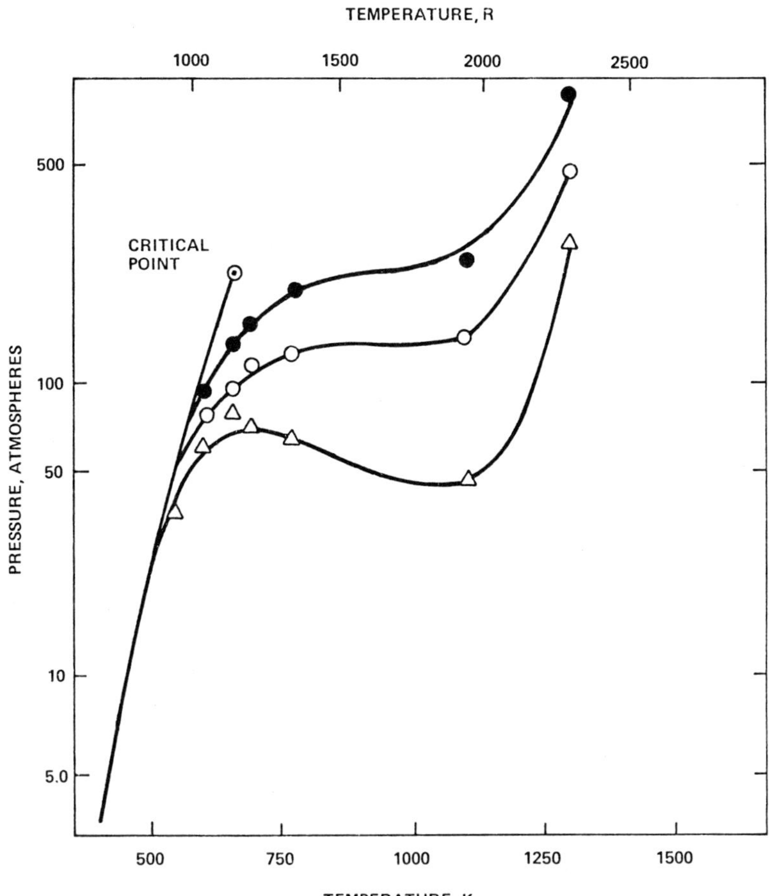

Figure 2. Accuracies of predicted enthalpies of gaseous water: (Δ), within ± 1.11 cal/g; (○), within ± 2.78 cal/g; (●), within ± 5.56 cal/g.

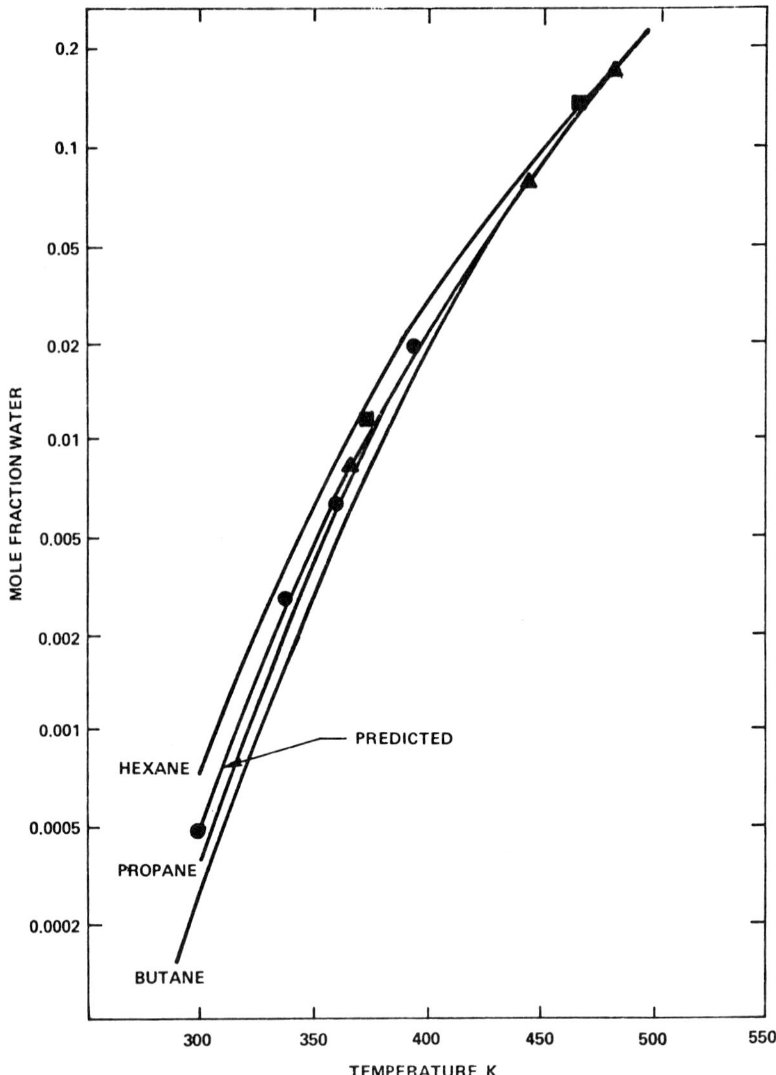

Figure 3. Water solubilities in paraffins. Predicted values: (●), butane; (▲), hexane; (■), octane. Lines for propane, butane, and hexane are adapted from the API technical data book.

Figure 3 shows the calculated solubilities of water in normal paraffins butane, hexane, and octane, as a function of temperature. For comparison, literature data (7) also are shown in Figure 3. The saturated water solubilities as calculated from the equation of state depend primarily on temperature, while literature data indicate a significant effect of hydrocarbon molecular size on the water solubilities.

Figure 4 shows the calculated phase relationships in the methane–butane–water ternary system (8) at two temperatures. The results shown in Figure 4 indicate that the equation of state predicts the bubble points well, but that the predicted dew points at the higher temperature may not be as accurate.

Figure 5 shows the VLE K-ratios of water in 1-butene, and also indicates the effect of small changes in the parameters on calculated

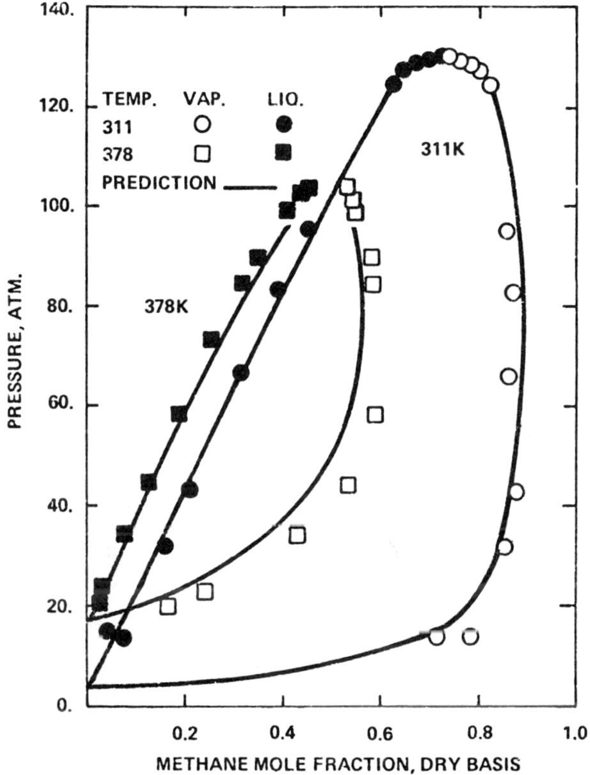

Figure 4. Predicted phase diagram of methane, butane water system at two temperatures (8)

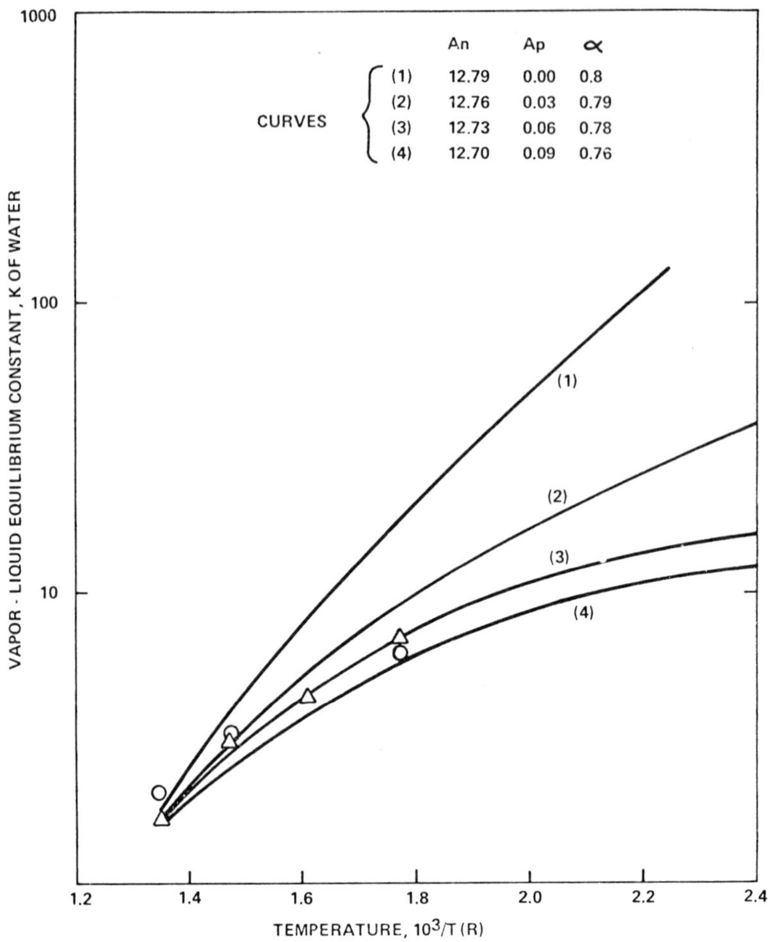

Figure 5. Hydrocarbon—rich vapor and liquid K— values for water-1 butene system. Calculated with equation-of-state parameters for 1-butene. Experimental data: (○), Leland et al. (9); (Δ), Wehe and McKetta (10).

results. The experimental data (*9, 10*) reported for this system scatter considerably. However, the results calculated with a small (0.06), but nonzero $A_p(T_c)$ for 1-butene, appear to agree best with the experimental data. Olefinic and other saturated hydrocarbons contain electron-rich π-bond which show polar character towards polar molecules.

The VLE calculated by the equation of state for the *n*-butane–1-butene–water system agrees fairly well with the data reported by Wehe and McKetta (*10*). The calculated water solubilities in benzene agree with literature data (*11*) from 320 K and higher. At lower temperatures the equation of state overestimates the water solubilities.

Methanol in Hydrocarbon-Rich Vapor and Liquid. The volumetric properties of methanol gas (*12*) and the second virial cross coefficients of methanol and light gases (*13*) were used to determine the pure-component parameters $A_p(T_c)$ and α for methanol. Table II shows the enthalpy departure of gaseous methanol from ideal gas at three temperatures and several pressures. For comparison, the experimental values (*14*) and the values calculated by the Soave equation (*1*) are also shown. Table II indicates that the Won modified equation of state predicts the enthalpy departure of methanol very well at low temperatures and fairly well at high temperatures, but that the original Soave equation considerably underestimates the enthalpy departure at all temperatures and pressures. Since the original Soave equation was meant to be applied only to hydrocarbons, we are not surprised at this result. Comparison of calculated and experimental second virial cross coefficients between methanol and methane (and also CO_2) is presented elsewhere (*15*).)

Figure 6 shows the VLE K-ratios of both methanol and *n*-hexane at 323 K at a methanol concentration ranging from infinite dilution to 25 mol %. Shown on the graph are experimental values (*16*), those calculated by the Won modified equation of state, values calculated by the original Soave correlation, and those calculated by the Soave correlation when using a a binary parameter k_{ij} chosen such that the calculated methanol K-ratio matches the experimental value at 12 mol % methanol. The Won modified equation of state was used with no binary parameter k_{ij}, but with the pure compound parameters $A_p(T_c)$ and α. Neither equation of state describes the strong composition dependence of the methanol K-ratio in the dilute region. However, the results calculated by the modified equation of state are very close to experimental data in other regions.

Mercaptans. Gas-phase volumetric data of methyl and ethyl mercaptans are very limited in the literature. The K-values of methyl mercaptan in ethane (*17*) at 211 K, and for ethyl mercaptan in propane (*18*) at 305 K were used to determine the values for $A_p(T_c)$ for these mercaptans. The parameter α was calculated from the acentric factor ω as suggested for nonpolar and slightly polar compounds by Soave (*1*).

Table III shows calculated K-values of methyl mercaptans in methane, ethane, propane, and *n*-butane, and of ethyl mercaptan in propane. Although the agreement with experimental data will be improved by introducing the adjustable parameter k_{ij} in Equation 14, it is interesting to note that good agreement is obtained anyway although only one data point was used in calculating the parameter.

Table II. Enthalpy

	Enthalpy Departure, H° − H *(cal/g)*			
	450 K			*478 K*
Pressure (atm)	*Original Soave*	*Modified Soave*	*Exp*[a]	*Original Soave*
6.8	4.9	7.8	8.3	4.3
13.6	10.3	17.1	17.8	8.9
20.4	16.5	27.9	28.3	14.2
27.2				19.8
40.8				34.3
54.4				
68				

[a] From Ref. *14*.

Discussion

Conventionally, the fugacity, f_1^L, of a Polar Solute 1 in liquid mixture is related to the activity coefficient, γ_1, by;

$$f_1^L = \gamma_1 x_1 f_1^R \exp \frac{[V_1^L (P - P_1^R)]}{RT} \tag{19}$$

where V_1^L is the partial molar volume of the polar solute in liquid mixture and f_1^R is a reference state fugacity at the system temperature and at the reference pressure, P_1^R, which is usually set at 0 or at the saturated vapor pressure of Solute 1.

Equation 19 shows that the activity coefficient method for high-pressure fluid-phase equilibria requires information for the partial molar volume, V_1^L, which is very difficult to obtain.

Substitution of Equation 19 for Equation 3 gives:

$$\phi_1^L = \gamma_1 f_1^R \exp \frac{[V_1^L (P - P_1^R)]}{RT} \Big/ P \tag{20}$$

Equation 20 shows that the fugacity coefficient in the liquid, ϕ_1^L, when properly evaluated, provides a self-sufficient thermodynamic quantity in the analysis of high-pressure fluid-phase equilibria.

Departure of Gaseous Methanol

Enthalpy Departure, $H^{\circ} - H$ (cal/g)				
478 K		*506 K*		
Modified Soave	*Exp*[a]	*Original Soave*	*Modified Soave*	*Exp*[a]
5.6	6.7	3.9	5.3	4.4
12.3	13.9	7.9	10.9	9.4
21.3	21.1	12.3	16.9	15.0
30.0	29.4	17.0	23.4	20.6
55	50	27.8	38.3	33.3
		41.7	58.3	53.9
		65.6	99.5	86.7

When the pressures are low and also when Component 1 is very dilute in Solvent 2, Equation 20 yields

$$(\phi_1^L)^{\infty} = \gamma_1^{\infty} f_1^R / P_2^s \tag{21}$$

Equation 21 shows that the infinite dilution fugacity coefficient in liquid, $(\phi_1^L)^{\infty}$ is identical to the relative volatility of infinitely dilute Solute 1 over Solvent 2.

At high pressures, the meaning of ϕ_1^L is not as clear as at low pressure. However, the use of a single equation of state for both liquid and gas mixtures has clear advantages over the conventional method, where activity coefficients, the gas-phase equation of state, and liquid partial molar volumes are required as a function of temperature (and pressure).

Conclusions

When volumetric data are available for a polar gas or for hydrocarbon gas mixtures containing polar species, the Won modification of the Soave equation of state can be used to determine the two pure-component parameters $A_p(T_c)$ and α for the polar compound. With these parameters, the Won modified equation of state provides good estimates of VLE K-ratios of trace polar compounds in hydrocarbon-rich mixtures.

The thermodynamic properties of the polar gas predicted by the Won modified equation of state are in excellent agreement with experimental data. This equation of state, however, is not recommended for liquid mixtures rich in the polar component.

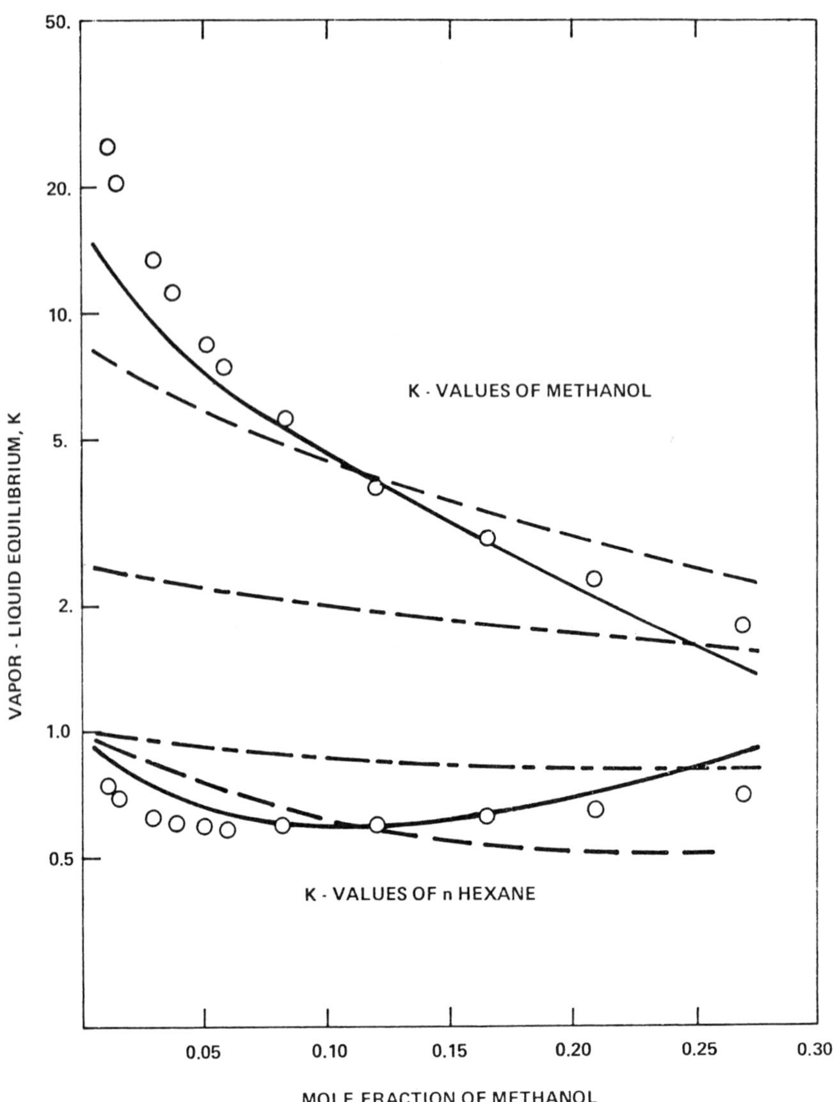

Figure 6. Vapor–liquid K-ratios of methanol and n*-hexane at 323.2 K: (—), predicted from modified Soave R–K; (— — —) predicted from Soave R–K; (– – –), calculated from Soave R–K with binary* $k_{ij} = 0.5$*; (○), experimental (16).*

Table III. Vapor–Liquid Equilibrium K-Values of Mercaptans in Light Hydrocarbons

Hydrocarbon Solvent	Temperature (K)	Pressure (atm)	K-Values Calculated	K-Values Experimental
Methyl Mercaptan Solute				
n-Butane	272	1.02	2.31	2.28[a]
Propane	255	2.6	0.45	0.54
	291	7.9	0.64	0.59
	339	23.5	0.79	0.75
Ethane	211	3.3	0.053	0.053
Methane	178	29.8	0.014	0.02
Ethyl Mercaptan Solute				
Propane	255	2.7	0.19	0.27[b]
	267	3.9	0.22	0.20
	278	5.5	0.26	0.30
	289	7.6	0.29	0.31
	300	10.1	0.32	0.29
	305	11.6	0.34	0.34
	311	13.3	0.37	0.37
	317	15.1	0.39	0.31

[a] From Ref. *17*.
[b] From Ref. *18*.

Acknowledgment

The authors are grateful for helpful discussions and contributions from F. T. Selleck, R. A. Chong, E. C. Hohmann, and R. Nielsen of Fluor Engineers and Constructors. Inc. and R. Goffredi of the University of Wisconsin. We wish to express our appreciation to the management of Fluor Engineers and Constructors, Inc. for permission to publish this work.

Glossary of Symbols

Notation

A = energy parameter in equation of state, atmosphere · $\left(\frac{\text{liter}}{\text{gram/mole}}\right)^2$

b = size parameter in equation of state, cubic centimeters per gram–mole

B = second virial coefficient, cubic centimeters per gram–mole

F = feed, gram–mole

f = fugacity, atmosphere

H = enthalpy, calories per gram
k = adjustable binary parameter
K = vapor–liquid equilibrium K-ratio
P = pressure, atmosphere
R = gas constant, 0.08206 atmosphere · liter/K
T = temperature, K
x, y = mole fraction in liquid and vapor
V = molar volume, cubic centimeters per gram–mole
v = vapor, gram–mole
z = mole fraction of feed

Greek Letter

α = defined by Equation 6
γ = activity coefficient
ϕ = fugacity coefficient
ω = acentric factor

Superscript and Subscript

c = critical property
i, j = components in mixture
L, V = liquid and vapor phases
n, p = nonpolar and polar contribution
r = reduced property
S = saturated state
∞ = state at inf .ite dilution

Literature Cited

1. Soave, G. *Chem. Eng. Sci.* **1972,** *27,* 1197.
2. Won, K. W. *69th Annual A.I.Ch.E. Meeting* **1976,** Tech. Paper 54A.
3. Prausnitz, J. M. "Molecular Thermodynamics of Fluid-Phase Equilibria;" Prentice-Hall: Englewood Cliffs, NJ 1969, 94, 99.
4. Rachford, H. H.; Rice, J. D. *J. Pet. Technol.* **1952,** *4,* 10, Sec. 1, 19, Sec. 2, 3.
5. Keenan, J. H.; Keyes, F. G.; Hill, P. G; Moore, J. G. "Steam Tables;" Wiley Interscience: New York, 1969.
6. Coan, C. R.; King, A. D. *J. Am. Chem. Soc.* **1971,** *93,* 1857.
7. "Technical Data Book, Petroleum Refining," 2nd ed.; American Petroleum Institute: Washington, DC, 1971; Chapter 9, p. 9.
8. McKetta, J. J.; Katz, D. L. *Ind. Eng. Chem.* **1948,** *40,* 853.
9. Leland, T. W.; McKetta, J. J.; Kobe, K. A. *Ind. Eng. Chem.* **1955,** *47,* 1265.
10. Wehe, A. H.; McKetta, J. J. *J. Chem. Eng. Data* **1961,** *6,* 167.
11. Thompson, W. H.; Snyder, J. R. *J. Chem. Eng. Data* **1964,** *9,* 516.
12. Ramsay; Young. In *Intern. Critical Tables,* **1928,** *3,* 436.
13. Hemmaplardh, B.; King, A. D. *J. Phys. Chem.* **1972,** *76,* 2170.
14. Storvick, T. S.; Smith, J. M. *J. Chem. Eng. Data* **1960,** *5,* 133.

15. Won, K. W. *Adv. Cryog. Eng.* **1978,** *23,* 544.
16. Wolff, H.; Hoppel, H. E. *Ber. Bunsenges. Phys. Chem.* **1968,** *72,* 710.
17. Zudkevitch, D.; Wilson, G. M. *Proc. 53rd Ann. Conv. GPA* **1974,** Tech. Sec. C, 101.
18. Hankinson, R. W.; Wilson, G. M. *Proc. 53rd Ann. Conv. GPA* **1974,** Tech. Sec. C, 98.

RECEIVED August 10, 1978.

14

Industrial Experience in Applying the Redlich-Kwong Equation to Vapor-Liquid Equilibria

R. D. GRAY, JR.

Exxon Research and Engineering Co., Florham Park, NJ 07932

This chapter provides a critical assessment of the strengths and weaknesses of methods which use the Redlich–Kwong equation of state to correlate and predict vapor–liquid equilibria. A brief review is given of well established strengths (effectiveness for high pressure and cryogenic light hydrocarbon systems) and weaknesses (inability to represent details of density dependence, problems with some versions of the Redlich–Kwong method for supercritical gases, and difficulties with polar compounds). A more detailed discussion is given for (cryogenic H_2-containing systems as well as for certain problems in representing details of paraffin–paraffin binaries for critical region behavior or for wide ranges of conditions and difficulties with heavy hydrocarbon systems). Based on this experience, a set of desirable criteria for the next generation of equation-of-state methods is provided.

This chapter draws on over eight years experience in using the Redlich–Kwong (RK) equation of state to correlate and predict vapor–liquid equilibrium (VLE) behavior in petroleum refining and petrochemical applications. The perspective is that of a thermodynamic data development and internal consulting group which takes responsibility for the accuracy of the predictions of the data methods it recommends. This kind of experience often leads one to uncover weaknesses in a data correlation. At the same time, one often is forced to extend a correlation far outside its intended region of validity, sometimes uncovering unexpected capabilities. Generally, one discovers things in application experience that are difficult to discover in any other way.

0-8412-0500-0/79/33-182-253$05.00/1

The objective of this chapter is to summarize the major patterns of strength and weakness which recur in the application of the RK equation of state to predict phase behavior. Some of these patterns are quite well established in previous literature, and so are summarized only briefly. Some less well-recognized strengths and weaknesses are explored in some detail. Based on this experience, some conclusions are drawn concerning desirable features of the next generation of methods for predicting phase-equilibrium behavior. It is hoped that this synopsis of industrial experience will prove useful not only to the many currently using RK methods, but also to those in the academic world who are engaged in developing new methods.

Chronology of RK Application to VLE

An abbreviated chronology of the development of RK methods is a useful way to begin this discussion. Although Wilson (*1*) first proposed the introduction of a temperature-dependent parameter to replace one of two constants in 1964, much of the popularity of the RK method stems from the extremely simple temperature dependence introduced by Soave (*2*) in 1972. In the intervening period, Joffe and Zudkevitch (*3, 4, 5*) in 1969 and 1970 and Chang and Lu (*6*) in 1970 had proposed making both constants temperature dependent, inspired in part by a series of papers by Chueh and Prausnitz (*7, 8*) which demonstrated that the RK equation can be adapted to predict both vapor and liquid properties. The more complex Joffe–Zudkevitch (JZ) method was not as widely adopted as the Soave procedure.

Two noteworthy recent developments are the Peng and Robinson equation (*9, 10, 11*) and the Graboski–Daubert version of the Soave method (*12*). The Peng–Robinson work is part of a systematic attack on phase behavior of interest in gas processing, with especially impressive treatment of critical region effects, while the Graboski–Daubert work provides a comprehensive application of the Soave method to a large variety of hydrocarbon and related systems of interest in refining applications.

In this chapter, most of the application experience is for the JZ method, which has been used at Exxon Research and Engineering Company since before its publication (*see* Acknowledgment Section). However, most of the conclusions apply to the Soave or other forms, with modifications as noted.

Summary of Procedures to Adapt the RK Equation to VLE Prediction

The RK equation of state, relating pressure P to molar volume V and temperature T is

$$P = \frac{RT}{V - b} - \frac{aT^{-1/2}}{V(V + b)} \tag{1}$$

In adapting the RK equation for vapor–liquid equilibrium calculations, pure-component parameters are adjusted to match vapor and liquid fugacity along the vapor pressure locus. In the Soave modification only the RK parameter a is temperature dependent, while for the JZ modification both a and b are temperature dependent.

Using the nomenclature introduced by Chueh and Prausnitz (*6, 7*), pure-component parameters are expressed in terms of dimensionless premultipliers Ω_{a_i} and Ω_{b_i}.

$$a_i = \Omega_{a_i} \frac{R^2 T_{c_i}^{2.5}}{P_{ci}} \tag{2}$$

$$b_i = \Omega_{b_i} \frac{R T_{ci}}{P_{ci}} \tag{3}$$

Here, $\Omega_{a_i}{}^\circ = 0.42747 \ldots$ and $\Omega_{b_i}{}^\circ = 0.08664 \ldots$ will be used to denote the values of the premultipliers used in the original RK equation.

In terms of this nomenclature, the Soave premultiplier, $\Omega_{a_i}{}^\circ$, generalized in terms of Pitzer's acentric factor ω_i and the reduced temperature Tr_i, is

$$\Omega_{a_i} = \Omega_a{}^\circ\, T_{r_i}{}^{1/2} \{1 + (0.48 + 1.574\, \omega_i - 0.176 \omega_i{}^2)(1 - T_{r_i}{}^{1/2})\}^2 \tag{4}$$

The Joffe–Zudkevitch (RKJZ) modification includes variations of both Ω_{a_i} and Ω_{b_i} with temperature to fit liquid density as well as to match vapor-to-liquid fugacity along the vapor pressure locus. (The original method proposed by Zudkevitch and Joffe (*4*) had matched liquid fugacity to a generalized vapor fugacity correlation, but the present study follows the method they adopted in a later publication (*5*).) By comparison with the Soave procedure, this method loses some accuracy in saturated vapor densities (although this causes very little loss of accuracy in vapor fugacity), while greatly improving the accuracy of saturated liquid density.

The principal disadvantage of the RKJZ method is the complex temperature dependence of Ω_{a_i} and Ω_{b_i}. Although Haman et al. (*13*) provided a generalized correlation for Ω_{a_i} and Ω_{b_i}, these quantities are not generalized in the form described here but are generated each time they are needed from liquid densities and vapor pressures. Various sources of liquid density and vapor pressure are used: versions of the Riedel correlations (*14, 15*); API 44 Antoine equations (*16*); petroleum fraction correlations; and curve fits of experimental information.

An advantage of the JZ procedure is that for subcritical components, Ω_{a_i} and Ω_{b_i} are used only as intermediate internal variables. That is, specifying vapor pressure and saturated liquid density provides all of the information needed to calculate a and b; critical properties are not needed to calculate Redlich–Kwong parameters of heavy solvents or petroleum fractions.

For supercritical components, users of the Soave method generally use Equation 4 extrapolated above the critical temperature, whereas the RKJZ method uses the limiting value of Ω_{a_i} and Ω_{b_i} at the critical temperature. Although the RKJZ procedure introduces a corner into the Ω_a and Ω_b temperature dependence (and thus into the temperature dependence of fugacity), it does have the advantage of preserving the physically reasonable high-temperature limit of the original RK method for supercritical gases. That is, the second virial coefficient tends asymptotically towards b, a small positive number. This limiting behavior is significant for the very wide temperature ranges in refining applications.

Mixing rules used in the present work follow the usual practice: the original molar average mixing rule for b, with an adjustable interaction parameter (C_{ij}) in the mixing rule for a:

$$a = \sum_i \sum_j y_i \, y_j \, a_{ij} \tag{5}$$

where

$$a_{ii} = a_i \qquad i = j \tag{6}$$

$$a_{ij} = (a_i \, a_j)^{1/2} \, (1 - C_{ij}) \qquad i \neq j \tag{7}$$

The major alternative to Equation 7 is to use the procedures of Chueh and Prausnitz to calculate a_{ij}, by calculating pseudocritical temperature, pressure, and volume in an intermediate step, with the k_{ij} parameter in the combining rule for pseudocritical temperature performing the function of Equation 7. This has the advantage of providing a connection with the considerable literature on k_{ij} parameters. The disadvantage for the RKJZ method is that it imports critical properties into mixing rules even though they are not required to define the pure-component parameters for subcritical compounds. For systems where critical properties are well-defined, one readily can transform from one set of mixing rules to the other, as noted by Kato, Chung, and Lu (*17*).

Once procedures for calculating pure-component parameters and mixing rules are established, the calculation of component fugacity coefficients ϕ_i for both vapor and liquid phases follows standard procedures (*see* e.g. (*4*)). For VLE calculations, the distribution of components between phases is expressed generally as the K-value—the vapor mole fraction divided by the liquid mole fraction—related to fugacity coefficients for each component by:

$$K_i = \hat{\phi}_i \text{ liquid}/\hat{\phi}_i \text{ vapor} \tag{8}$$

Generally Understood Capabilities of RK Methods

For refining and gas-processing applications, the benchmark general-purpose VLE method is that of Chao and Seader (*18*), generally used with the modifications of Grayson and Streed (*19*). By comparison with the Chao–Seader method for these applications, either the RKJZ or Soave version has the following capabilities, which are relatively well understood: (a) scope—wider than Chao–Seader, extending closer to the mixture critical and to cryogenic temperatures; (b) accuracy—generally better than Chao–Seader, even with a rough estimate of the C_{ij} parameter (or with $C_{ij} = 0$ for light hydrocarbons); and (c) flexibility—the interaction parameters provide easy adjustment of the method for specific systems.

One example will serve to underscore the reason for the advantage over Chao–Seader at high pressure. Figure 1 shows the convergence of RKJZ *K*-values to unity as the mixture critical pressure is approached, for a temperature and composition on the mixture critical locus for the methane–ethane–butane ternary (*20*). This mixture was chosen in order to check RKJZ apparent critical pressure vs. the 1972 corresponding-states correlation of Teja and Rowlinson (*21*), which presumably has a better theoretical basis than the RKJZ method. In these comparisons, the Teja and Rowlinson correlation uses two interaction parameters per binary pair, based primarily on fits to binary critical loci; the RKJZ method uses $C_{ij} = 0$ for all binaries, based on binary VLE data.

Note that the RKJZ method not only predicts qualitatively the approach to mixture critical conditions; it is also quantitatively superior to the Teja and Rawlinson procedure in this instance. The ability of the RKJZ method to sense the approach to mixture critical conditions has been a great advantage in its application, by comparison with the Chao–Seader method, which has a stated limitation of pressure less than 0.8 times the true critical pressure.

Generally Understood Limitations of RK Methods

Limitations of the RK methods which have been mentioned (or assumed) in previous literature include: (a) poor second virial coefficient prediction, especially for compounds having nonzero acentric factors; (b) poor prediction of component liquid densities (this is a disadvantage only of the Soave form; the RKJZ method is fit to component liquid densities); and (c) inability to represent all PVT properties at the component critical simultaneously; the Soave form fails to reproduce the critical density while the RKJZ form gives nonzero values of $(\partial P/\partial V)_T$ and $(\partial^2 P/\partial V^2)_T$ at the critical point.

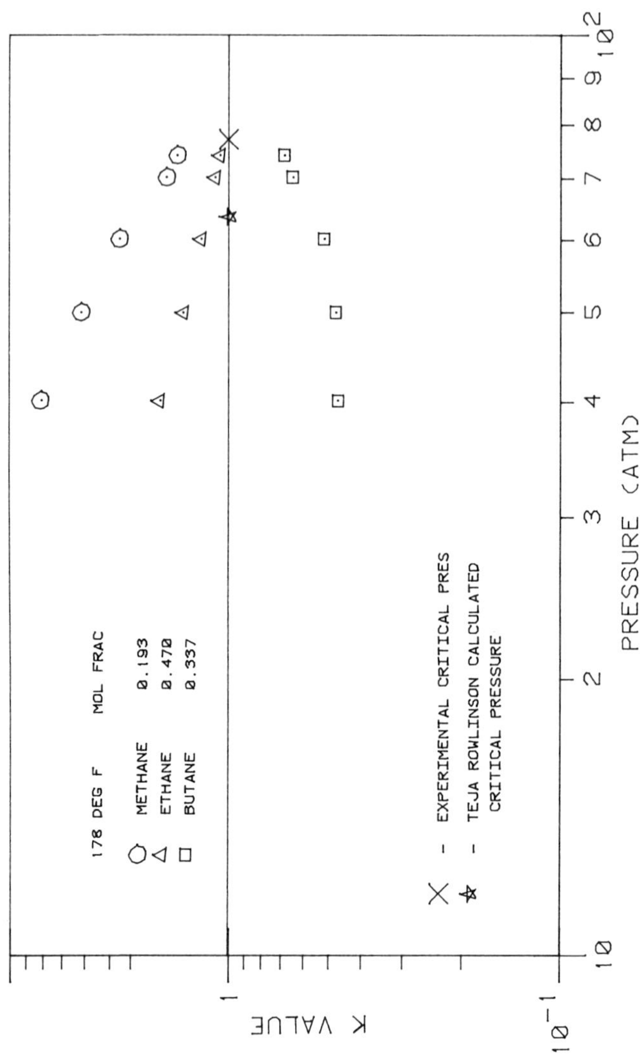

Figure 1. Convergence of RKJZ to P_c *using experimental* T_c

Other limitations include: (d) poorly defined basis for extrapolating *a* and *b* parameters above component critical; (e) inability to represent polar/nonpolar systems in detail; e.g., using an average C_{ij} to represent a low-pressure isotherm for an alcohol/hydrocarbon system gives maximum errors of perhaps a factor of two in *K*-value, whereas a method based on activity coefficients would fit the same data within a few percent; and (f) an inability to represent other derived properties (enthalpy, etc.) for nonpolar systems to the same degree of accuracy as VLE.

Some General Observations on Redlich–Kwong Methods

The thermodynamics community was rather slow to accept the early modified RK methods for VLE prediction. This writer, despite being an interested observer during the development of the RKJZ method, only became an enthusiast for the method after considerable applications experience. The reason for this early skepticism was the feeling that it was asking too much of the simple volume dependence of the RK equation to represent the fugacity functionality in both phases with sufficient accuracy.

It is worthwhile to ask the question: "Why is the RKJZ (or the Soave method) better than one would expect?" Answering this question in any depth is beyond the scope of this chapter, although it is explored for specific systems below. However, there are two general observations one can make.

The first observation is that, because of compensating errors, the vapor fugacity predictions of the RK equation are relatively insensitive to the adjustment of the constants necessary to fit vapor pressure. Thus, once component fugacity is matched along the vapor pressure locus, the effect of pressure and temperature on vapor fugacity is reasonably well represented. Further, the effect of pressure on liquid fugacity usually does not depend on highly accurate liquid densities. However, the RKJZ method is generally more accurate then the Soave method in representing the effect of pressure for light gas–heavy solvent systems because of its better representation of liquid volumetric behavior.

The second observation is that the success of the RKJZ and Soave methods may be attributed to the mixing rules (which are so important in fugacity prediction). These mixing rules essentially incorporate the van der Waals one-fluid (VDW-1) approximation favored in recent corresponding-states work (*21*). That is, if one interprets the *b* parameter as proportional to the critical volume V_c, then the *a* parameter is proportional to T_cV_c, and the mixing rules for *a* and *b* are equivalent to determining T_cV_c and V_c for the reference substance (although it should be noted that the value of *b* for the RKJZ method is more closely proportional to critical volume than it is for the Soave method). In a sense, one might regard the RK methods as a good mixing rule coupled with only the most rudimentary reference substance. Nevertheless, the accuracy of VLE predictions with these models is quite competitive with those of corresponding-states models incorporating much more elaborate reference fluids.

Some Surprising Successes with RK Methods

Our early application experience with the RKJZ method, apart from high-pressure systems, tended to be for systems where it was the method of last resort—where nothing else then available in our collection of computer programs could work. Thus, often its successes were doubly surprising.

One of the most surprising early successes was with cryogenic H_2–hydrocarbon systems. Here what was so surprising was not just that the method could be made to work, but that it worked so easily once the procedure for incorporating H_2 developed by Chueh and Prausnitz (*7*) was adopted. Results for H_2–hydrocarbon systems have been explored in more detail in a recent study (*22*); for the present discussion it is sufficient to note that a single C_{ij} value correlates H_2 and hydrocarbon K-values over substantial ranges of temperature and pressure for cryogenic systems.

Another system for which success is better than might be expected is the CO_2–methane binary shown in Figure 2. Here very accurate data over a wide range of conditions have recently become available (*23, 24*), for which Professors Kidnay and Kobayashi and their students deserve special praise. This system is very nonideal (owing to the CO_2 quadrupole), and is shown on a very expanded scale; rms errors in K-value at each temperature, except for the lowest, were less than 3%, and the smooth trend in C_{ij} exhibited for data from two different sources demonstrates really remarkable consistency of results between two laboratories. The points below $-$ 89°C are essentially for CO_2 well below its triple point near infinite dilution in methane. Although CO_2 liquid properties were extrapolated carefully into this region to obtain RKJZ parameters, the apparent S-curve in C_{ij} may be an artifact of the extrapolation.

Water–hydrocarbon systems, shown in Figure 3, comprise another class of systems which, rather surprisingly, can be handled accurately enough for many purposes. This work with the RKJZ method parallels similar studies by Heidemann (*25*) with the Soave method and by Peng and Robinson with their equation (*10*). As in their work, only fugacities in the hydrocarbon-rich liquid phases are fit by the model directly; if liquid water is present, it is assumed to be pure, since the C_{ij} fitting the

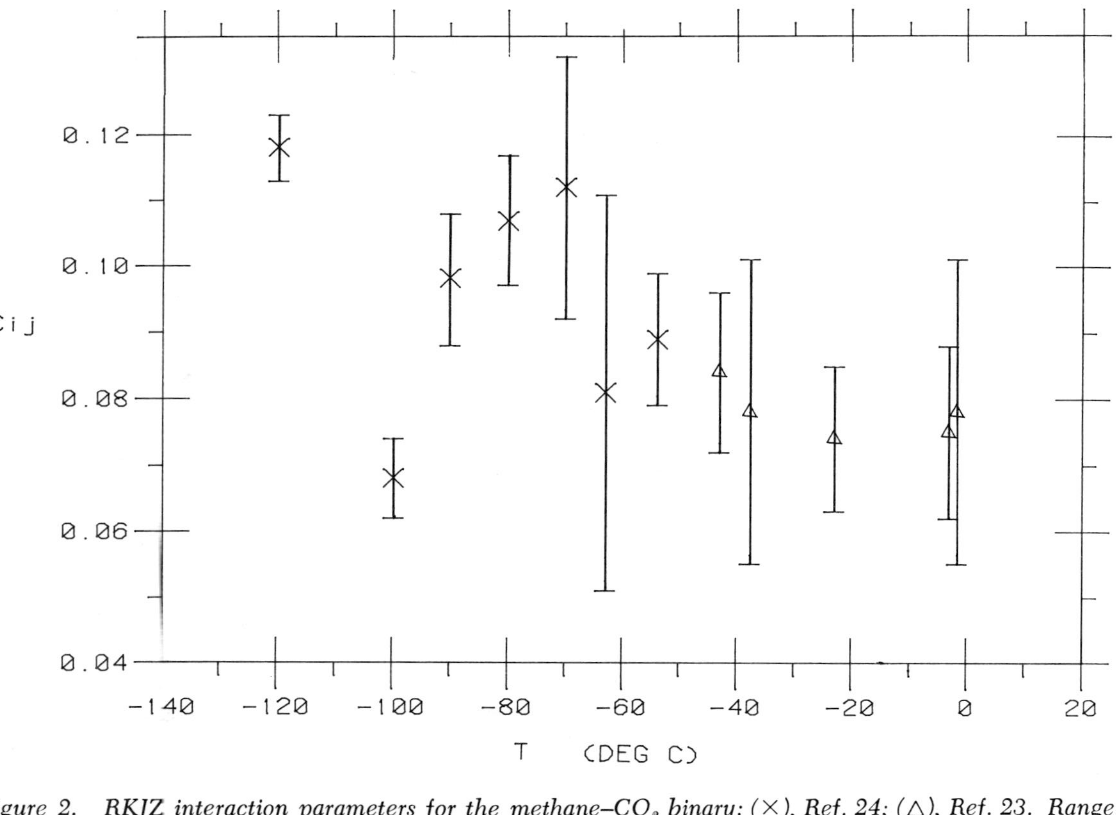

Figure 2. RKJZ interaction parameters for the methane–CO_2 binary: (×), Ref. 24; (△), Ref. 23. Range: rms deviation up to 1% above minimum.

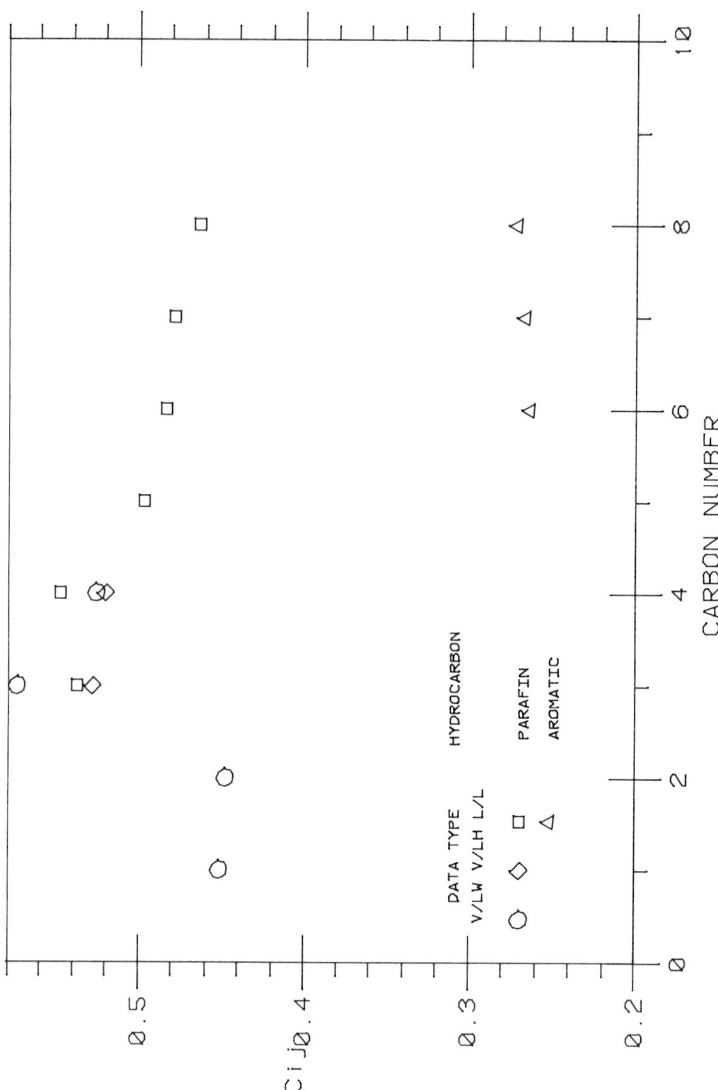

Figure 3. **RKJZ interaction parameters for the H_2O–hydrocarbon systems**

water fugacity in the hydrocarbon-rich phases yields hydrocarbon solubilities in liquid water which are in error by many orders of magnitude. Figure 3 shows C_{ij} data vs. carbon number based on the data from several sources (*26–31*).

What is striking here is not just that one can "fudge" the RKJZ method to include water by introducing large C_{ij} values of about 0.4. More importantly, if one looks at C_{ij}'s for a variety of systems, they fall into recognizable patterns. Note that C_{ij}'s for paraffins, taken from vapor–liquid and liquid–liquid data, are reasonably consistent, but that correlating data for aromatics requires sharply lower C_{ij}'s because of the higher water solubility.

One should, of course, be cautious in extrapolating predictions based on C_{ij}'s from a narrow temperature range. The available data indicate some temperature dependence in C_{ij} for water, but the data are not accurate enough or available over a wide range of conditions to support temperature dependence for most systems.

In general, within the sometimes stringent limitation of temperature dependence of C_{ij} one can map the infinite dilution fugacity of any polar compound into hydrocarbon systems. Further, if the infinite dilution behavior follows known patterns with hydrocarbon type, this can be made the basis for a correlation of C_{ij}. This ability to incorporate polar compounds over narrow ranges of concentration is extremely useful in refining and hydrocarbon processing applications.

Some Unexpected and/or Unexplored Limitations

Although RK methods are surprisingly versatile, they have a variety of limitations. This discussion will concentrate on those which might be considered unexpected, or which serve to define the boundaries of the known region of validity.

There is one kind of limitation which, although difficult to summarize concisely, should be mentioned, since it might come as an unpleasant surprise to the uninitiated. That is that the RKJZ or Soave models cannot represent certain details of light hydrocarbon systems at subambient temperatures to anything close to experimental accuracy. This kind of limitation was difficult to distinguish from systematic experimental error before the development of more accurate experimental techniques, notably by Kobayashi and co-workers and Kahre, who have measured low-temperature phase behavior for methane binaries (*32, 33, 34, 35, 36*).

If one looks at deviation trends for these systems, it is apparent that no adjustment of C_{ij} will allow one to fit the heavy-component *K*-value, while maintaining reasonable accuracy for methane at temperatures less than 50° to 75°C above the methane critical temperature (that is, below about $-$ 50°C) at moderate to high pressures (say, above 10 or 20 atm). These systematic deviations (of 15 to 40%) seem to indicate problems in characterizing simultaneously the extremes of mixing effects (or possibly, systematic experimental error) in expanded solvents and dense vapor, in the region where the lightest compound is only slightly supercritical or is subcritical. Once this effect was recognized as a source of confusion in correlating experimental data, it was not very important in the application experience reported here. The Soave method is slightly better than the RKJZ method in representing this region, and it appears that the Peng–Robinson method is substantially better.

Most other workers treating these data have assumed that these discrepancies are simply manifestations of the common problem of systematic experimental error for heavy-component *K*-value found in the older experimental data for light hydrocarbons. Furthermore, these deviations are masked by the common practice of reporting deviations in bubble-point pressure and in absolute differences in vapor mole fraction, which are both relatively insensitive to deviations in heavy-component *K*-value. Nevertheless, if the high precision of the newer experimental results is to be believed, there are systematic errors in these methods which might be important in some applications.

A related limitation of the RKJZ or Soave methods, which again is difficult to quantify, is also worth mentioning because it is not obvious from published results. The accuracy of these methods deteriorates if one attempts to fit too wide a range of conditions even for normal fluids. This effect is often masked by the generally spotty quality of the experimental data base, but one generally can discern substantial systematic deviation trends by careful examination of accurate data. Thus, ultimately the simplified volume dependence of the RK equation does place limitations on its accuracy. Consequently, one should keep in perspective the claims of Zudkevitch and Joffe (*3*), to represent hydrocarbon systems based on C_{ij}'s determined from one or two data points, or of Soave (*2*) to represent hydrocarbon systems with $C_{ij} = 0$. These claims are quite true in the context in which they were made, demonstrating the generality of the RK methods. Nevertheless, when highest accuracy is required, one should be reconciled to different C_{ij}'s for different regions.

Another limitation in a region of apparent strength is in the representation of critical-region effects at very high pressures, such as those occurring in mixtures of methane with moderately heavy hydrocarbons. An extreme example is shown in Figure 4, which shows calculated and experimental phase boundaries at 100°F for the pseudobinary mixture of methane with Kensol-16, a narrow-boiling oil with carbon number in the C_{15} to C_{20} range. This mixture was studied in 1950 by Rzasa and Katz (*37*). The calculated-phase envelopes show the extreme sensitivity of

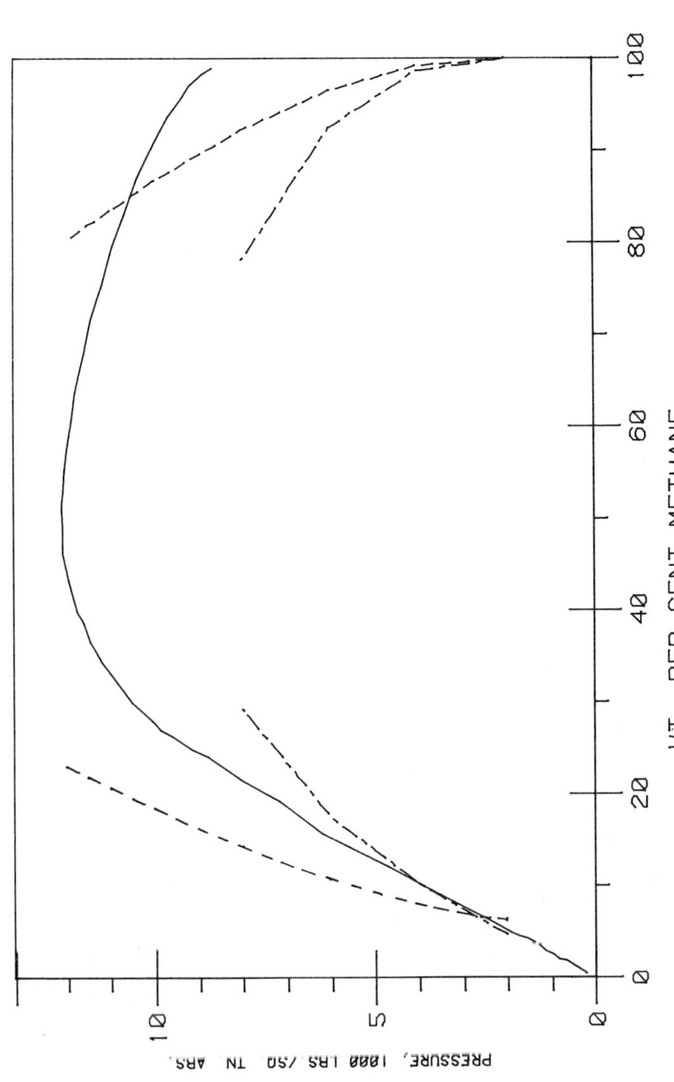

Figure 4. Phase boundaries for the methane–kensol system: (——), *data of Rzasa and Katz;* (— — —), $c_{ij} = 0.0$; (- - -), $c_{ij} = 0.05$.

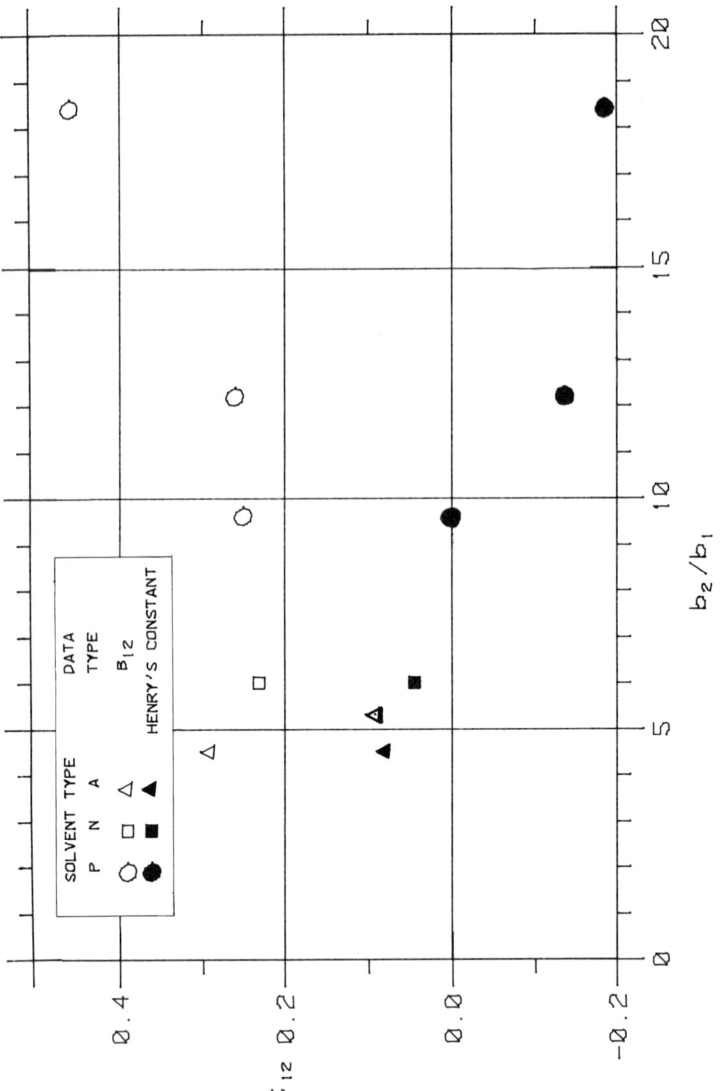

Figure 5. RKJZ interaction parameters for methane–C_{10}^{+} hydrocarbons

mixture critical pressure (calculated by the RKJZ method) to C_{ij}: for $C_{ij} = 0$, the calculated critical pressure is well below the experimental value of about 12000 psia, while for $C_{ij} = 0.05$, it is far above it. This demonstrates that extreme care is necessary to represent critical-region effects for mixtures with large differences in molecular size.

If molecular-size differences are too large, it is not only the critical region which should be of concern. Note in Figure 4 that the C_{ij} which represents the critical point would not represent the dew-point line very well. More striking is Figure 5, in which the C_{ij}'s necessary to correlate methane–heavy hydrocarbon behavior (recently measured by Prausnitz and co-workers (*38, 39, 40*)) are plotted vs. the ratio of RKJZ parameters b_2/b_1 (where 2 designates the solvent and 1 the methane); this ratio is essentially equivalent to the ratio of critical volumes. Note the divergent trends in C_{ij} necessary to correlate these highly asymmetric systems: as solvent molecular weight increases, decreasing C_{ij}'s are needed to correlate Henry's constant for methane; but increasing C_{ij}'s are needed to correlate second virial cross-coefficients (and thus dew-point composition at high pressure). These trends are for the RKJZ method; similar trends (although considerably displaced) occur for the Soave version. Clearly the trends demonstrated in Figure 5 impose limitations on the use of RK methods for asymmetric systems at high pressures. For many applications, the method will be satisfactory provided one understands the nature of the limitations.

Finally, it is worthwhile noting that the strengths of the RKJZ method—its capability to represent critical-region behavior, as well as nonideal mixing—can in practice impose limitations. As noted by Deiters and Schneider (*41*), this capability makes it possible to represent phase behavior of remarkably complex topology, especially for high-pressure systems. This complex phase behavior is always potentially present, even though flash or distillation algorithms do not account for it. Consequently, algorithms which work well for the Chao–Seader method (which represents nonideal mixing but not critical phenomena) or one of the methods based on the BWR equation of state (which represents critical phenomena but is not used usually for systems exhibiting nonideal mixing), may develop new pathology when used with a RK method. In practice, maintaining a RK method can be expected to require more advanced expertise in both programming and applied thermodynamics.

The Next Generation of Equation-of-State Methods

A logical conclusion for this chapter is, based on this applications experience, to reflect on what features in a new generation of equations of state would represent desirable improvements over the RK methods for phase-equilibrium calculations. Such reflections are of general interest since the strengths and weaknesses of the RK methods seem to be shared by any of the corresponding states methods using the VDW-1 approximation. Here is a personal, but by no means original, list of features:

(a) the accuracy (and flexibility) of RKJZ or Soave methods in region where these methods are satisfactory;

(b) good representation of density of both phases, including good representation of second virial coefficients;

(c) simultaneous representation of pure-component critical properties (ρ_c, $(dP/dT)_c$, T_c, P_c);

(d) well-defined supercritical extrapolation of temperature-dependent parameters;

(e) well-understood extrapolation to high molecular weights;

(f) independently adjustable infinite dilution fugacities for each component in a binary;

(g) group contribution features built into pure-component and/or mixture parameters; and

(h) special modifications for polar compounds in all of the above.

Many of these features are already in some of the emerging methods, and all are at least under study somewhere. No one has yet been ambitious enough to include all in a single method. Obviously, one cannot expect any modifications of the simple RK methods to combine all of these features, although some of them can be introduced by various add-on artifices. Thus, one can expect the RK methods eventually to be largely supplanted in application work where these features are important by the more elaborate methods now under development.

It is outside the scope of this chapter to assess the potential of these emerging methods, except to comment that for those of us interested in a wide range of molecular size, the perturbed hard-chain model of Donohue and Prausnitz (*42*) appears to come closest to combining all the features of interest. Regardless of which of the new methods ultimately find wide use in industrial applications, the process of selecting, adapting, and testing them for this purpose will take years. During this period, the RK methods will provide the benchmark by which the emerging methods are judged.

Glossary of Symbols

a, b = parameters in Redlich–Kwong equation of state, Equation 1
a_{ij} = interaction parameter used in calculating mixture parameter in Equation 5
C_{ij} = binary interaction parameter defined by Equation 7

K_i = ratio of vapor mole fraction to liquid mole fraction of Component i for vapor and liquid in equilibrium
P = system pressure
P_{c_i} = critical pressure of Component i
R = gas constant
T = system temperature
V = system volume
V_c = critical volume
T_{c_i} = critical temperature of Component i
$T_{r_i} = T/T_{c_i}$ = reduced temperature of Component i
Ω_{a_i} = premultiplier to determine Redlich–Kwong a parameter
Ω_{b_i} = premultiplier to determine Redlich–Kwong b parameter
ω_i = acentric factor for Component i
$\hat{\phi}_i$ = fugacity coefficient of Component i in mixture
ρ_c = critical density

Acknowledgments

The author thanks Exxon Research & Engineering Company for permission to publish this work. J. Joffe and D. Zudkevitch developed the original version of the RKJZ programs. G. A. Lyon coded most of the adaptations used in this work, and H. E. Newman supplied the computer graphics. M. S. Graboski and T. E. Daubert generously provided information concerning their version of the Soave method in advance of publication.

Literature Cited

1. Wilson, G. M. *Adv. Cryog. Eng.* **1964,** *9,* 168.
2. Soave, G. *Chem. Eng. Sci.* **1972**, *27,* 1197.
3. Zudkevitch, D.; Joffe, J., paper presented at the New Orleans meeting of AIChE, March 1969.
4. Zudkevitch, D.; Joffe, J. *AIChE J.* **1970,** *16*(1), 112.
5. Joffie, J.; Schneder, G. M.; Zudkevitch, D. *AIChE J.* **1970,** *16*(3), 496.
6. Chang, S. D.; Lu, B.C.-Y. *Can. J. Chem. Eng.* **1970,** *46,* 21.
7. Chueh, P. L.; Prausnitz, J. M. *Ind. Eng. Chem., Fundam.* **1967,** *6,* 492.
8. Chueh, P. L.; Prausnitz, J. M. *AIChE J.* **1967,** *13,* 1099.
9. Peng, D.–Y.; Robinson, D. B. *Ind. Eng. Chem., Fundam.* **1976,** *15,* 59.
10. Peng, D.–Y.; Robinson, D. B. *Can. J. Chem. Eng.* **1976,** *54,* 595.
11. Peng, D.–Y.; Robinson, D. B. *AIChE J.* **1977,** *23,* 137.
12. Graboski, M. S.; Daubert, T. E. *Ind. Eng. Chem., Process Des. Dev.* **1978,** *17,* 443.
13. Haman, S. E. M.; Chung, W. K.; Elshaxal, I. M.; Lu, B. C.–Y. *Ind. Eng. Chem., Process Des. Dev.* **1977,** *16,* 51.
14. Riedel, L. *Chem.–Ing.–Tech.* **1954,** *26,* 83.
15. Riedel, L. *Chem.–Ing.–Tech.* **1954,** *26,* 259.
16. *Am. Pet. Inst. Proj. 44,* loose leaf sheets extant 1977.
17. Kato, M.; Chung, W. K.; Lu, B. C.-Y. *Chem. Eng. Sci.* **1976,** *31,* 733.
18. Chao, K. C.; Seader, J. D. *AIChE J.* **1961,** *7,* 598.
19. Grayson, H. G.; Streed, C. W. *World Pet. Congr. Proceedings, 6th* **1963,** *Sec. III,* 234.
20. Cota, H. M.; Thodos, G. *J. Chem. Eng. Data* **1962,** *7,* 62.

21. Teja, A. S.; Rowlinson, J. S. *Chem. Eng. Sci.* **1973,** *28,* 529.
22. Gray, R. D., paper presented at the New York meeting of AIChE, 1977.
23. Davalos, J.; Anderson, W. R.; Phelps, R. E.; Kidnay, A. J. *J. Chem. Eng. Data* **1976,** *21,* 81.
24. Mraw, S. C.; Hwang, S. C.; Kobayashi, R. *J. Chem. Eng. Data* **1978,** *23,* 135.
25. Heidemann, R. A. *AIChE J.* **1974,** *20,* 847.
26. Olds, R. H.; Sage, B. H.; Lacey, W. N. *Ind. Eng. Chem.* **1942,** *34*(10), 1223.
27. Rigby, M.; Prausnitz, J. M. *J. Phys. Chem.* **1968,** *72,* 330.
28. Coan, C. R.; King, A. D. *J. Am. Chem. Soc.* **1971,** *93*(8), 1857.
29. Sage, B. H.; Lacey, W. N. "Some Properties of Lighter Hydrocarbons in H_2S, and CO_2," Monograph on API Research Project 32, 1955.
30. Kobayashi, R.; Katz, D. L. *Ind. Eng. Chem.* **1953,** *45,* 440.
31. Polak, J.; Lu, B. C.-Y. *Can. J. Chem. Eng.* **1973,** *51,* 4018.
32. Wichterle, I.; Kobayashi, R. *J. Chem. Eng. Data* **1972,** *17,* 4.
33. Kahre, L. C. *J. Chem. Eng. Data.* **1974,** *19,* 67.
34. Elliot, D. G.; Chen, R. J. J.; Chappelear, P. S.; Kobayashi, R. *J. Chem. Eng. Data* **1974,** 19, 71.
35. Chu, I.-C.; Chu, T.–C.; Chen, R. J. J.; Chappelear, P. S.; Kobayashi, R. *J. Chem. Eng. Data* **1976,** *21,* **41.**
36. Kahre, L. C. *J. Chem. Eng. Data* **1975,** *20,* 363.
37. Rzasa, M. J.; Katz, D. L. *Trans. Am. Inst. Min., Metall. Pet. Eng.* **1950,** *189,* 119.
38. Chappelow, C. C.; Prausnitz, J. M. *AIChE J.* **1974,** *20,* 1097.
39. Cukor, P. M.; Prausnitz, J. M. *J. Phys. Chem.* **1972,** *76,* 598.
40. Kaul, B. K.; Prausnitz, J. M. *AIChE J.* **1978,** *24,* 223.
41. Deiters, U.; Schneider, G. M. *Ber. Bunsenges. Phys. Chem.* **1976,** *80,* 1316.
42. Donohue, M. M.; Prausnitz, J. M. *AIChE J.* **1978,** *24,* 849.

RECEIVED October 5, 1978.

15

Determination of the Critical Exponent δ for a Binary Mixture Using a Centrifugal Field

RODOLFO SALINAS, H. S. HUANG,[1] and JACK WINNICK[2]

University of Missouri, Columbia, MO 65201

A determination of the critical exponent δ has been made for the liquid–liquid system, n-*decane–β,β′ dichloroethyl ether (chlorex). Utilizing schlieren photographs of the system in an ultracentrifuge at a temperature slightly below the critical solution point, the density gradient was obtained as a function of radius. These gradients, used in conjunction with sedimentation theory, provided a means for calculating values for the exponent δ. The values thus obtained are consistent with accepted values for the exponents β and γ in two-fluid systems. They are, however, smaller than those found for pure fluids.*

The behavior of binary solutions near their critical solution point has been studied extensively in centrifugal fields (*1, 2, 3*). This behavior is similar to that of a pure component at its gas–liquid critical point, now known to be governed by exponential relations of the form

$$\xi = \lim_{T \to T_c} \frac{d \ln |Y - Y_c|}{d \ln |T - T_c|} \qquad (1)$$

where ξ is the critical exponent for property Y as the critical temperature is approached (*4*). The critical isotherm is described by

$$\delta = \lim_{V \to V_c} \frac{d \ln (P - P_c)}{d \ln (V - V_c)} \qquad (2)$$

[1] Present address: Argonne National Laboratory, Argonne, IL 60439.
[2] Present address: Georgia Institute of Technology, Atlanta, GA 30332.

0-8412-0500-0/79/33-182-271$05.00/1

Similar behavior has been found for binary solutions at their critical point. Simple transformations allow Equation 1 to be used (5). In particular, the exponent δ describes the chemical potential, μ, along the critical isotherm:

$$| \mu_1 - \mu_c | = C \, | x_1 - x_c |^{\delta} \text{ at } T = T_c, P = P_c \tag{3}$$

Experimental values for δ can be used to test the universality of modern critical point theory. However, its value in binary mixtures has not been determined with much precision (6). There are, in fact, no experimental measurements reported which can be used to ascertain δ to better than about $\pm$ 20%.

Theory

Attempts to evaluate δ for binary fluids have, until now, required vapor pressure measurements (7). These measurements, performed along several isotherms, allow determination of the Gibbs free energy as a function of mole fraction, x, and temperature, T. Fitting these data to a semi-empirical four-parameter equation permits derivatives to be taken analytically. The difference between $\partial G/\partial x$ near and at the critical solution composition at any temperature defines the quantity, Δ:

$$\Delta = \frac{\partial G}{\partial x}(x, T) - \frac{\partial G}{\partial x}(x_c, T) \tag{4}$$

From scaling (8), however:

$$\Delta = \Delta x \, | \Delta x |^{\delta - 1} \, h(Y) \tag{5}$$

where $\Delta x = (x - x_c)/x_c$ and $Y = \epsilon/\Delta x^{1/\beta}$ and $\epsilon = (T - T_c)/T_c$.

A value of δ is selected, and, using the values of Δ calculated from Equation 4, $h(Y)$ is determined at each temperature. The function $h(Y)$ so determined should be independent of temperatures near T_c. Since the derivatives required by Equation 4 magnify any errors in the data, accurate assessment of δ by this technique is difficult. Therefore we have used a method which avoids this problem.

Measurement of the concentration distribution in a centrifugal field leads directly (9) to the gradient of the chemical potential:

$$\left(\frac{\partial \mu_1}{\partial x_1}\right) \frac{dx_1}{dr} = (M_1 - \rho V_1) \, \omega^2 r \tag{6}$$

where M_1 is the molecular weight of Component 1, ρ the density of the solution, V_1 the partial molal volume of 1, and ω the speed of rotation. If Equation 5 is differentiated and combined with the Gibbs–Duhem Equation

$$\frac{\partial \mu_1}{\partial x_1} = \left(\frac{1 - x_1}{x_c}\right) | \Delta x |^{\delta - 1} \left\{ \delta \cdot h(Y) - \frac{Y\, h'(Y)}{\beta} \right\} \quad (7)$$

where $h'(Y)$ is the derivative of $h(Y)$ and $h(Y)$ may be represented approximately by (8)

$$h(Y) = E_1 \left(\frac{Y + Y_o}{Y_o}\right) \left\{ 1 + E_2 \left(\frac{Y + Y_o}{Y_o}\right)^{2\beta} \right\}^{\frac{\gamma - 1}{2\beta}} \quad (8)$$

Combining this result with (*6*):

$$E_1 \left(\frac{1 - x_1}{x_c}\right) | \Delta x |^{\delta - 1} \left\{ \delta\, h(Y) - \frac{Y\, h'(Y)}{\beta} \right\} \frac{dx_1}{dr} = (M_1 - \rho V_1)\, \omega^2 r \quad (9)$$

where E_1 converts the dimensionless LHS to the dimensions of the RHS. This equation is used, along with values of dx/dr which are determined experimentally, to find the value for δ which best represents the binary system under investigation. It must be noted that ϵ may vary with r. This is true since the critical temperature generally will be a function of pressure and pressure varies in a known manner with radius:

$$\frac{d\epsilon}{dr} = \left(\frac{\partial \epsilon}{\partial T_c}\right) \left(\frac{dT_c}{dP}\right) \left(\frac{\partial P}{\partial r}\right) \quad (10)$$

This derivative can be evaluated at any radius if dT_c/dP is known.

Experimental

The system investigated was *n*-decane–β-β' dichloroethyl ether. This system was chosen because of its convenient critical solution temperature, 26.5°C (*10*). Coexistence curve and index of refraction data have been reported previously by Chu (*10, 11*).

Samples. The *n*-decane used in this study was of research grade, obtained from Phillip 66 Petroleum Chemical Company. The purity was 99.95% as determined by the manufacturer from a chromatographic analysis. No treatment (*6*) other than degassing was performed on it. β, β' dichloroethyl ether was obtained from K&K Chemical Laboratory and was purified by gas chromatography (GC) following the procedures described by Chu (*11*).

A pycnometer of approximately 50 mL was cleaned with hot cleaning solution for use in preparing the mixtures. The rubber cap, syringes, and needles were rinsed with research-grade acetone. Weighings of the pycnometer with and without the samples were performed in a temperature- and humidity-controlled room using an analytical balance with a precision of 0.0002 g. The sample used was one at 0.3952 ± 0.0001 mol fraction *n*-decane, the reported composition at the critical solution point (*10*). The phase separation temperature was 26.9°C.

Equipment. The Model E analytical ultracentrifuge, manufactured by Beckman Instruments, Inc., was equipped with a schlieren optical system and a photographic unit to take pictures of the schlieren patterns at various rotation speeds, up to the maximum of 60,000 rpm. Rotor temperature was maintained constant to within ± 0.02°C (*12*) and rotor speed was controlled electronically to within ± 0.2% of any selecting setting. The pictures were measured using a Gaertner M2060 microcomparator. Both the ultracentrifuge and the microcomparator are housed in a temperature-controlled room.

The optical cell used for this investigation was supplied by Beckman. The cell has the following basic components: a centerpiece which contains the sample of investigation, two sapphire windows that seal the open ends of the centerpiece, window holders, window liners and gaskets, and a cell housing that contains all components mentioned above. A screw ring holds these components in the cell housing.

The centerpiece is the central component of the cell assembly. It is shaped to minimize convection. The thickness of the centerpiece represents the thickness of the fluid column through which the light will pass. An aluminum centerpiece with a thickness of 1.2 cm and a 2½° sector angle was used in this study.

Determination of dT_c/dP. The pressure dependence of the critical temperature was determined experimentally using a small sample (1 mL) near critical composition. This was inserted into a glass ampule along with a magnetic stirring bar. This was inserted into a glass ampule along with a magnetic stirring bar. The opening of the ampule was connected by a length of nylon tubing to a cylinder of nitrogen. The ampule was suspended in a constant-temperature bath, maintained to better than ± 0.001°C. Temperature was monitored with a quartz thermometer.

Stirring was done with a hand-held magnet on a long rod. The temperature corresponding to appearance of two phases was measured as nitrogen pressure was applied incrementally, up to 1.85 atm.

Results

The schlieren patterns displayed in the viewing screen and the photographic plates represent the refractive index gradient distribution across the optical cell (*see* Figure 1). The height of the schlieren line from the base line, H, is related to the refractive index gradient dn/dr by the following expression (*13*);

$$\frac{dn}{dr} = \frac{\mathrm{H}\tan\theta}{\mathrm{Lam_1m_2}} \tag{11}$$

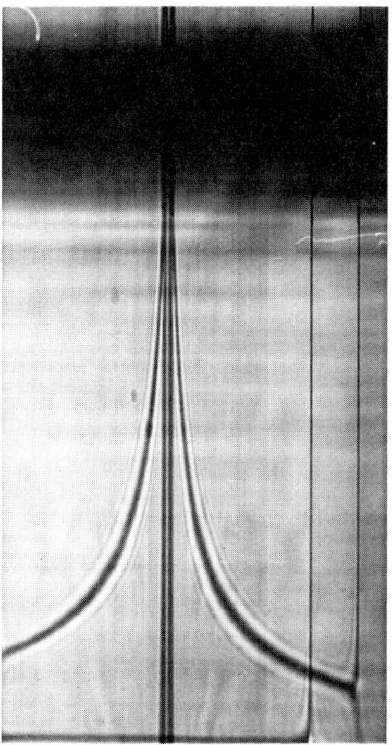

Figure 1. Actual Schlieren photograph of interace for n-*decane–chlorex (β,β dichloroethyl ether). Liquid–liquid interface is double line to left of center. The other two vertical lines are the pure decane reference and mixture vapor–liquid interfaces.*

where m_1 and m_2 are instrument constants, found from calibration: $m_1 = 2.15$; $m_2 = 3.37$; H is the value of the index of refraction gradient as measured from the base line in centimeters; θ is the phase plate angle (60.00°); a the cell thickness (1.20 cm); and L is the optical lever arm (59.0 cm).

The index of refraction n of the mixture is related to composition through (*14*)

$$\frac{(n^2 - 1)}{(n^2 + 2)} V = x_1 \frac{(n_1{}^2 - 1)}{(n_1{}^2 + 2)} V_1 + x_2 \frac{(n_2{}^2 - 1)}{(n_2{}^2 + 2)} V_2 \tag{12}$$

Here n_1 and n_2 are the refractive indices of pure n-decane and chlorex. These have been determined by Chu (*11*) at the wavelength used (546 nm) to be 1.4100 and 1.4596, respectively. The mixture molar volume is:

$$V = x_1 V_1 + x_2 V_2 + V^{\mathrm{E}} \tag{13}$$

Here V is the mixture molar volume, $V_{1,2}$ the pure component molar volume, and V^E the molar excess volume. No excess volumes were available for this system (*11*), requiring these measurements to be made as part of this study.

Because the excess volume for this system was expected to be small (*10*), and any effect is second order (*15*), only a few determinations were deemed necessary. A standard pycnometric technique was used (*16*). The densities at 29°C of three samples of known mole fraction, in addition to the pure components, were found. The excess volumes calculated from these results are shown as Table I. For the accuracy required by the calculations, a value for V^E of 1.0 $cm^3\ mol^{-1}$, independent of composition over the range of x considered, was sufficient. The effect of error in V^E is discussed later. Differentiation of Equation 12 then leads to

$$\frac{dn}{dx} = \left[\left.\frac{dn}{dx}\right|_0 - \frac{(n^2-1)(n^2+1)}{6n[xV_1 + (1-x)V_2]}\frac{dV^E}{dx}\right]\left[1 - \frac{V^E}{V}\right] \quad (14)$$

where $dn/dx|_0$ is the gradient for the ideal mixture, e.g., with $V^E = 0$ and $dV^E/dx = 0$;

$$\left.\frac{dn}{dx}\right|_0 = \frac{(n^2+2)^2\left[\frac{(n_1^2-1)}{(n_2^2+2)}V_1 - \frac{(n_2^2-1)}{(n_2^2+2)}V_2 - \frac{(n^2-1)}{(n^2+2)}(V_1 - V_2)\right]}{6n[xV_1 + (1-x)V_2]} \quad (15)$$

and $x = x_1$.

Equation 14, along with the measured values of dn/dr, yields dx/dr:

$$dx/dr = (dn/dr)/(dn/dx) \quad (16)$$

This is substituted into Equation 9.

Exact values for V and dV^E/dx could not be determined from the rough excess volume results. However, the effect of the second term in Equation 14 is very small, as shown later. The pure molar volume, V_1, was used in place of V_1 in Equation 9. Here the effect will be seen only

Table I. Excess Volumes at 29°C for *n*-Decane–Chlorex

X_1	*Density (g/mL)*	V^E *(mL/mol)*
0.00	$1.2090 \pm 5 \times 10^{-5}$	0
0.2486	$1.0287 \pm 5 \times 10^{-5}$	0.98 ± 0.02
0.4013	$0.9462 \pm 5 \times 10^{-5}$	0.96 ± 0.02
0.6925	$0.8197 \pm 5 \times 10^{-5}$	1.04 ± 0.02
1.00	$0.7222 \pm 5 \times 10^{-5}$	0

in the value of E_1 as determined in the least-square fitting. The density of the mixture, ρ, in Equation 9 was calculated using the experimental excess volume results.

Solution of Equation 9 requires values for x and n as $f(r)$. Since these are not known precisely at any r, values must be guessed and an integration of Equation 16 performed across the cell. In practice initial guesses were made at both sides of the meniscus. Numerical integration was performed across the entire cell, followed by a mass balance. Since the total sample is visible in the schlieren picture, the average mole fraction calculated must match that of the sample as inserted.

For each guessed value for the left-side interfacial composition, the right-side value was iterated until the mass balance was satisfied. This pair of interface compositions allowed regression of Equation 9 to find the best values of E_1 and δ, provided an exact value for x_c was known. The values for β and γ used in Equation 8 were 0.35 (the most recent measurements (*21*) indicate $\beta = 0.32$) and 1.25, respectively. As discussed later, the exact values used do not have a significant effect on the value of δ that is found. Since the samples prepared here may have contaminants different from those in other laboratories, and these samples show a change in x_c with time (*10*), we allowed x_c to float in this computation. The values obtained were always within 0.005 of Chu's one-atmosphere results.

The entire procedure is repeated with new left-side interface guesses, providing a set of $E_1 - \delta$ pairs. The fit of Equation 9 to the experimental results was used for the final determination of E_1 and δ. In doing this fitting, Equation 10 must be integrated along the radius to obtain ϵ which is needed in Equation 9. To do this integration a value of ϵ must be known at any radius. Since the temperature was not measured accurately, the decane-rich interfacial compositions were placed on the line through Chu's data (*see* Figure 2), and the corresponding value of ϵ obtained was used. The experimental value for dT_c/dP is 0.0258°C/atm.

In this final fitting, points which showed deviation of more than two standard deviations were excluded. These points were always very near the meniscus, where accurate measurement was difficult because of the steepness of the schlieren trace. Only about two of the approximately 100 data points per set were excluded. The best fit was that which gave the lowest sum of the squares difference between measured and calculated heights, H and H'. The rms difference between calculated and experimental points is 0.02 cm., approximately the precision obtainable experimentally.

The final results are shown in Figures 2 and 3. Figure 2 is a plot of the rms of the differences between the experimental and calculated heights vs. δ. Figure 3 shows how these residuals vary with the radius

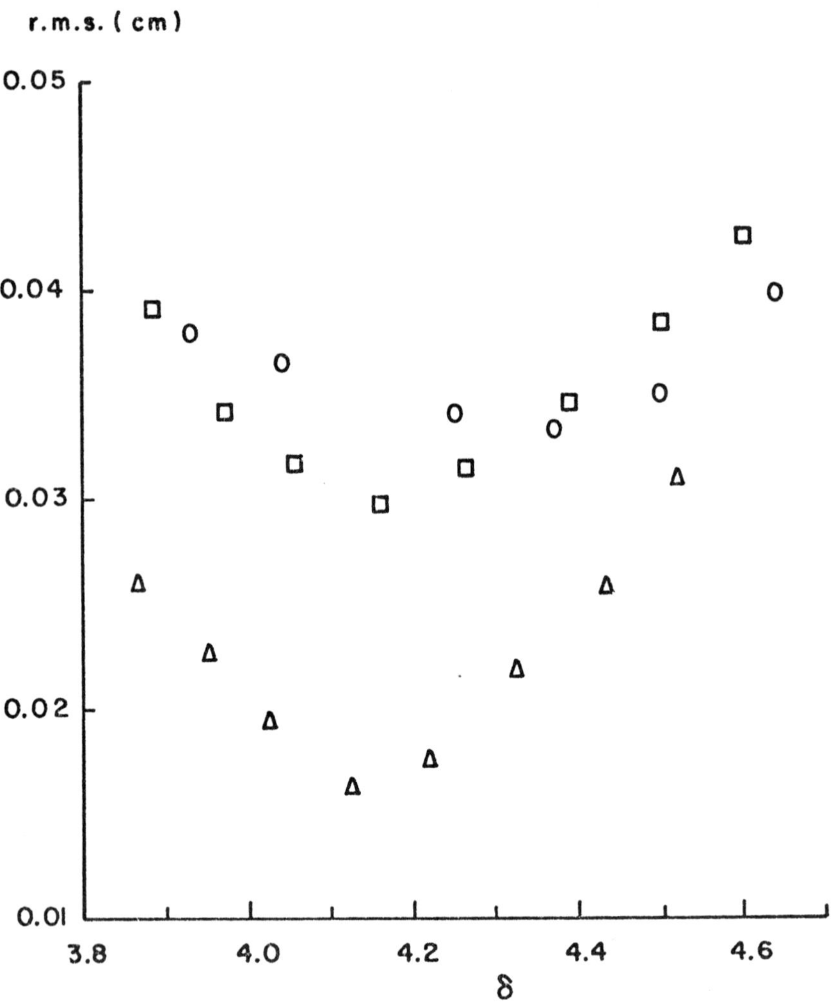

Figure 2. A plot of rms vs. δ: (○), 1; (□), 2; and (△), 3.

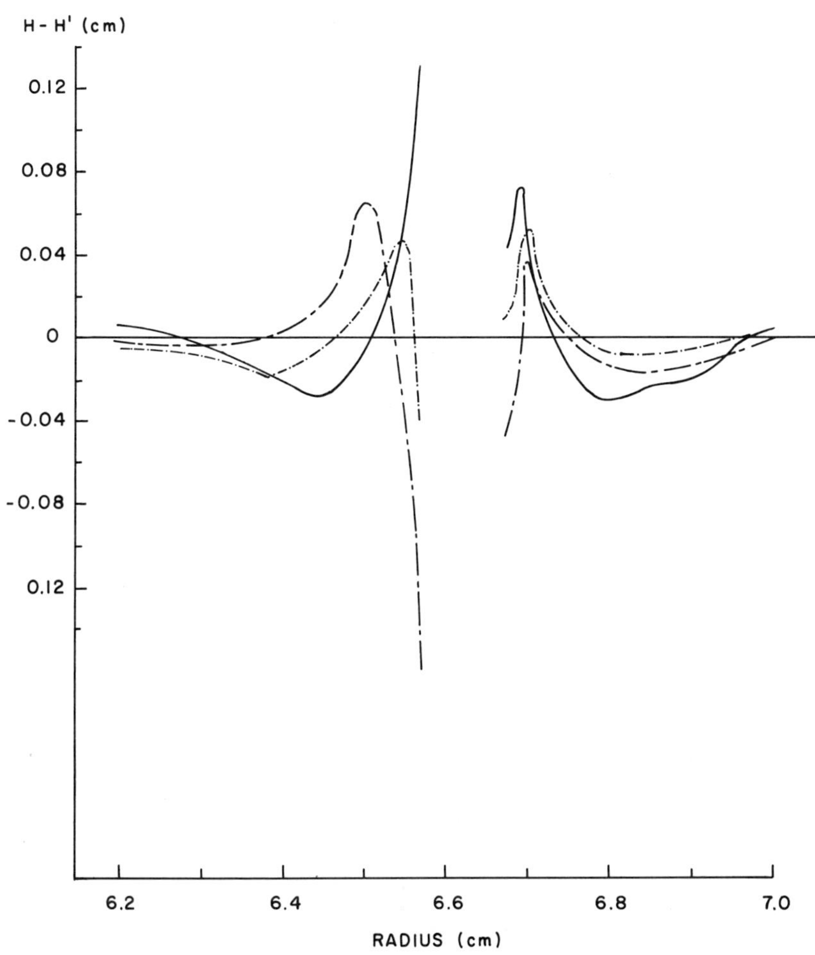

Figure 3. H–H′ *vs.* r *for various values of* δ*: (— · —), 4.17; (– – – –), 4.44; and (——), 3.85.*

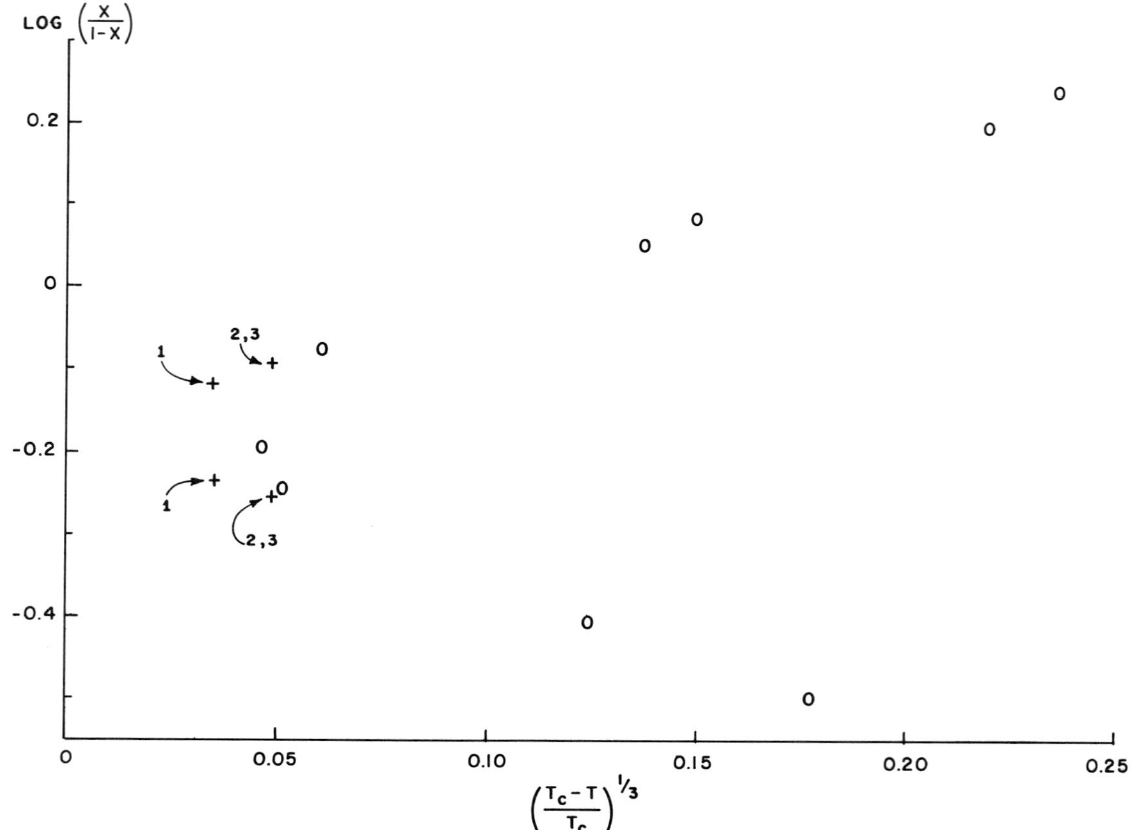

Figure 4. Coexistence curve for n-*decane–chlorex:* (○), *Chu;* (+), *present.*

along the cell for three different values of δ. Note that close to the meniscus a change in δ of 0.3 causes a highly significant deviation between calculated and measured values.

Complete results are shown in Table II. Column 2 lists the speed; Columns 3 and 4 show the results of the fitting technique described above. Columns 5 and 6 indicate the interfacial compositions, Column 7 lists the root-mean square difference between calculated and experimentally measured heights, and the final column lists the ε values for each run.

Table II. Experimental Results

Run	*Speed*	δ	E_1	X_{LM}	X_{RM}	*rms*[a]	$\epsilon \times 10^5$
1	10,000	4.33	3.93×10^5	.4308	.3680	.033	4.35
2	15,000	4.17	2.90×10^5	.4460	.3560	.030	11.7
3	15,000	4.19	2.78×10^5	.4453	.3554	.019	11.7

[a] The term rms $= \sqrt{\frac{\sum^{N} (H_i - H_i')^2}{N}}$.

Note that for Runs 2 and 3 at 15,000 rpm the separation of the components is greater than for Run 81 at 10,000, as is expected. Also note that for Runs 2 and 3 the interfacial compositions are about the same.

Figure 4 is a reproduction of the coexistence curve for this system at one atmosphere pressure as presented by Chu (*10*). The experimental compositions from the present study at both sides of the interface, deduced by the calculation procedure described above, are also plotted.

Discussion

The sensitivity of the calculations to experimental error and theoretical simplifications is examined now.

Equation 8 has been shown to fit critical region data for gases and magnetic substances (*9, 17*). Although different values for E_2 have been suggested, no effect on our results was seen when different values of this parameter were used. In particular the E_2 values tested were (*17*): $E_2 = .32$; $E_2 = .37$; $E_2 = .25$; and (*10*) $Y_o = 0.076$.

The sensitivity of the calculations to the values of β and γ were tested by allowing β and γ to take their classical values ($\beta = 0.5$, $\gamma = 1.0$). The minimum sum of the squares in Run 1 (rms = 0.041) was obtained for a higher value of δ ($\delta = 4.54$) than when the best values of β or δ were used. Clearly no tendency for δ to approach its classical value was seen.

Equation 12 has been tested for a large variety of systems (*14*) and found to be an approximation adequate enough for our purposes if the true molar volume is used, including the excess volume on mixing. The most noticeable of the individual effects of V^E on the calculations is seen in dn/dx, Equation 14. The effect of $V^E = 1.0$ cm^3 mol^{-1} on dn/dx is less than 1%. The effect of $\partial V^E/\partial x = 0.5$ cm^3 mol^{-1} is less than 0.25%. Their combined effect is less than 3% on δ.

The effect of the pressure gradient, caused by rotation, on the density of the mixture is even less than that of the excess volume. At the highest speed of rotation, 15,000 rpm, the pressure at the meniscus is about 8 atm. Judging from the pure-component compressibilities, this will cause a density increase of about 0.08%, or about 10% of the excess volume effect. Since we have no data on the mixture compressibilities, and since the effect is so slight, we have ignored it in our calculations.

The error in measurement of the height, H, of the Schlieren trace is estimated to be about $\pm$ 0.01 cm. Two readings must be taken for each height, one for the mixture and one for the reference line (*see* Figure 1). The total error per datum is then $\pm$ 0.02 cm. Other errors (e.g., temperature, impurities in the sample, and rotational speed fluctuations) that are difficult to estimate may add somewhat to this value. It is, however, approximately the same as the rms error shown in Table III.

The thermodynamic consistency of the final results can be ascertained from the inequality:

$$\gamma \geq \beta(\delta - 1) \tag{17}$$

The best values for γ and β are: $\gamma = 1.23 \pm 0.02$; $\beta = 0.32 \pm 0.01$. Equation 17 then implies $\delta = 4.84 + 0.18$. Thus, the present results lie slightly below the expected ones.

The method described here seems to have the potential for a more precise determination of δ for binary systems. Work is currently in progress on the system: *n*-hexane–perfluoro–*n*-hexane. This system is more stable over periods of time than the chlorex–decane system yet it has an equally convenient critical temperature, 22.6°C (*18, 19, 20*).

Glossary of Symbols

E = constant in scaling equation
G = Gibbs free energy
h = scaling function defined by Equation 8
M = molecular weight
n = index of refraction
P = pressure

r = radius in centrifugal field
T_c = critical temperature
V = mixture molar volume
V_1 = molar volume of Component 1
$\bar{V}_1$ = partial molar volume of Component 1
V_c = mole fraction of Component 1
x_1 = mole fraction of Component 1
x_c = critical mole fraction
Δx = dimensionless composition variable, defined below Equation 5
Y = dimensionless variable defined below Equations 5 or, in Equation 1, generic property.

Greek

β, γ, δ = critical exponents
Δ = difference in full energy derivative, Equation 4
ϵ = dimensionless temperature
ξ = generic critical exponent
μ = chemical potential
ρ = mass density
ω = rotor speed

Acknowledgment

The authors with to thank C. M. Knobler and R. L. Scott for their helpful comments and the use of their laboratories. This work was performed with the financial aid of the National Science Foundation under Grant GK-36514. R. Salinas wishes to thank CONACYT of Mexico for his support during the course of this research.

Literature Cited

1. Dickinson, E., et al. *Phys. Rev. Lett.* **1975,** *34,* 180.
2. Fannin, A. A.; Knobler, C. M. *Chem. Phys. Lett.* **1974,** *25,* 92.
3. Hildebrand, J. H., et al. *J. Phys. Chem.* **1954,** *58,* 577.
4. Stanley, H. E. "Introduction to Phase Transitions and Critical Phenomena"; Oxford University Press: London, 1971.
5. Rowlinson, J. S. "Critical Phenomena," *Natl. Bur. Stand. (U.S.), Spec. Publ.* **1966,** *273,* 9.
6. Scott, R. L. "Critical Exponents for Binary Fluid Mixtures," *Special Report on Thermodynamics,* in press.
7. Simon, M.; Fannin, A. A.; Knobler, C. M. *Ber. Bunsenges. Phys. Chem.* **1972,** *76,* 321.
8. Vicentini-Missoni, M.; Levelt Sengers, J. M. H.; Green, M. S. *J. Res. Natl. Bur. Stand., Sect. A* **1969,** *73,* 563.
9. Guggenheim, E. A. "Thermodynamics"; North-Holland: Amsterdam, 1949.
10. Chu, B.; Kao, W. P. *J. Chem. Phys.* **1969,** *50,* 3986.

11. Chu, B. *J. Chem. Phys.* **1964**, *41*, 226.
12. Gropper, L.; Boyd, W. *Anal. Biochem.* **1965**, *11*, 328–245.
13. Longworth, L. G. *Ind. Eng. Chem., Anal. Ed.* **1946**, *18*, 219.
14. Hubbard, J. C. *Z. Phys. Chem.* **1910**, *74*, 207.
15. Cullinan, J. J.; Lenczyk, J. P. *Ind. Eng. Chem., Fundam.* **1969**, *8*, 819.
16. Winnick, J.; Kong, J. *Ind. Eng. Chem., Fundam.* **1974**, *13*, 292.
17. Vicentini-Missoni, M.; Joseph, R. I.; Green, M. S.; Levelt Sengers, J. M. H. *Phys. Rev.*, in press.
18. Bedford, R. G.; Dunlap, R. D. *J. Am. Chem. Soc.* **1957**, *80*, 282.
19. Gaw, W. J.; Scott, R. L. *J. Chem. Thermodyn.* **1971**, *31*, 337–345.
20. Knobler, C. M., private communications.
21. Greer, S. *Phys. Rev., Sect. A* **1976**, *14*, 1770.

Received August 8, 1978.

16

State-of-the-Art Determination of the Second Virial Coefficient of Ethylene for Temperatures from 0° to 175°C

M. WAXMAN and H. A. DAVIS

Thermophysics Division, Center for Thermodynamics and Molecular Science, National Bureau of Standards, Washington, DC 20234

Values of the second virial coefficient of ethylene for temperatures between 0° and 175°C have been determined to an estimated accuracy of 0.2 cm^3/mol or less from low-pressure Burnett PVT *measurements. Our values, from −167 to −52 cm^3/mol, agree within an average of 0.2 cm^3/mol with those recently obtained by Douslin and Harrison from a distinctly different experiment. This close agreement reflects the current state of the art for the determination of second virial coefficient values. The data and error analysis of the Burnett method are discussed.*

Over a span of many decades, the virial coefficients associated with the volumetric behavior of a gas have acquired, perhaps in a fortuitous manner, scientific significance. The determination of virial coefficients, either experimentally or theoretically, still presents scientific challenges. Its historical background, which we have summarized from Mason and Spurling (*1*), suggests how the subject developed.

Historical. About a century ago, experimentalists and theoreticians independently popularized the use of various forms of polynomial expressions for the mathematical description of physical behavior, e.g., the volumetric behavior of a gas. The early use was based solely on convenience, especially for interpolation of data. Theoretical significance was later established for many cases. For the volumetric behavior of a gas, each term of an infinite series expansion in density was given specific significance in terms of intermolecular forces.

The first major contribution was made in 1901 by Kamerlingh Onnes, who described the isothermal volumetric behavior of a gas by the finite polynomial series

$$Pv = A'' + \frac{B''}{v} + \frac{C''}{v^2} + \frac{D''}{v^4} + \frac{E''}{v^6} + \frac{F''}{v^8},$$

where P designates the gas pressure, v its specific volume, and A'', B'', C'', D'', E'', and F'' designate the coefficients of the series. His comments concerning the application of both finite and infinite series will always be relevant. Kamerlingh Onnes appreciated the fact that the coefficients of such a finite series might differ from the corresponding terms of an infinite series, particularly for terms of higher power than quadratic. Kamerlingh Onnes suggested the name "virial coefficients" for the series coefficients and further recommended that this name be reserved for the coefficients of the infinite series

$$Pv = RT\left(1 + \frac{B^{\infty}}{v} + \frac{C^{\infty}}{v^2} + - - - + \frac{M^{\infty}}{v^M} \cdots\right).$$

where R denotes the gas constant and T is the absolute temperature. The name virial probably was associated with the classical virial theorem of Clausius, in which the time average kinetic energy of a system of particles is equal to a function of the position and the total force acting on each particle. Clausius called this function the virial of the system. For a closed system of gas molecules, the function could be modified to include only intermolecular forces. If this modified function were expanded as a power series in density, the coefficient of the leading term would be equivalent to the second virial coefficient, B^{∞}. Other significant contributions of Kamerlingh Onnes concerned the expression of the equation of state in a reduced, hence universal, form using critical constants and the correlation of virial coefficients based on the principle of corresponding states. He suggested that the accuracy with which the coefficients could be evaluated from experiments reduced with increasing powers to such an extent that only the first few terms of the series were significant and the rest were to be considered collectively as a "virial remainder" to adjust the fit to the experimental data. This is because of the high correlation between the series coefficients.

In spite of the inherent limitations of the infinite density series, its theoretical significance has grown. In fact, at the present time, theoretical calculations of the second virial coefficient, e.g., for argon from an interatomic potential, are purported to be more accurate than values obtained from the best experimental data. For fluids of more complicated molecules, the accuracy of the experimental values still exceeds that of calculations based on theory; hence accurate data are needed to guide theory in its development. For any gas, the accuracy to be expected from state-of-the-art determinations of the second virial coefficient is about 0.2 cm^3/mol for values ranging from -50 to -200 cm^3/mol. Such accuracy is illustrated for the second virial coefficient of ethylene by our recent experimental determinations, reported here by us, and by the independent results of Douslin and Harrison (*2*). Industrial demands presented the impetus for these ethylene determinations.

Ethylene Thermodynamic Program. During the early 1970's, the inadequacy of available thermodynamic information for industrial processing of ethylene became apparent to concerned people in this country as well as in the United Kingdom, West Germany, the Soviet Union, and elsewhere. Cognizant of this need, the Office of Standard Reference Data at NBS, in cooperation with U.S. industry, established a comprehensive program to obtain high-quality experimental data where they were required and from these data as well as from the data to obtain a definitive and wide-range correlation with other thermodynamic properties for ethylene. Several different laboratories with proven experimental capability were asked to participate in this program. Because of the suitability of our Burnett apparatus for the accurate determination of the equation of state of gases from low to moderate pressure measurements at moderate temperatures, we were asked to determine the second virial coefficients of ethylene. In this chapter, we discuss the results of our measurements. Our discussion includes the pertinent aspects of the Burnett experimental method and the analysis of data obtained with such a method for the determination of virial coefficients. From the analysis of our ethylene Burnett data, we present values of the ethylene second virial coefficient for temperatures from 0° to 175°C and we compare these results with other state-of-the-art results (*2*).

Burnett Method

Methodology. In general, the determination of the four gas variables—pressure P, volume V, quantity n, and temperature T—is necessary for the calculation of the compressibility factor Z defined by the gas equation

$$Z = PV/nRT.$$

In the Burnett method, only two of these variables need to be measured (*3*). Here, P and T variables are measured on an isotherm before and after stepwise expansion of the gas contained in the sample volume into an evacuated expansion volume. A schematic Burnett apparatus representative of such a process is given in Figure 1.

The use of the conservation of the quantity of gas during an expansion and the gas equation leads to a relationship between the ratio of the measured pressures and the product of the ratios of the volumes $(V_{I} + V_{II})V/_{I}$ and the compressibility factors. Similarly, from the conservation of the quantity of gas, a relationship results between the ratio of the densities and the ratio of the volumes. The unknowns are the compressibility factors, the volume ratio, and the gas densities. Their evaluation is

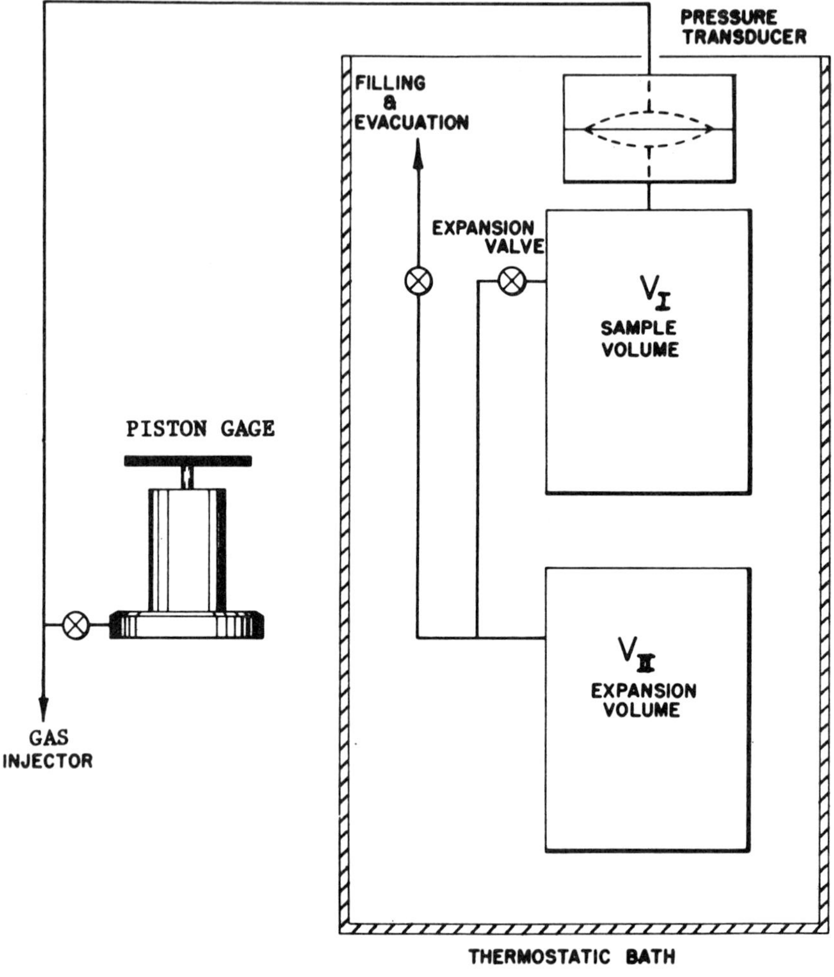

Figure 1. Schematic of Burnett PVT *apparatus*

Glossary of Symbols for Figure 4

Symbol	Definition
●	= results for N treated as a single parameter for all sequences of expansions
○	= results for N treated as a single parameter or as a parameter for each sequence of expansions
$P1$	= maximum pressure, 3 MPa
$P2$	= maximum pressure, 3.7 MPa
G	= results for the Gulf sample
M	= results for Matheson sample
C	= results for cubic series
Q	= results for quadratic series
N_{constant}	= N as constant
I	= standard deviation of the individual quantities N, B_2, and Z
ΔN	= $N_{\text{parameter}} - N_{\text{constant}}$
ΔB_2	= $B_2\ (N_{\text{parameter}}) - B_2\ (N_{\text{cosntant}})$
ΔZ	= $Z\ (N_{\text{parameter}}) - Z\ (N_{\text{cosntant}})$

made tractable by the representation of the compressibility factor as a truncated virial series in pressure or density. This leads to two nonlinear equations for constant temperature

$$\frac{P_{r-1}}{P_r} = \frac{N_r\left[1 + \sum_{i=2}^{I} b_i P_{r-1}{}^{i-1}\right]}{1 + \sum_{i=2}^{I} b_i P_r{}^{i-1}} \tag{1}$$

in terms of a truncated virial expansion in pressure, and

$$P_r = \frac{\rho_0}{N_1, N_2 \ldots N_r}\, RT\left[1 + \sum_{j=2}^{J} B_j \left(\frac{\rho_0}{N_1, N_2 \ldots N_r}\right) \rho^{j-1}\right] \tag{2}$$

in terms of a finite virial expansion in density. In Equations 1 and 2 the symbol ρ denotes the gas density, N_r the volume ratio $(V_{\text{I}} + V_{\text{II}})/V_{\text{I}}$ for the expansion r, b_i the pressure virials, and B_j the density virials. The subscript o denotes the initial condition of the sample gas, r its condition after the rth expansion in the sequence of expansions extending from $r = \text{o}$ to $r = \mathcal{R}$, and $r - 1$ its condition just before the rth expansion. Further, N_r is defined as N, the value of the ratio of pressures P_{r-1} and P_r in the limit of zero pressure plus a slight correction for any distortion of the Burnett volumes owing to pressure. The unknown parameters in Equation 1 are the pressure virials (b_i) and the volume ratio (N); in

Equation 2 they are the density virials (B_j), the volume ratio (N), and the initial density (ρ_0). The theory of the Burnett method is presented in considerably more detail in Refs. *3, 4, 5, 6,* and *7.*

Data Analysis. Generally, several sequences of expansions are necessary to determine the volumetric behavior of a gas. Either generalized equation for the appropriate pressure and temperature ranges may be applied to experimental data to form an overdetermined set of nonlinear equations which, when combined with the least-squares constraint, can be evaluated for the unknown parameters and the true pressures. The procedure followed in such nonlinear evaluations involves linearizing the equations by means of a first-order Taylor series expansion about estimates of the parameters and, optionally, the variables, and then treating the resulting equations as a linear regression problem to obtain first-order corrections for the estimates (*6–11*). The process is iterated until the least-squares principle is satisfied, i.e., until the values of the derivatives of the sum of the squares of the residuals with respect to each parameter converge to zero. This, obviously, means that the sum of squares should have a minimum value. The computer algorithm used for the evaluation of the density-series equations readily permits the using least-squares convergence criterion; its inclusion in the algorithm for the evaluation of the pressure equations is far more formidable. For algebraic simplicity, it is common practice to use the smallest value of the sum of the squares for the convergence criterion. However, the use of only this criterion can lead to erroneous results. Convergence requires that both the derivative and the sum of the squares criteria be satisfied. We include both criteria in our Burnett algorithms. Our helium volume-ratio values, reported herein, probably represent the first published results for which convergence has been proven in the evaluation of pressure-series equations.

The results themselves have a subtlety associated with their interpretation owing to the presence of the volume-ratio parameter and, optionally, the initial density parameter. The Burnett equations have more flexibility to fit Burnett data than only a density series to *PVT* data. The statistical uncertainties reflect the quality of the experimental data relative to the particular model used to describe the experiment. The estimation of accuracy for Burnett results is necessarily somewhat subjective since the effect of systematic errors on parameter values is not explicit in nonlinear equations, such as the Burnett equations. Accuracy, however, can be estimated from a study of the effects of systematic errors in computer model calculations and from the magnitude of the change in the volume-ratio value determined with nonideal and nearly ideal gases. For these reasons, we include such information along with our virial coefficient results for ethylene.

Experimental Apparatus

The apparatus used in this experiment is basically the same as the Burnett *PVT* apparatus shown in Figure 2 and described in our earlier publications (*12, 13*). The only noteworthy modification stems from a need for refinishing the surfaces exposed to the sample gas because of a chemical reaction between the ethylene sample and the transducer backing surface during our first ethylene measurements. In this section we describe the apparatus, discuss the nature of the reaction, and report the purity analyses for the gases—ethylene and helium—used in this experiment.

The gas being studied was confined to the Burnett volumes and separated from the pressure-measuring instrumentation by a differential pressure transducer. This arrangement allowed all of the volumes containing the sample gas to be thermostated together in a forced-flow liquid bath to within 0.002 K of the desired isotherm temperature. The isotherm temperature, as defined by the International Practical Temperature Scale of 1968, was determined from the resistance measurement of a calibrated capsule-type platinum thermometer located in the wall of the Burnett vessel. The measured pressure was that of the counterbalancing gas used to null the transducer. The transducer itself contributed an error of less than 1 ppm in the transfer pressure and in the constancy of the sample volume. The pressure measurement was determined by using two primary gauges: a controlled-clearance piston gauge with the counterbalancing gas as its pressure fluid and a mercury barometer. The pressure range, 0.3–3.7 MPa, was selected for optimum accuracy and precision (better than 0.003% and 0.001%). The standard deviation of the analytical fits to our pressure data was comparable with this measurement precision.

As time progressed, our initial isothermal ethylene measurements exhibited a slowly decreasing or increasing pressure trend whose magnitude depended on the isotherm temperature. For the first measured isotherm at 0°C, the trend barely exceeded our measurement precision; in contrast, for the next isotherm at 25°C, the trend increased by a factor of two to five, to 50 ppm. Consequently, we rejected the 25°C data and disassembled the apparatus to determine the cause of the reaction.

Visual inspection suggested that, owing to the presence of ethylene, the backing surface of the transducer had corroded slightly although this surface, as well as the other gas-volume surfaces, had previously been gold plated. This backing surface, which is slightly concave to allow only limited deflection of the transducer diaphragm, had been optically lapped before the gold-plating process. Apparently the abrasive material used in the lapping process had become embedded in the backing surface and later, during the experiment, reacted with the ethylene. Subsequently, we undertook a limited study of possible ethylene chemisorption on metal surfaces. We exposed gold-plated and unplated stainless steel and nickel sheets to instrument-grade ethylene at 150°C and 6 MPa. The sheets then were examined by photospectral techniques. These qualitative analyses, conducted by N. Erickson of NBS, indicated the presence of a carbon peak considerably larger than that appearing on surfaces only exposed to the atmosphere (*14*). With this background information, the need for passivating the surfaces of the gas volumes by prior exposure to

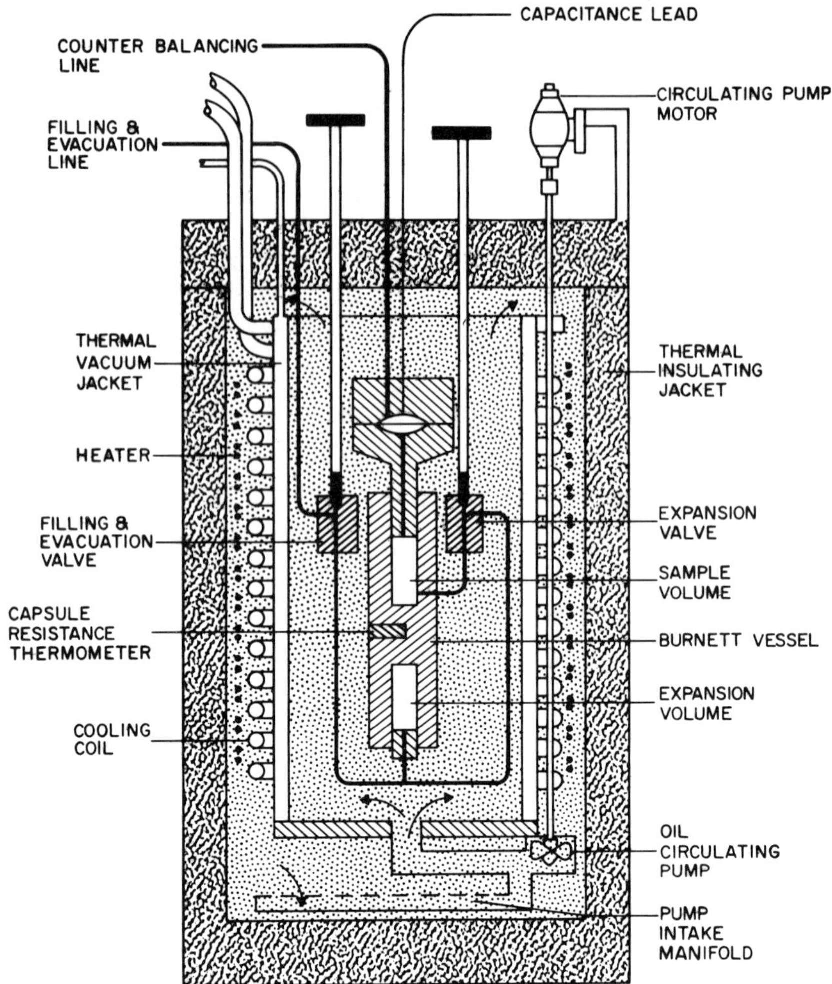

Figure 2. Burnett PVT *apparatus*

ethylene became evident. The time required to passivate the surfaces increased with increasing temperature from a few hours at 25°C to a few days at 175°C. Prior to the reassembly of the apparatus, all of the surfaces to be exposed to the sample gas were refinished to remove all visible traces of surface contaminants; they then were gold plated.

Two different sources for the ethylene sample gas were used. The Gulf Oil Company–U.S., a participant in the OSRD ethylene program, contributed a cylinder of high purity ethylene for this program from its Cedar Bayou olefin plant. This source was used for the isotherms from 0° to 100°C, which were determined in order of increasing temperature. For the higher temperature isotherms and for comparative information at 25° and 100°C, the source was changed to a cylinder of chemically pure ethylene obtained from Matheson. The change occurred after gas chromatography (GC) for analyzing high-purity ethylene was developed at NBS in the framework of the OSRD program. The analyses by percentage volume for the principal impurities in our two supplier were:

	Gulf Oil Company		*Matheson*
	Gulf	*NBS (average)*	*NBS (average)*
CH_4	0.004_5	0.007_0	0.002_2
CH_2H_6	$.003_0$	$.003_9$	$.002_5$
Estimated purity	99.99	99.98_6	99.99_3

Results obtained at 25° and 100°C for both sample sources indicated that although the impurity variation had little effect on the compressibility factor—0.005% at 25°C and 0.006% at 100°C for our range of pressures—its effect was systematic. Impurities had a more dramatic effect at conditions characteristic of the critical region. This is illustrated in Figure 3 for ethylene density results obtained at 25°C by us and others (*2, 15*). For samples with reported purities of 99.99% or better, the

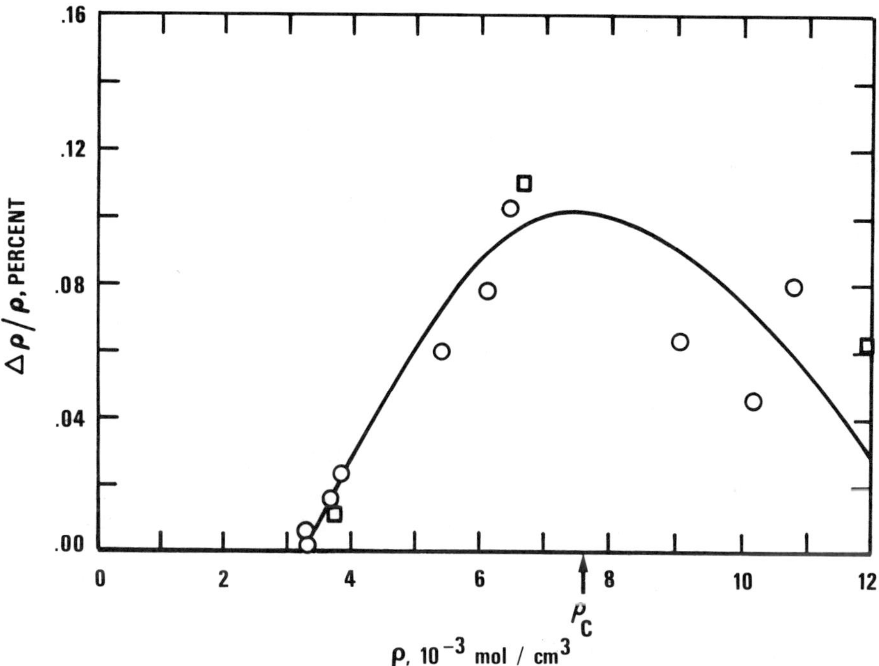

Figure 3. Ethylene density comparison at 25°C, near the critical density: $\Delta\rho$ = ρ(NBS) − ρ(Douslin) and ρ_c = critical density. All gas samples involved in this comparison had reported purities of 99.99% or better. NBS: (○), Hastings and Levelt Sengers; (□), Waxman and Davis.

Principal Impurities by Percentage Volume for Figure 3

	This Work	*Hastings (15)*	*Douslin (2)*
CH_4	0.002_2(av)	0.005	n.l.
C_2H_2	< .0001	< .0001	0.0005
C_2H_6	$.002_5$(av)	$.003_9$(av)	< .001
CO_2	n.d.	< .0002	.002

densities vary as much as 0.1% near the critical density. However, if the 0.01% impurity had been entirely a volatile impurity, such as methane, the effect would be only about 0.05%. This disagreement probably resulted from the introduction of additional impurities into the actual gas samples used rather than from errors in the purity analyses. To preserve the sample quality during the experiment, the apprartus was evacuated and flushed with the sample gas prior to the initial filling of the sample volume.

Helium gas of 99.999% purity was used to determine the value of the volume-ratio parameter for each isotherm.

Results

Multiple sets of Burnett data were obtained for each isotherm—three sets for ethylene and two sets for helium. Each set consisted of data from a series of four consecutive expansions from the highest to the lowest pressure compatible with our optimum accuracy and precision. The initial pressure for each set was selected so as to intersperse the data from all of the sets over the entire pressure range of interest, 0.3 MPa to 3.7 MPa. Consistent with the extent of the nonideal behavior of the gas, the density-series generalized equation was applied to the ethylene data and the pressure-series generalized equation was applied to the helium data. The parameters in the resulting overdetermined sets of equations then were evaluated using the least-squares constraint.

The highest power of the series terms chosen to define each isotherm reflected the extent of the nonideality. For the ethylene isotherms, a cubic series was used for temperatures from 0° to 25°C and a quadratic series was used for temperatures from 75° to 175°C. At 50°C, a quadratic series as well as a cubic series were used. For the helium isotherms, a quadratic series was used with the virial coefficient of the quadratic term treated as a constant obtained from published values rather than as a parameter. The other parameters were evaluated more accurately with the quadratic coefficient treated as a constant rather than as a parameter since the contribution of this term was so small for our range of pressures. The term functioned only as a virial remainder. In the helium data analyses, the parameters were common to all of the data. For the ethylene data analyses, only the virial coefficient parameters were common to all of the data; an initial density parameter was required for each sequence of expansions and the volume-ratio parameter was treated in three different ways in an attempt to assess accuracy. In the first case, the volume-ratio parameter was constrained to be the value determined from the helium data; in the second, it was a single parameter common to all of the ethylene data; and in the third, it was considered to be an adjustable parameter for each sequence of expansions. In all of these analyses, the parameters were evaluated simultaneously along with estimates of the true pressures.

In Table I we present our ethylene Burnett results, and in Figure 4 we compare results for N treated as a parameter with respect to those for N treated as a constant. To simplify the comparison, average results for the different conditions are used where differences are negligible. The majority of the results follow a consistent pattern. That is, they have a standard deviation less than 10 ppm in pressure along with variations among the different N treatments less than 0.13 cm^3/mol in the second virial coefficient and less than 0.004% in the compressibility factor. Further, the differences between the results for N treated as a single parameter for all of the data and N treated as an adjustable parameter for each sequence of expansions are negligible, namely, less than 0.03 cm^3/mol in the second virial coefficient. For this situation, little ambiguity arises in the selection of values and we can estimate accuracies based on our knowledge of the effects of systematic errors. There are some results, e.g., illustrated by those for the 50°C isotherm, which have larger variations although the pressure data are fitted with an excellent standard deviation of less than 10 ppm. In this case, we prefer the results that have been obtained with N treated as a constant for two reasons: first, we are confident that the effect of an error in the N value on the virial coefficients is smaller than these larger variations; second, the information contained in the sample gas data is used exclusively to determine the properties of the gas, and not of the apparatus property as well. Even if N were fixed, some uncertainty still has to be associated with the results, as the variations do exist and usually are unexplainable. The relative independence of the precision of fit to the measured pressures with the different treatments of N illustrates the point we made previously that the Burnett density models have excessive flexibility in fitting the pressure data.

The effects of systematic errors are best studied by analyses of accurate Burnett data with superimposed simulated errors. For a relative pressure offset of 0.003%, which is comparable with the accuracy of piston gauges, the ethylene second virial coefficient of −167 cm^3/mol changes only by 0.02 cm^3/mol. Thus, this type of error is largely cancelled in the Burnett method. An offset in N of 11 ppm, which is comparable with the N variation we expect, changes the same second virial coefficient by 0.1 cm^3/mol. Errors resulting from truncation of the series

Table I. Ethylene

Temperature (°C)	N *Status (Case)*	*$10^5\sigma$ (ΔP/P)*	N *(helium)*	*10^5ΔN*	*$10^5\sigma$(N)*	*Number of Series Terms*
0	1	0.42	1.77725		0.6	3
	2	.44		−1	1.5	
	3 (set 1)	.33		−1	1.1	
	(set 2)			−1	1.1	
	(set 3)			0	1.1	
	1	0.83	1.77725		0.6	3
	2	.66		3	1.3	
	3 (set 1)	.60		2	1.4	
	(set 2)			3	2.2	
	(set 3)			3	2.2	
25	1	0.62	1.78070		0.5	3
	2	.57		4	2.5	
	3 (set 1)	.66		4	3.5	
	(set 2)			4	3.2	
	(set 3)			4	3.1	
	1	0.74	1.78070		0.5	3
	2	.77		2	3.3	
	3 (set 1)	.80		0	4.3	
	(set 2)			0	3.8	
	(set 3)			−1	3.8	
50	1	0.58	1.78075		0.6	3
	2	.60		−2	3.4	
	3 (set 1)	.57		−5	3.8	
	(set 2)			−4	3.4	
	(set 3)			−4	3.6	
	1	0.73	1.78075		0.6	2
	2	.65		3	1.6	
	3 (set 1)	.70		3	1.8	
	(set 2)			3	1.8	
	(set 3)			3	1.8	
	1	0.61	1.78075		0.6	2
	2	.62		−4	3.5	
	3 (set 1)	.66		−3	4.1	
	(set 2)			−3	4.2	
	(set 3)			−3	4.2	
75	1	0.77	1.78078		0.8	3
	2	.70		−7	4.0	
	3 (set 1)	.78		−5	5.4	
	(set 2)			−6	5.0	
	(set 3)			−5	5.4	
	1	0.81	1.78078		0.8	2
	2	.84		1	2.1	
	3 (set 1)	.83		2	2.2	
	(set 2)			1	2.2	
	(set 3)			1	2.2	
100	1	0.78	1.78079		0.6	2
	2	.56		5	1.5	
	3 (set 1)	.48		5	1.3	
	(set 2)			4	1.0	
	(set 3)			5	1.3	

Burnett Results[a, b, c]

B_2 (cm³/mol)	ΔB_2 (cm³/mol)	$\sigma(B_2)$	ρ_{max} (10^{-3} mol/cm³)	Z	$10^5\Delta Z$	Gas Source
−167.67		0.02	1.88	0.71350		Gulf
	0.04	.08			2	
	.03	.05			2	
−167.75		0.02	2.74	0.60265		
	−0.13	.05			−4	
	−.10	.05			−2	
			(1.88)	(0.71350)		
					(−5)	
					(−4)	
−139.69		0.03	1.86	0.76550		Gulf
	−0.18	.12			−7	
	−.20	.16			−7	
−139.82		0.03	1.86	0.76546		Matheson
	−0.09	.16			−3	
	.01	.19			1	
−117.59		0.04	1.58	0.83054		Gulf
	0.11	.18			4	
	.11	.20			11	
			(0.88)	(0.90155)		
					(4)	
					(9)	
−117.69		0.02	1.58	0.83053		
	−0.10	.04			−6	
	−.10	.04			−6	
			(0.88)	(0.90154)		
					(−6)	
					(−6)	
−117.62		0.03	0.88	0.90155	7	
	0.17	0.19			6	
	0.14	0.21				
−99.53		0.06	1.40	0.87209		Gulf
	0.43	.25	1.40		16	
	.34	.33	1.40		13	
−99.61		0.02	1.40	0.87208		
	−0.04	.08	1.40		−2	
	.02	.08	1.40		−2	
−84.55		0.02	1.26	0.90179		Gulf
	−0.20	.06	1.26		−10	
	.19	.06	1.26		−10	

Table I.

Temperature (°C)	N Status (Case)	$10^5\sigma$ (ΔP/P)	N (helium)	$10^5\Delta$N	$10^5\sigma$(N)	Number of Series Terms
	1	0.79	1.78079		0.6	2
	2	.85		0	2.7	
	3 (set 1)	.76		1	2.5	
	(set 2)			0	2.5	
125	1	0.59	1.78080		0.5	2
	2	.57		2	1.5	
	3 (set 1)	.57		2	1.5	
	(set 2)			2	1.6	
	(set 3)			2	1.6	
150	1	0.62	1.78078		0.3	2
	2	.65		2	1.8	
	3 (set 1)	.70		2	2.0	
	(set 2)			2	1.9	
	(set 3)			2	1.9	
175	1	0.66	1.78077		0.7	2
	2	.69		−1	1.9	
	3 (set 1)	.70		−1	1.9	
	(set 2)			−1	2.0	
	(set 3)			−1	1.9	

[a] 0°C data were obtained before the apparatus was refinished. In the data reduction, the maximum pressure was varied to determine whether the effect of sorption was significant.

[b] For Case 1, N is a constant determined from Burnett helium data; for Case 2, N is a single parameter; and for Case 3, N is a parameter for each designated set (sequence of expansions).

depend on the extent of the nonideal behavior of the gas. In an earlier analysis of Burnett carbon dioxide data at 0°C, an increase in the power of the series from quadratic to cubic with N constant or as a parameter changed the second virial coefficient of −150 cm³/mol by no more than 0.04 cm³/mol (*8*). For ethylene at 75°C, a similar series modification for N constant changes the second virial coefficient of −99 cm³/mol by 0.08 cm³/mol. However, the change for N treated as an adjustable parameter is large in comparison, being 0.4 cm³/mol.

For this experiment, the treatment of N as a constant results in the best approximation of the series coefficients as virial coefficients. The most accurate as a true virial coefficient is, of course, the second (linear) coefficient since the experiment was optimized for the determination of the linear behavior and not for the nonlinear behavior of the sample gas. However, the role of the nonlinear terms is more than that of virial remainders terms to accommodate only the analytical fit to the data. The nonlinear coefficients are characteristic of the true virial coefficients to the extent suggested to us by the variations of their values and by the definitiveness of their temperature behavior. Our listed values of the third virial coefficient agree within 2% with those determined by Douslin and Harrison (*2*). In Table II, we present what we consider to be our best values for the virial coefficients for N constant. Where several groups of

Continued

B_2 (*cm³/mol*)	ΔB_2 (*cm³/mol*)	$\sigma(B_2)$	ρ_{mxa} (*10⁻³ mol/cm³*)	Z	$10^5 \Delta Z$	*Gas Source*
−84.63		0.03	1.26	0.90174		Matheson
	0.01	.12	1.26		0	
	−.03	.12	1.26		-0_5	
−72.11		0.02	1.14	0.92407		
	−0.09	.07	1.14		4	Matheson
	−.10	.07	1.14		5	
−61.34		0.02	1.06	0.93992		Matheson
	−0.01	.10	1.06		0	
	−.02	.11	1.06		−1	
−52.19		0.02	0.980	0.95282		Matheson
	0.02	.11	0.980		1	
	.04	.11	0.980		2	

[c] $\Delta Z = Z\ (N_{parameter}) - Z\ (N_{constant})$; $\Delta N = N_{parameter} - N_{constant}$; $\Delta B_2 = B_2\ (N_{parameter}) - B_2\ (N_{constant})$; $\sigma('')$ = Standard deviation of (''); and ΔP = measured pressure − "true" pressure.

results are given for an isotherm in Table I, we have chosen, except for the 50°C isotherm, the values from the group with the smallest variations. For the 50°C isotherm, the values are averaged ones based on all of the listed results.

Our estimate of accuracy for the second virial coefficient is influenced by the pattern and magnitude of the variations for each isotherm. For isotherms with variations no greater than 11 ppm in N and 0.04 cm³/mol in B_2, we believe that the model is characteristic of the data to the extent that coefficients of a nonorthogonal series can be determined experimentally. For this situation, the uncertainty of the volume ratio determined from helium data contributes the largest error to the value of the second virial coefficient. We estimate this error to be 0.05 cm³/mol or less for the typical N uncertainty of 4 ppm. Any error caused by truncation of the series should be no greater than the largest variation of 0.04 cm³/mol. Our calculated estimate of the effect of the sample impurities has the same trend as that indicated by results for the two different sample sources at 25° and 100°C. However, both the magnitude and uncertainty of the calculated estimate are negligible, at most 0.01 ± 0.005 cm³/mol, as compared with 0.1 ± 0.1 cm³/mol for the change in the experimental value of the second coefficient from one sample source to the other. The large experimental uncertainty is associated with the

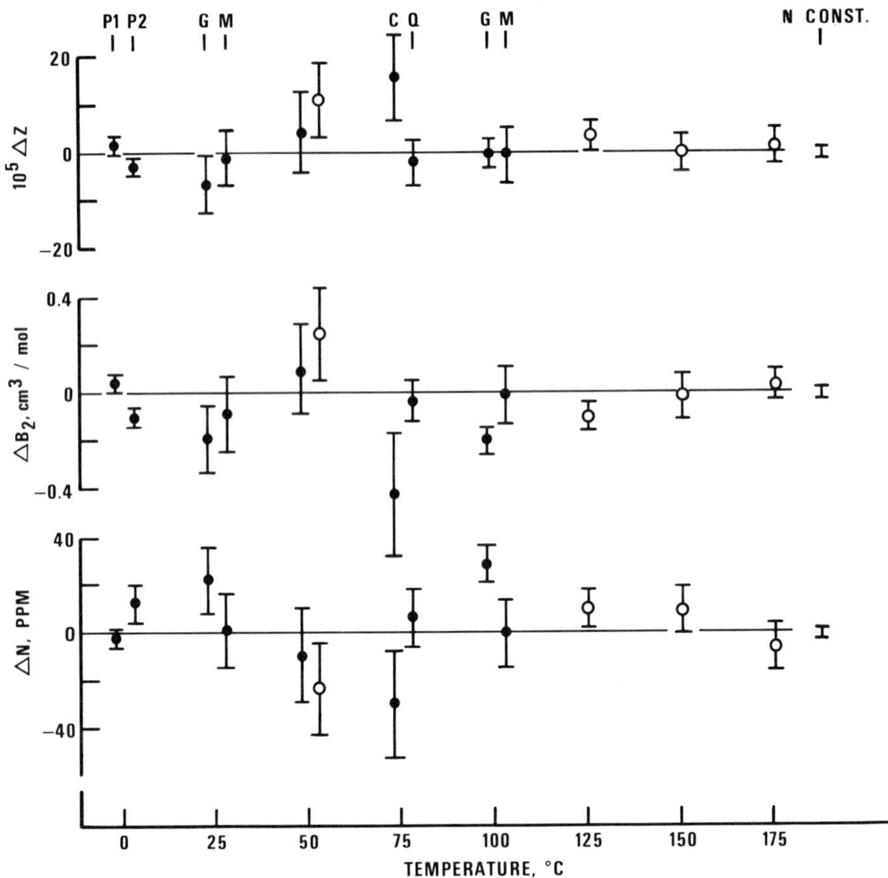

Figure 4. Comparison of Ethylene Burnett results for N *treated as a parameter to those for* N *treated as a constant*

presence of excess variations in the results for the Gulf source at 25° and 100°C. The error analyses just discussed are applicable to the 0°, 75°, 100°, 150°, and 175°C isotherms.

An estimate of errors for the other isotherms is not as straightforward because of the large magnitude and inconsistent pattern of the variations. For the 50°C isotherm, the value of the second virial coefficient for N treated as a constant changes by 0.1 cm^3/mol depending on the series power and the maximum density of the data. In contrast, the value of the compressibility factor remains essentially unchanged, 0.00001 or less. Each of the different fits for the 50°C ethylene data has at least one undesirable feature. For the cubic fit, the contribution of the cubic power term to the compressibility factor is too small, being at most 0.00025, to allow for a realistic evaluation of its coefficient. This slight overfit could

Table II. Ethylene Virial Coefficients[a, b, c, d]

Temperature (°C)	B_2 *(cm³/mol)*	B_3 *(cm⁶/mol²)*	B_4 *(cm⁹/mol³)*	ρ_{max} *(10^{-3} mol/cm³)*
0.00_0	−167.67	7,775	186,000	1.88
25.00_0	−139.82	7,165	114,000	1.86
50.00_8	−117.63 av −117.69[d]	6,560 av 6,000[d]		1.58
75.01_1	−99.61	5,885		1.40
100.01_5	−84.61	5,280		1.26
125.01_9	−72.11	4,820		1.14
150.02_3	−61.34	4,410		1.06
175.02_7	−52.19	4,140		0.98

[a] The overall accuracy for the B_2 values is estimated to be 0.2 cm³/mol or less, for B_3 values it is 4% or less, and for B_4 values it is about 30%. The listed B_3 values agree within 2% with those obtained by Douslin and Harrison (*2*).

[b] The number of significant figures for B_2, B_3, and B_4 is not intended to reflect an associated accuracy; rather, it only serves to retain our computational accuracy for the compressibility factor to 0.001%.

[c] The 0°C values correspond to the results for a maximum density of 0.00188 cm³/mol.

[d] Value corresponds to the results with the smallest standard deviations of B_2 and B_3 for all treatments of N.

adversely affect the other coefficients. The quadratic fit for the same maximum density of 0.00158 mol/³ has smaller standard deviations of the parameters; however, the variations are large. For the quadratic fit with a reduced maximum density, the standard deviations of the parameters as well as the variations are large. Because the choice of best values is not evident here, we have used average values and have included the average deviation of 0.04 cm³/mol as an additional error contribution to the value of the second virial coefficient at 50°C. This type of error is attributed to the slight inadequacy of data quality relative to the model. It also is included in the overall error for the values of the second virial coefficient at 25° and 125°C. Reflecting our understanding of the effects of systematic errors, we estimate the overall error in our values of the second virial coefficient to be less than 0.2 cm³/mol at all listed temperatures.

In Figure 5, we present a comparison of our preferred values for the ethylene second virial coefficient with comparable state of the art results obtained by Douslin and Harrison (*2*). The experimental method and data analysis used by Douslin are independent from ours. In Douslin's experiment, all of the variables required for the calculation of the compressibility factor are measured, whereas in the Burnett method only two variables are measured. Aslo, in this experiment the same sample of gas is retained for the entire experiment; in the Burnett isothermal method, the sample is changed for each sequence of measurements. Furthermore,

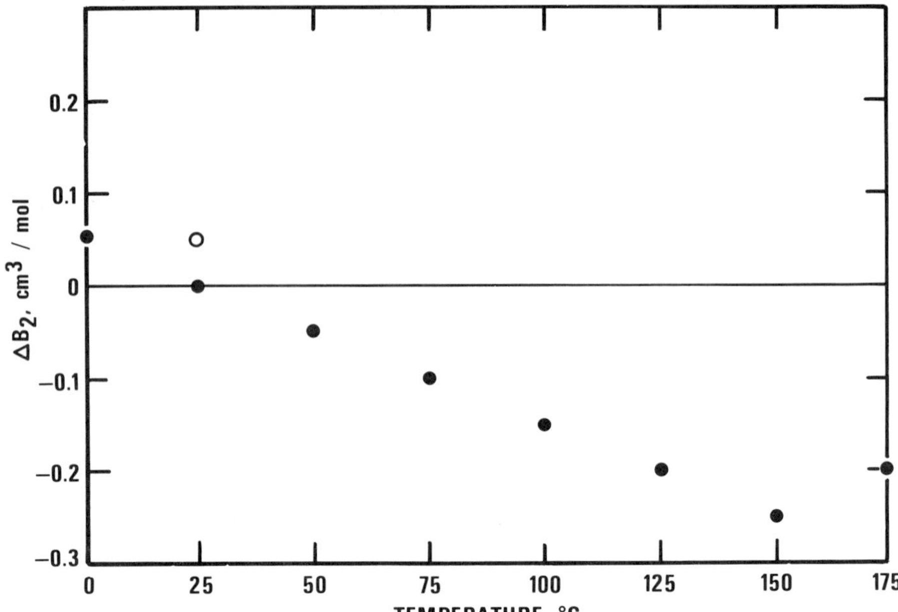

Figure 5. Ethylene second virial coefficient comparison. $\Delta B_2 = B_2$ *(Douslin)* $-$ B_2 *(this chapter): (●), Douslin's graphical value; (○), least-squares value from Douslin's data.*

errors in the variables propagate quite differently in the data analyses used in the two approaches. Douslin obtains values for the virial coefficients graphically from the relationship

$$B_j^*(\rho) = \left[\left(\frac{p}{\rho RT} - 1\right) - \sum_{j=2}^{J-1} B_j \rho^{j-1}\right] / \rho^J, \; B_J = \lim_{\rho \to 0} B_j^*$$

by imposing the constraint of linear extrapolation on the function B_j^*, which curves strongly at low densities if the wrong choice of B_j is made. Just as in a least-squares analysis, these graphical virial coefficients are highly correlated. We analyzed Douslin's data for moderate to high densities at 25°C by least squares. The resulting value of the second virial coefficient and the graphical value agree to within the uncertainty of the least-squares analysis, 0.1 cm³/mol. This close agreement may not reflect the comparative merit of either data analysis used; rather, it may only be indicative of the excellent quality of Douslin's data. The difference between the values of the second virial coefficients obtained from the two experiments increases linearly with increasing temperature from 0.05 cm³/mol at 0°C to −0.25 cm³/mol at 150°C. At 175°C, the difference decreases slightly to -0.20 cm^3/mol. Perhaps the temperature trend is to be expected as effects of chemisorption and polymerization on both experiments, although different, would be more pronounced at the higher temperatures. Nevertheless, we know of no case in the literature where two independent sets of second virial coefficients obtained by distinctly different methods agree to such an extent over such a wide range of temperatures.

Glossary of Symbols

A'', B'', C'', E'', and F'' = coefficients of a series used by Kamerlingh Onnes
B^{∞}, C^{∞}, . . . M^{∞} = coefficients of an infinite series used by Kamerlingh Onnes
b_i = coefficients (virial) of a pressure series to define the compressibility factor
B_j = coefficients (virial) of a density series to define the compressibility factor
n = quantity of gas
N = Burnett volume ratio for the limit of zero pressure
N_r = Burnett volume ratio for the Burnett rth expansion
P = pressure of the gas
r = Burnett expansion number
R = gas constant
T = absolute temperature of the gas
v = specific volume of the gas
V = volume of the gas
V_{I} = Burnett sample volume
V_{II} = Burnett expansion volume
Z = gas compressibility factor
ρ = density of the gas
ρ_0 = initial density of the Burnett sample gas
ρ_r = density of the Burnett sample gas for the rth expansion
σ = standard deviation

Acknowledgments

We acknowledge the benefit we have received from past discussions with numerous people who have been active in nonlinear statistical analysis and Burnett data reduction. Also we would like to express our thanks to M. Klein for his contributions to this experiment as part of the OSRD ethylene program, and to M. Moldover and J. M. H. Levelt Sengers for their helpful comments on this chapter.

In order to describe materials and experimental procedures adequately, it is occasionally necessary to identify commercial products by manufacturer's name. In no instance does such identification imply endorsement by the National Bureau of Standards, nor does it imply that the particular product is necessarily the best available for that purpose.

Literature Cited

1. Mason, E. A.; Spurling, T. H. "The Virial Equation of State"; Pergamon: Elmsford, 1969; Chapter 1.
2. Douslin, D. R.; Harrison, R. H. *J. Chem. Thermodyn.* **1976**, *8*, 301–330.
3. Burnett, E. S. *J. Appl. Mech.* **1936**, *3*, A136–A140.
4. Krammer, G. M.; Miller, J. G. *J. Phys. Chem.* **1957**, *61*, 785–788.
5. Burnett, E. S. U.S. Dept. of the Interior, Bur. of Mines Report of Investigation 6267, 1963.
6. Waxman, M.; Hastings, J. R.; Chen, W. T. *Symp. Thermophys. Propr., 5th* **1970**, 248–261.
7. Hall, K. R.; Canfield, F. B. *Physica (The Hague)* **1970**, *47*, 99–108.
8. Waxman, M.; Davis, H. A.; Hastings, J. R. *Symp. Thermophys. Prop., 6th* **1973**, 245–255.
9. Levelt Sengers, J. M. H., personal communication on the matrix formulation of the Deming algorithm.
10. Deming, W. E. "Statistical Adjustment of Data"; Wiley: New York.
11. Britt, H. I.; Luecke, R. H. *Technometrics* **1973**, *15*(2), 233–247.
12. Waxman, M.; Hilsenrath, J.; Chen, W. T. *J. Chem. Phys.* **1973**, *58*(9), 3692–3701.
13. Waxman, M.; Hastings, J. R. *J. Res. Nat. Bur. Stand., Sect. C* **1971**, *C75C*, 165–176.
14. Erickson, N. E., personal communication.
15. Hastings, J. R.; Levelt Sengers, J. M. H.; Balfour, F. W., unpublished data.

RECEIVED September 21, 1978.

17

Thermodymanic Properties of Refrigerant 500

A. P. KUDCHADKER[1]

Thermodynamics Research Center and Department of Chemical Engineering, Texas A&M University, College Station, TX 77843 and
Department of Chemical Engineering, Indian Institute of Technology, Kanpur, U. P., India

For design calculations involving Refrigerant 500, a minimum-boiling azeotrope of 39.4 mol % of 1,1-difluoroethane and 60.4 mol % of difluorodichloromethane, reliable real gas thermodynamic properties are required which have been calculated from 0.2 to 100 bar and from 220 to 540 K using the recently proposed Boublik–Adler–Chen–Kreglewski equation of state and the PVT *data reported in the literature. This equation of state has 21 universal constants and only five adjustable constants which have been calculated for R-500 from the* PVT *data, saturated vapor pressure and liquid density, and the critical constants. In order to calculate the absolute values of the real gas properties, the reference state properties, which are also reported here, are required. All properties are given in SI units.*

Recent applications of minimum-boiling azeotropes to the vapor compression refrigeration systems have indicated their superiority over their pure constituents under identical conditions (*1, 2*). While retaining the merits of pure component refrigerants, the use of minimum-boiling azeotropes has resulted in increased capacity, attainment of lower temperatures, and, consequently, improved coefficients of performance. Refrigerant 500 (R-500) is a minimum-boiling azeotrope of 26 wt % (39.4 mol %) of 1,1-difluoroethane (R-152a) and 74 wt % (61.6 mol %) of difluorodichloromethane (R12).

[1] Present address: Department of Chemical Engineering, Indian Institute of Technology Bombay, Powai, Bombay 400 076, India.

0-8412-0500-0/79/33-182-305$05.00/1

Table 1. Compression factor, fugacity coefficient, and residual thermodynamic functions of real fluid Refrigerant 500

The compression factor is calculated from the Boublik-Alder-Chen-Kreglewski equation of state:

$$\frac{PV}{RT} = Z = Z^h + Z^a; \qquad Z^h = \frac{1 + (3\alpha - 2)\xi + (3\alpha^2 - 3\alpha + 1)\xi^2 - \alpha^2\xi^3}{(1 - \xi)^3}; \qquad Z^a = \sum_{n=1}\sum_{m=1} mD_{nm}\left(\frac{u}{kT}\right)^n\left(\frac{V^o}{V}\right)^m; \qquad \text{where } \xi = 0.74048V^o/V;$$

$$V^o = V^{oo}[1 - C\exp(-3u^o/kT)]^3; \qquad \frac{u}{k} = \frac{u^o}{k}\left(1 + \frac{\eta}{kT}\right);$$

D_{nm} are universal constants, V^{oo}, C, α, u^o/k, and η/k are characteristic constants, R, V, P, T are, respectively, the gas constant, the molar volume ($cm^3\ mol^{-1}$), the pressure (bar), and the temperature (K).

The thermodynamic functions for the real fluid less the functions for the ideal gas at the same temperature and pressure or volume (*the residual functions*) are calculated from the following equations.

The fugacity coefficient $\phi = f(T,P)/P$ and the residual Gibbs energy are:

$$\ln\phi = \frac{G - G^{id}}{RT} = (\alpha^2 - 1)\ln(1 - \xi) + \frac{(\alpha^2 + 3\alpha)\xi - 3\alpha\xi^2}{(1 - \xi)^2} + \sum_{n=1}\sum_{m=1} D_{nm}\left(\frac{u}{kT}\right)^n\left(\frac{V^o}{V}\right)^m + Z - 1 - \ln Z.$$

The residual enthalpy is:

$$\frac{H - H^{id}}{RT} = \frac{T}{\xi}\left(\frac{\partial\xi}{\partial T}\right)_v(1 - Z) + y\sum_{n=1}\sum_{m=1} nD_{nm}\left(\frac{u}{kT}\right)^n\left(\frac{V^o}{V}\right)^m + Z - 1;$$

where:

$$\frac{T}{\xi}\left(\frac{\partial\xi}{\partial T}\right)_v = \frac{T}{V^o}\left(\frac{\partial V^o}{\partial T}\right)_v = -\frac{9Cu^o}{kT}[\exp(3u^o/kT) - C]^{-1}; \quad y = \frac{1 + 2\eta/kT}{1 + \eta/kT}.$$

The residual heat capacity at constant pressure is:

$$\frac{C_p - C_p^{id}}{R} = \frac{C_v - C_v^{id}}{R} - 1 - \frac{T}{R}\left(\frac{\partial P}{\partial T}\right)_v^2\left(\frac{\partial P}{\partial V}\right)_T^{-1}; \qquad \frac{C_v - C_v^{id}}{R} = -\left[2T\left(\frac{\partial\xi}{\partial T}\right)_v + T^2\left(\frac{\partial^2\xi}{\partial T^2}\right)_v\right]\frac{(Z^h - 1)}{\xi}$$

$$- T^2\left(\frac{\partial\xi}{\partial T}\right)_v^2\left[\frac{4\alpha^2 + 6\alpha + (2\alpha^2 - 6\alpha)\xi}{(1 - \xi)^4} - \frac{(\alpha^2 - 1)}{(1 - \xi)^2}\right] - Z^a\left[\frac{2T}{V^o}\left(\frac{\partial V^o}{\partial T}\right)_v - \left(\frac{T}{V^o}\right)^2\left(\frac{\partial V^o}{\partial T}\right)^2 + \frac{T^2}{V^o}\left(\frac{\partial^2 V^o}{\partial T^2}\right)_v\right] + T\frac{\partial y}{\partial T}\sum_{n=1}\sum_{m=1} nD_{nm}\left(\frac{u}{kT}\right)^n\left(\frac{V^o}{V}\right)^m$$

$$- y\sum_{n=1}\sum_{m=1} n(n - 1)D_{nm}\left(\frac{u}{kT}\right)^n\left(\frac{V^o}{V}\right)^m + \frac{T}{V^o}\left(\frac{\partial V^o}{\partial T}\right)_v\left[2y\sum_{n=1}\sum_{m=1} nmD_{nm}\left(\frac{u}{kT}\right)^n\left(\frac{V^o}{V}\right)^m - \frac{T}{V^o}\left(\frac{\partial V^o}{\partial T}\right)_v\sum_{n=1}\sum_{m=1} m^2D_{nm}\left(\frac{u}{kT}\right)^n\left(\frac{V^o}{V}\right)^m\right];$$

where:

$$\left(\frac{\partial^2\xi}{\partial T^2}\right)_v = \left(\frac{\partial^2 V^o}{\partial T^2}\right)_v = \frac{1}{T}\left(\frac{\partial\xi}{\partial T}\right)_v\left\{\frac{3u^o}{kT}\left[1 - \frac{2C}{\exp(3u^o/kT) - C}\right] - 2\right\}.$$

$$\left(\frac{\partial P}{\partial T}\right)_v = \frac{R}{V}\left\{Z + T\left(\frac{\partial \xi}{\partial T}\right)_v \frac{[1 + 3\alpha + (6\alpha^2 - 2)\xi - (3\alpha - 1)\xi^2]}{(1 - \xi)^4} - y\sum_{n=1}\sum_{m=1} nmD_{nm}\left(\frac{u}{kT}\right)^n\left(\frac{V^o}{V}\right)^m + \frac{T}{V^o}\left(\frac{\partial V^o}{\partial T}\right)_v \sum_{n=1}\sum_{m=1} m^2 D_{nm}\left(\frac{u}{kT}\right)^n\left(\frac{V^o}{V}\right)^m\right\}.$$

$$\left(\frac{\partial P}{\partial V}\right)_T = -\frac{RT}{V^2}\left[\frac{1 + (6\alpha - 2)\xi + (9\alpha^2 - 6\alpha + 1)\xi^2 - 4\alpha^2\xi^3 + \alpha^2\xi^4}{(1 - \xi)^4} + Z^a + \sum_{n=1}\sum_{m=1} m^2 D_{nm}\left(\frac{u}{kT}\right)^n\left(\frac{V^o}{V}\right)^m\right].$$

Universal Constants, D_{nm}		
D_{11} = −8.8043	D_{25} = −133.70055	D_{43} = +12.231796
D_{12} = +4.1646270	D_{26} = +860.25349	D_{44} = −12.110681
D_{13} = −48.203555	D_{27} = −1535.3224	
D_{14} = +140.43620	D_{28} = +1221.4261	**Characteristic Constants**
D_{15} = −195.23339	D_{29} = −409.10539	V^{oo} = 41.80 cm³ mol⁻¹
D_{16} = +113.51500	D_{31} = −2.8225	C = 0.12
D_{21} = +2.9396	D_{32} = +4.7600148	α = 1.045
D_{22} = −6.0865383	D_{33} = +11.257177	u^o/k = 353.81 K
D_{23} = +40.137956	D_{34} = −66.382743	η/k = 45 K
D_{24} = −76.230797	D_{35} = +69.248785	
	D_{41} = +0.3400	
	D_{42} = −3.1875014	

Notes: The constants D_{nm} were obtained by fitting to the second virial coefficients and PVT data for argon.

At $1/V = 0$ all the residual functions are equal to zero and $Z = \phi = 1$. In the following tables, the values below the critical temperature are given for the liquid and the vapor phase. For the liquid, the relative errors are about five times larger than in other regions. The equations are valid for argon at least up to the reduced temperature $T/T^c = 7$ and pressure $P/P^c = 20$ and similar ranges of validity are anticipated for other substances. Below the critical temperature, the equations are valid for the vapor at $P \geqslant 0$ and for the liquid at $P > 1$ bar; however, not below $T/T^c = 0.55$.

The Boublik–Alder–Chen–Kreglewski (henceforth referred to as BACK) equation so far has been successfully applied to calculate the residual properties of sulfur dioxide, hydrogen sulfide, benzene, naphthalene, tetralin, *cis*- and *trans*-decalin, furan, and tetrahydrofuran (*3, 4*).

In the present investigation, the compression factor, fugacity coefficient, and residual thermodynamic functions of the real fluid R-500 have been calculated using the BACK equation of state. For this purpose, the experimental data, i.e., critical constants, saturated liquid density, saturation vapor pressure, and the *PVT* data have been utilized.

The *PVT* data for R-500 as reported by Sinka and Murphy (*5*) extends from 322 to 478 K, 14 to 58 bar, and 17 to 149 $m^3\ kmol^{-1}$, and from 298.15 to 413.15 K, 0.2 to 7 bar, and 3 to 80 $m^3\ kmol^{-1}$ as stated by Prasad and Kudchadker (*6*).

ASHRAE (*7*) has reported the saturated and superheated properties of R-500 from 188 to 530 K and from 0.04 to 55 bar. These values are based upon the measurements of Sinka and Murphy (*5*). For the present calculations all temperatures were converted to IPTS-68 and all derived properties are reported in SI units.

Basic Data

Critical constants (*5*): $T_c = 378.65$ K; $P_c = 44.26$ bar; $V_c = 0.200$ $m^3\ kmol^{-1}$; $\rho_c = 497\ kg\ m^{-3}$.

Saturated liquid density (*5*): the data were fitted to the modified Guggenheim equation as proposed by Haynes and Hiza (*8*).

$$(\rho - \rho_c)/m^3\ kmol^{-1} = 932.468\ (1 - T_r)^{0.35} + 1481.05\ (1 - T_r) - 2421.08\ (1 - T_r)^{4/3} + 1442.92\ (1 - T_r)^{5/3}$$

Saturation vapor pressure (*5*): the data were fitted to the Wagner equation (*9*).

$$\ln P_r = \begin{array}{l} -7.04046[(1 - T_r)/T_r] + 1.47146[(1 - T_r)^{1.5}/T_r] \\ -2.86953[(1 - T_r)^3/T_r] + 13.7355[(1 - T_r)^7/T_r] \\ -38.2370[(1 - T_r)^9/T_r] \end{array}$$

where $T_r = T/T_c$.

Derived Properties

The BACK equation of state used in this work contains 24 universal constants and five characteristic constants as reported in Table I. The equation with 24 universal constants fitted the experimental data on R-500 much better than the equation with 20 constants for high-boiling liquids (*3*). The five characteristic constants were obtained using the above data and the nonlinear regression program from the Thermodynamics Research Center computer program file. The saturated and superheated properties were computed from 0.2 to 100 bar and from 220 to 540 K using the formulae given in Ref. *4* and are reported in Tables II and III. The properties are the compression factor Z, the fugacity coefficient ϕ, the residual enthalpy $(H - H^{id})/RT$, and residual entropy $(S - S^{id})/R$. The ideal gas thermodynamic properties required to calculate the absolute values of enthalpy and entropy are reported in Table IV. These are calculated for R-500 using the following standard formulae for a mixture from its pure component data:

$$(H^{id} - H°_0)_m = y_A(H^{id} - H°_0)_A + y_B(H^{id} - H°_0)_B$$

$$(S^{id} - S°_0)_m = y_A S°_A + y_B S°_B - y_A R \ln x_A - y_B R \ln x_B$$

$$C°_{pm} = y_A C°_{pA} + y_B C°_{pB}$$

where m is R-500, A is 1,1-difluoroethane, B is difluorodichloromethane, y is the mole fraction and $C_p°$ is heat capacity at constant pressure. The ideal gas thermodynamic properties of 1,1-fluoroethane and difluorodichloromethane were taken from Ref. *3*. The reference state is the ideal gas at 1.01325 bar and 0 K.

Table II. Properties of Saturated Liquid and Vapor of R-500 [a]

T	P	*Volume, m^3 $kmol^{-1}$*		$(H - H^{id})/RT$		$(S - S^{id})/R$	
(K)	*(bar)*	*Liquid*	*Vapor*	*Liquid*	*Vapor*	*Liquid*	*Vapor*
220	0.3922	0.0717	45.844	−11.75	−0.057	−11.81	−0.040
230	0.6495	0.0731	28.721	−10.95	−0.082	−11.02	−0.057
233.15	0.7539	0.0736	25.011	−10.72	−0.091	−10.78	−0.064
239.67	1.01325	0.0746	19.004	−10.25	−0.111	−10.31	−0.078
240	1.0272	0.0746	18.765	−10.23	−0.112	−10.28	−0.079
250	1.5610	0.0762	12.706	−9.563	−0.150	−9.607	−0.105
260	2.291	0.0778	8.870	−8.950	−0.196	−8.975	−0.137
270	3.259	0.0796	6.357	−8.380	−0.249	−8.384	−0.176
273.15	3.621	0.0802	5.751	−8.210	−0.268	−8.205	−0.189
280	4.512	0.0816	4.659	−7.851	−0.312	−7.828	−0.220
290	6.098	0.0837	3.479	−7.355	−0.386	−7.304	−0.278
298.15	7.672	0.0857	2.774	−6.973	−0.454	−6.896	−0.323
300	8.068	0.0862	2.638	−6.889	−0.470	−6.806	−0.335
310	10.47	0.0890	2.025	−6.448	−0.568	−6.331	−0.408
320	13.37	0.0922	1.568	−6.027	−0.682	−5.874	−0.494
330	16.83	0.0960	1.220	−5.621	−0.816	−5.429	−0.598
340	20.90	0.101	0.948	−5.223	−0.976	−4.990	−0.725
350	25.68	0.107	0.731	−4.824	−1.175	−4.548	−0.888
360	31.25	0.116	0.552	−4.406	−1.435	−4.085	−1.109
370	37.74	0.130	0.392	−3.921	−1.826	−3.552	−1.456
378.65	44.26	0.200	0.200	−2.884	−2.884	−2.469	−2.469

[a] This assumes that the azeotropic composition is invariant with temperature and pressure.

Table III. Properties of

T (K)	Z	ϕ	$\frac{H - H^{id}}{RT}$	$\frac{S - S^{id}}{R}$
		P (bar) = 0.2		
220.00	.9914	.9915	−.0289	−.0203
230.00	.9926	.9927	−.0246	−.0172
233.15	.9929	.9930	−.0235	−.0164
239.67	.9935	.9936	−.0213	−.0148
240.00	.9936	.9936	−.0212	−.0148
250.00	.9944	.9944	−.0183	−.0127
260.00	.9951	.9951	−.0160	−.0111
270.00	.9956	.9957	−.0141	−.0097
273.15	.9958	.9958	−.0135	−.0093
280.00	.9961	.9961	−.0124	−.0086
290.00	.9965	.9965	−.0110	−.0076
298.15	.9968	.9968	−.0101	−.0069
300.00	.9969	.9969	−.0099	−.0068
310.00	.9972	.9972	−.0089	−.0061
320.00	.9975	.9975	−.0080	−.0055
330.00	.9977	.9977	−.0072	−.0049
340.00	.9979	.9979	−.0066	−.0045
350.00	.9981	.9981	−.0060	−.0041
360.00	.9982	.9982	−.0055	−.0037
370.00	.9984	.9984	−.0051	−.0034
378.65	.9985	.9985	−.0047	−.0032
380.00	.9985	.9985	−.0047	−.0032
400.00	.9987	.9987	−.0040	−.0027
420.00	.9989	.9989	−.0034	−.0024
440.00	.9991	.9991	−.0030	−.0021
460.00	.9992	.9992	−.0026	−.0018
480.00	.9993	.9993	−.0023	−.0016
500.00	.9994	.9994	−.0021	−.0015
520.00	.9995	.9995	−.0018	−.0013
540.00	.9995	.9995	−.0017	−.0012
		P (bar) = 1.0		
220.00	.0039	.4181	−11.7504	−10.8783
230.00	.0038	.6924	−10,9529	−10.8582
233.15	.0038	.8023	−10.7175	−10.4972
239.67	.9667	.9679	−.1097	−.0771
240.00	.9669	.9680	−.1092	−.0767
250.00	.9712	.9721	−.0942	−.0659
260.00	.9748	.9754	−.0819	−.0570
270.00	.9777	.9782	−.0718	−.0498
273.15	.9786	.9790	−.0689	−.0478
280.00	.9802	.9806	−.0633	−.0437
290.00	.9824	.9827	−.0561	−.0387
298.15	.9839	.9842	−.0511	−.0351

Liquid and Vapor of R-500[a]

Z	ϕ	$\frac{H - H^{id}}{RT}$	$\frac{S - S^{id}}{R}$
		P (*bar*) = *0.5*	
.0020	.8345	−11.7515	−11.5706
- - -	- - -	- - -	- - -
.9812	.9816	−.0624	−.0438
.9821	.9824	−.0594	−.0417
.9837	.9840	−.0538	−.0376
.9838	.9840	−.0535	−.0374
.9858	.9860	−.0463	−.0323
.9876	.9877	−.0404	−.0280
.9890	.9891	−.0354	−.0245
.9894	.9895	−.0340	−.0235
.9902	.9903	−.0313	−.0216
.9913	.9914	−.0278	−.0191
.9920	.9921	−.0253	−.0174
.9922	.9922	−.0248	−.0170
.9930	.9930	−.0223	−.0152
.9936	.9937	−.0201	−.0137
.9942	.9943	−.0182	−.0124
.9947	.9948	−.0165	−.0113
.9952	.9952	−.0151	−.0103
.9956	.9956	−.0138	−.0094
.9960	.9960	−.0127	−.0086
.9963	.9963	−.0118	−.0081
.9963	.9963	−.0117	−.0080
.9969	.9969	−.0100	−.0068
.9973	.9973	−.0086	−.0059
.9977	.9977	−.0075	−.0052
.9980	.9980	−.0066	−.0046
.9983	.9983	−.0058	−.0041
.9985	.9985	−.0052	−.0036
.9987	.9987	−.0046	−.0033
.9988	.9988	−.0041	−.0030
		P (*bar*) = *1.01325*	
.0040	.4126	−11.7504	−10.8652
.0039	.6833	−10.9528	−10.5720
.0038	.7918	−10.7174	−10.4840
.0038	1.0573	−10.2513	−10.3070
.0038	1.0723	10.2284	−10.2982
- - -	- - -	- - -	- - -
.9708	.9717	−.0955	−.0668
.9744	.9751	−.0831	−.0578
.9774	.9780	−.0727	−.0505
.9783	.9788	−.0699	−.0484
.9800	.9804	−.0641	−.0443
.9821	.9825	−.0569	−.0392
.9837	.9839	−.0518	−.0356

Table III.

T (K)	Z	ϕ	$\frac{H - H^{id}}{RT}$	$\frac{S - S^{id}}{R}$
300.00	.9842	.9845	−.0501	−.0344
310.00	.9858	.9860	−.0449	−.0308
320.00	.9872	.9873	−.0404	−.0277
330.00	.9884	.9885	−.0366	−.0250
340.00	.9894	.9895	−.0332	−.0227
350.00	.9904	.9904	−.0303	−.0207
360.00	.9912	.9913	−.0277	−.0189
370.00	.9919	.9920	−.0254	−.0174
378.65	.9925	.9925	−.0237	−.0162
380.00	.9926	.9926	−.0234	−.0160
400.00	.9937	.9937	−.0200	−.0137
420.00	.9946	.9946	−.0173	−.0119
440.00	.9954	.9954	−.0150	−.0104
460.00	.9960	.9960	−.0132	−.0092
480.00	.9965	.9965	−.0116	−.0081
500.00	.9970	.9970	−.0103	−.0073
520.00	.9973	.9973	−.0092	−.0066
540.00	.9977	.9977	−.0083	−.0059
		P *(bar)* = *2.0*		
220.00	.0078	.2099	−11.7482	−10.1869
230.00	.0076	.3475	−10.9508	−9.8938
233.15	.0076	.4027	−10.7154	−9.8058
239.67	.0075	.5376	−10.2494	−9.6288
240.00	.0075	.5453	−10.2265	−9.6200
250.00	.0073	.8166	−9.5625	−9.3598
- - -	- - -	- - -	- - -	- - -
260.00	.9480	.9507	−.1690	−.1185
270.00	.9543	.9564	−.1475	−.1029
273.15	.9561	.9580	−.1415	−.0986
280.00	.9596	.9613	−.1296	−.0901
290.00	.9641	.9654	−.1146	−.0794
298.15	.9672	.9683	−.1042	−.0720
300.00	.9679	.9689	−.1020	−.0704
310.00	.9712	.9720	−.0912	−.0628
320.00	.9740	.9747	−.0820	−.0564
330.00	.9765	.9770	−.0741	−.0508
340.00	.9786	.9791	−.0672	−.0461
350.00	.9805	.9809	−.0612	−.0419
360.00	.9822	.9825	−.0559	−.0383
370.00	.9837	.9840	−.0513	−.0351
378.65	.9849	.9851	−.0477	−.0327
380.00	.9850	.9852	−.0472	−.0323
400.00	.9873	.9875	−.0403	−.0276
420.00	.9892	.9893	−.0347	−.0239
440.00	.9907	.9908	−.0302	−.0209
460.00	.9919	.9920	−.0264	−.0184

Continued

Z	ϕ	$\frac{H - H^{id}}{RT}$	$\frac{S - S^{id}}{R}$
.9840	.9843	−.0507	−.0349
.9856	.9858	−.0455	−.0312
.9870	.9872	−.0410	−.0281
.9882	.9884	−.0371	−.0253
.9893	.9894	−.0337	−.0230
.9902	.9903	−.0307	−.0210
.9911	.9911	−.0281	−.0192
.9918	.9919	−.0258	−.0176
.9924	.9924	−.0240	−.0164
.9925	.9925	−.0237	−.0162
.9936	.9936	−.0203	−.0139
.9945	.9946	−.0175	−.0120
.9953	.9953	−.0152	−.0105
.9959	.9959	−.0134	−.0093
.9965	.9965	−.0118	−.0083
.9969	.9969	−.0105	−.0074
.9973	.9973	−.0093	−.0066
.9976	.9976	−.0084	−.0060

P *(bar)* = *30.*

Z	ϕ	$\frac{H - H^{id}}{RT}$	$\frac{S - S^{id}}{R}$
.0118	.1405	−11.7461	−9.7832
.0115	.2326	−10.9487	−9.4901
.0114	.2695	−10.7134	−9.4021
.0112	.3598	−10.2474	−9.2252
.0112	.3649	−10.2246	−9.2164
.0110	.5464	−9.5608	−8.9563
.0108	.7854	−8.9484	−8.7069
.9296	.9345	−.2277	−.1600
.9324	.9370	−.2182	−.1530
.9380	.9418	−.1993	−.1393
.9450	.9480	−.1757	−.1223
.9499	.9524	−.1593	−.1106
.9510	.9534	−.1559	−.1082
.9561	.9580	−.1392	−.0963
.9605	.9620	−.1249	−.0862
.9643	.9655	−.1127	−.0776
.9676	.9686	−.1020	−.0702
.9705	.9713	−.0928	−.0637
.9731	.9738	−.0847	−.0582
.9754	.9759	−.0776	−.0533
.9771	.9776	−.0722	−.0495
.9774	.9779	−.0714	−.0490
.9809	.9812	−.0608	−.0418
.9837	.9839	−.0524	−.0361
.9860	.9861	−.0455	−.0315
.9879	.9880	−.0398	−.0278

Table III.

T *(K)*	Z	ϕ	$\frac{H - H^{id}}{RT}$	$\frac{S - S^{id}}{R}$
480.00	.9930	.9930	−.0233	−.0164
500.00	.9939	.9939	−.0207	−.0146
520.00	.9947	.9947	−.0185	−.0132
540.00	.9953	.9954	−.0166	−.0119
		P *(bar) = 4.0*		
220.00	.0157	.1058	−11.7439	−9.4973
230.00	.0153	.1751	−10.9466	−9.2041
233.15	.0152	.2029	−10.7114	−9.1162
239.67	.0150	.2708	−10.2455	−8.9393
240.00	.0149	.2747	−10.2226	−8.9305
250.00	.0146	.4113	−9.5590	−8.6705
260.00	.0144	.5912	−8.9469	−8.4213
270.00	.0142	.8198	−8.3796	−8.1809
273.15	.0141	.9025	−8.2092	−8.1067
- - - - -	- - - - -	- - - - -	- - - - -	- - - - -
280.00	.9152	.9223	−.2729	−.1920
290.00	.9251	.9306	−.2398	−.1678
298.15	.9319	.9365	−.2169	−.1513
300.00	.9334	.9378	−.2121	−.1479
310.00	.9405	.9439	−.1889	−.1312
320.00	.9465	.9493	−.1692	−.1172
330.00	.9518	.9540	−.1523	−.1052
340.00	.9563	.9581	−.1378	−.0950
350.00	.9603	.9618	−.1251	−.0862
360.00	.9638	.9650	−.1141	−.0785
370.00	.9669	.9679	−.1044	−.0719
378.65	.9693	.9702	−.0970	−.0667
380.00	.9697	.9705	−.0959	−.0660
400.00	.9744	.9749	−.0816	−.0562
420.00	.9781	.9785	−.0702	−.0485
440.00	.9812	.9815	−.0609	−.0423
460.00	.9838	.9840	−.0533	−.0372
480.00	.9860	.9861	−.0470	−.0330
500.00	.9878	.9879	−.0416	−.0294
520.00	.9894	.9894	−.0371	−.0265
540.00	.9907	.9907	−.0332	−.0239
		P *(bar) = 10.0*		
220.00	.0391	.0433	−11.7308	−8.5914
230.00	.0382	.0717	−10.9341	−8.2982
233.15	.0379	.0830	−10.6991	−8.2104
239.67	.0373	.1108	−10.2339	−8.0338
240.00	.0373	.1124	−10.2110	−8.0250
250.00	.0366	.1682	−9.5485	−7.7657
260.00	.0359	.2416	−8.9377	−7.5174

Continued

Z	ϕ	$\frac{H - H^{id}}{RT}$	$\frac{S - S^{id}}{R}$
.9895	.9896	−.0351	−.0246
.9909	.9909	−.0311	−.0220
.9920	.9921	−.0278	−.0198
.9930	.9930	−.0249	−.0179
	P (bar) = 5.0		
.0196	.0849	−11.7417	−9.2759
.0191	.1406	−10.9445	−9.9827
.0190	.1629	−10.7093	−8.8948
.0187	.2175	−10.2436	−8.7179
.0187	.2206	−10.2207	−8.7092
.0183	.3302	−9.5573	−8.4493
.0180	.4746	−8.9454	−8.2002
.0177	.6581	−8.3783	−7.9600
.0176	.7246	−8.2081	−7.8859
.0175	.8840	−7.8503	−7.7270
.9041	.9131	−.3071	−.2163
.9132	.9206	−.2771	−.1943
.9151	.9221	−.2708	−.1898
.9243	.9299	−.2405	−.1678
.9322	.9366	−.2149	−.1494
.9389	.9425	−.1931	−.1339
.9448	.9477	−.1744	−.1206
.9499	.9523	−.1582	−.1093
.9544	.9563	−.1441	−.0994
.9583	.9599	−.1318	−.0909
.9614	.9627	−.1223	−.0843
.9618	.9632	−.1209	−.0834
.9678	.9687	−.1027	−.0709
.9726	.9732	−.0882	−.0611
.9765	.9769	−.0765	−.0532
.9797	.9800	−.0669	−.0467
.9825	.9827	−.0589	−.0414
.9847	.9849	−.0522	−.0369
.9867	.9868	−.0465	−.0332
.9884	.9884	−.0416	−.0300
	P (bar) = 20.0		
.0781	.0225	−11.7090	−7.9155
.0762	.0372	−10.9131	−7.6223
.0756	.0431	−10.6786	−7.5346
.0745	.0575	−10.2143	−7.3584
.0745	.0583	−10.1915	−7.3496
.0729	.0872	−9.5308	−7.0914
.0716	.1252	−8.9221	−6.8445

Table III.

T (K)	Z	ϕ	$\frac{H - H^{id}}{RT}$	$\frac{S - S^{id}}{R}$
270.00	.0354	.3349	−8.3719	−7.2781
273.15	.0352	.3687	−8.2021	−7.2044
280.00	.0350	.4498	−7.8454	−7.0464
290.00	.0347	.5872	−7.3527	−6.8203
298.15	.0345	.7162	−6.9726	−6.6388
300.00	.0345	.7475	−6.8887	−6.5977
310.00	.8328	.8589	−.5360	−.3839
320.00	.8524	.8727	−.4715	−.3353
330.00	.8687	.8847	−.4184	−.2959
340.00	.8825	.8952	−.3740	−.2633
350.00	.8942	.9045	−.3364	−.2361
360.00	.9044	.9127	−.3043	−.2129
370.00	.9132	.9200	−.2766	−.1932
378.65	.9199	.9256	−.2555	−.1782
380.00	.9209	.9265	−.2524	−.1761
400.00	.9338	.9376	−.2127	−.1482
420.00	.9440	.9466	−.1815	−.1266
440.00	.9522	.9541	−.1566	−.1095
460.00	.9590	.9603	−.1363	−.0958
480.00	.9646	.9655	−.1196	−.0845
500.00	.9694	.9700	−.1057	−.0752
520.00	.9734	.9738	−.0939	−.0673
540.00	.9768	.9770	−.0839	−.0607

P *(bar)* = *30.0*

T (K)	Z	ϕ	$\frac{H - H^{id}}{RT}$	$\frac{S - S^{id}}{R}$
220.00	.1170	.0156	−11.6869	−7.5269
230.00	.1141	.0258	−10.8920	−7.2337
233.15	.1132	.0298	−10.6579	−7.1462
239.67	.1115	.0398	−10.1946	−6.9704
240.00	.1115	.0403	−10.1718	−6.9617
250.00	.1091	.0603	−9.5128	−6.7043
260.00	.1071	.0865	−8.9062	−6.4588
270.00	.1054	.1198	−8.3451	−6.2232
273.15	.1050	.1319	−8.1769	−6.1508
280.00	.1040	.1607	−7.8240	−5.9961
290.00	.1030	.2097	−7.3378	−5.7758
298.15	.1024	.2557	−6.9641	−5.6003
300.00	.1023	.2669	−6.8818	−5.5608
310.00	.1021	.3321	−6.4512	−5.3488
320.00	.1024	.4049	−6.0408	−5.1367
330.00	.1033	.4847	−5.6445	−4.9202
340.00	.1052	.5703	−5.2535	−4.6919
350.00	.1088	.6603	−4.8511	−4.4361
360.00	.6091	.7337	−1.3134	−1.0037
370.00	.6692	.7582	−1.1044	−.8277

Continued

Z	ϕ	$\frac{H - H^{id}}{RT}$	$\frac{S - S^{id}}{R}$
.0705	.1735	−8.3588	−6.6071
.0702	.1910	−8.1897	−6.5341
.0696	.2329	−7.8350	−6.3778
.0690	.3039	−7.3457	−6.1548
.0687	.3706	−6.9690	−5.9764
.0686	.3869	−6.8859	−5.9361
.0685	.4814	−6.4504	−5.7193
.0689	.5869	−6.0335	−5.5006
.0697	.7022	−5.6273	−5.2738
.7203	.7882	−.9108	−.6728
.7564	.8078	−.7911	−.5776
.7850	.8249	−.6974	−.5050
.8084	.8399	−.6216	−.4471
.8256	.8515	−.5665	−.4057
.8280	.8532	−.5586	−.3998
.8592	.8757	−.4598	−.3271
.8828	.8939	−.3860	−.2738
.9013	.9088	−.3289	−.2333
.9161	.9213	−.2836	−.2016
.9282	.9317	−.2470	−.1762
.9382	.9406	−.2168	−.1556
.9466	.9481	−.1917	−.1385
.9537	.9546	−.1706	−.1241

P *(bar)* = *40.0*

Z	ϕ	$\frac{H - H^{id}}{RT}$	$\frac{S - S^{id}}{R}$
.1557	.0122	−11.6646	−7.2559
.1518	.0201	−10.8707	−6.9627
.1507	.0232	−10.6370	−6.8753
.1484	.0310	−10.1746	−6.6999
.1483	.0314	−10.1519	−6.6912
.1452	.0469	−9.4946	−6.4347
.1424	.0672	−8.8899	−6.1905
.1401	.0931	−8.3311	−5.9566
.1395	.1024	−8.1636	−5.8847
.1382	.1248	−7.8124	−5.7314
.1367	.1628	−7.3292	−5.5138
.1359	.1984	−6.9582	−5.3408
.1357	.2071	−6.8767	−5.3020
.1353	.2576	−6.4503	−5.0941
.1354	.3142	−6.0456	−4.8878
.1362	.3762	−5.6573	−4.6796
.1381	.4429	−5.2787	−4.4642
.1416	.5133	−4.8994	−4.2325
.1481	.5860	−4.4982	−3.9637

Table III.

T *(K)*	Z	ϕ	$\frac{H - H^{id}}{RT}$	$\frac{S - S^{id}}{R}$
378.65	.7071	.7766	−.9759	−.7232
380.00	.7122	.7793	−.9586	−.7093
400.00	.7730	.8141	−.7572	−.5516
420.00	.8153	.8418	−.6204	−.4482
440.00	.8468	.8644	−.5201	−.3743
460.00	.8713	.8830	−.4433	−.3189
480.00	.8908	.8986	−.3827	−.2758
500.00	.9067	.9118	−.3338	−.2415
520.00	.9198	.9321	−.2935	−.2135
540.00	.9308	.9328	−.2600	−.1904
		P *(bar) 50.0*		
220.00	.1944	.0101	−11.6422	−7.0493
230.00	.1895	.0167	−10.8493	−6.7561
233.15	.1880	.0193	−10.6160	−6.6688
239.67	.1852	.0257	−10.1545	−6.4937
240.00	.1850	.0261	−10.1318	−6.4850
250.00	.1811	.0389	−9.4761	−6.2294
260.00	.1776	.0557	−8.8733	−5.9863
270.00	.1746	.0771	−8.3165	−5.7539
273.15	.1738	.0848	−8.1498	−5.6827
280.00	.1722	.1033	−7.8003	−5.5307
290.00	.1702	.1347	−7.3198	−5.3154
298.15	.1690	.1642	−6.9515	−5.1448
300.00	.1688	.1714	−6.8705	−5.1065
310.00	.1680	.2132	−6.4480	−4.9025
320.00	.1679	.2600	−6.0483	−4.7010
330.00	.1686	.3113	−5.6666	−4.4996
340.00	.1704	.3667	−5.2977	−4.2943
350.00	.1737	.4253	−4.9342	−4.0792
360.00	.1794	.4862	−4.5643	−3.8431
370.00	.1900	.5480	−4.1615	−3.5600
378.65	.2110	.6004	−3.7147	−3.2046
380.00	.2172	.6083	−3.6204	−3.1233
400.00	.5357	.6893	−1.6817	−1.3096
420.00	.6551	.7390	−1.2274	−.9250
440.00	.7262	.7776	−.9760	−.7244
460.00	.7761	.8088	−.8060	−.5939
480.00	.8137	.8348	−.6811	−.5005
500.00	.8431	.8566	−.5849	−.4301
520.00	.8667	.8751	−.5084	−.3750
540.00	.8861	.8910	−.4462	−.3308

Continued

Z	ϕ	$\frac{H - H^{id}}{RT}$	$\frac{S - S^{id}}{R}$
.1638	.6586	−3.9952	−3.5776

Z	ϕ	$\frac{H - H^{id}}{RT}$	$\frac{S - S^{id}}{R}$
.5276	.6984	−1.6564	−1.2974
.5418	.7024	−1.6019	−1.2487
.6692	.7523	−1.1374	−.8528
.7399	.7902	−.8954	−.6600
.7884	.8206	−.7343	−.5366
.8245	.8455	−.6170	−.4492
.8525	.8663	−.5274	−.3838
.8748	.8838	−.4565	−.3330
.8931	.8988	−.3992	−.2924
.9082	.9116	−.3520	−.2594

P *(bar)* = *100.0*

Z	ϕ	$\frac{H - H^{id}}{RT}$	$\frac{S - S^{id}}{R}$
.3859	.0061	−11.5284	−6.4360
.3758	.0101	−10.7399	−6.1422
.3728	.0116	−10.5082	−6.0551
.3669	.0155	−10.0510	−5.8813
.3666	.0157	−10.0286	−5.8727
.3583	.0233	−9.3805	−5.6207
.3509	.0333	−8.7863	−5.3827
.3444	.0458	−8.2391	−5.1567
.3426	.0504	−8.0756	−5.0878
.3389	.0613	−7.7334	−4.9414
.3342	.0797	−7.2645	−4.7356
.3310	.0970	−6.9067	−4.5741
.3304	.1013	−6.8282	−4.5381
.3275	.1258	−6.4210	−4.3481
.3256	.1533	−6.0392	−4.1640
.3246	.1836	−5.6798	−3.9849
.3248	.2164	−5.3396	−3.8092
.3262	.2515	−5.0157	−3.6353
.3292	.2884	−4.7049	−3.4614
.3339	.3267	−4.4039	3.2853
.3399	.3607	−4.1488	−3.1290
.3411	.3660	−4.1092	−3.1041
.3660	.4452	−3.5221	−2.7129
.4148	.5210	−2.9162	−2.2642
.4915	.5885	−2.3313	−1.8012
.5749	.6458	−1.8627	−1.4254
.6474	.6937	−1.5204	−1.1548
.7063	.7342	−1.2691	−.9602
.7537	.7687	−1.0788	−.8158
.7924	.7983	−.9302	−.7050

Table III. Continued

T(K)	Z	ϕ	$\frac{H - H^{id}}{RT}$	$\frac{S - S^{id}}{R}$
		P (bar) = 44.26		
220.00	.1722	.0112	−11.6551	−7.1618
230.00	.1679	.0184	−10.8616	−6.8686
233.15	.1666	.0213	−10.6280	−6.7812
239.67	.1641	.0284	−10.1661	−6.6059
240.00	.1640	.0288	−10.1434	−6.5972
250.00	.1605	.0430	−9.4867	−6.3411
260.00	.1574	.0617	−8.8829	−6.0974
270.00	.1548	.0854	−8.3249	−5.8641
273.15	.1541	.0939	−8.1578	−5.7926
280.00	.1527	.1145	−7.8073	−5.6398
290.00	.1510	.1493	−7.3253	−5.4232
298.15	.1501	.1819	−6.9554	−5.2513
300.00	.1499	.1899	−6.8741	−5.2127
310.00	.1493	.2362	−6.4495	−5.0065
320.00	.1493	.2881	−6.0470	−4.8024
330.00	.1501	.3449	−5.6616	−4.5972
340.00	.1520	.4061	−5.2874	−4.3864
350.00	.1554	.4709	−4.9155	−4.1624
360.00	.1615	.5379	−4.5297	−3.9097
370.00	.1742	.6055	−4.0841	−3.5824
378.65	.2812	.6609	−2.8836	−2.4694
380.00	.3995	.6665	−2.2298	−1.8241
400.00	.6170	.7257	−1.3411	−1.0204
420.00	.7050	.7684	−1.0287	−.7652
440.00	.7624	.8022	−.8336	−.6132
460.00	.8040	.8298	−.6956	−.5090
480.00	.8360	.8528	−.5917	−.4325
500.00	.8613	.8721	−.5105	−.3737
520.00	.8818	.8886	−.4453	−.3272
540.00	.8987	.9027	−.3919	−.2895

[a] Dashed lines separate gas and liquid phases.

Discussion

The enthalpies of vaporization, ΔH_v, at selected temperatures have been calculated using the residual enthalpy data in Table II. These are compared with the ASHRAE values (*10*) in Table V. The uncertainty in our calculated values is estimated to be $\pm$ 0.2 kJ mol^{-1} at lower temperatures increasing to $\pm$ 1.0 kJ mol^{-1} near the critical. The ΔH_v value at 239.67 K (bp 1.01325) calculated from the Clapeyron equation, the vapor pressure data, and the compression data is 20.12 kJ mol^{-1}. This value is designated as the experimental (expt.) value in Table V. The

Table IV. Ideal Gas Thermodynamic Properties of R-500 (J mol^{-1} K^{-1})

			(J mol^{-1} k^{-1})		
T *(K)*	$C°_P$	S°	$(G° - H°_0)$ / T	$(H° - H°_0)$ / T	$H° - H°_0$
200	56.39	273.83	−232.53	41.30	8259
298.15	70.66	299.08	−250.38	48.69	14518
300	70.90	299.52	−250.69	48.83	14648
400	82.78	321.62	−265.72	55.89	22356
500	92.00	341.12	−278.89	62.23	31114
600	99.08	358.54	−290.73	67.81	40684
700	104.61	374.25	−301.60	72.65	50851
800	109.02	388.52	−311.55	76.96	61571
900	112.59	401.58	−320.86	80.73	72654
1000	115.53	413.59	−329.52	84.07	84066
1100	117.97	424.70	−337.66	87.04	95742
1200	120.41	435.09	−345.39	89.71	107647
1300	121.78	444.75	−352.63	92.12	119757
1400	123.38	453.90	−359.60	94.30	132014
1500	124.62	462.40	−366.13	96.28	144416

agreement between the ΔH_v values at the normal boiling point is satisfactory as shown in Table V. However, the difference between our values and those of ASHRAE increases as the critical point is approached but is well within our estimated uncertainty. In the superheated region, the difference between ASHRAE and our enthalpy values at extreme conditions of temperature and pressure is about 5%, our values being greater.

It is felt that the thermodynamic properties reported here are closer to the true values, especially at low pressures. Also, the BACK equation of state, with only five adjustable constants, has now been used for the refrigerants. Calculations for 1,1-difluoroethane and difluorodichloromethane, the pure components making up R-500, are in progress.

Table V. Comparison of Enthalpy of Vaporization (ΔH_v) of R-500

	ΔH_v *(kJ mol^{-1})*		
T *(K)*	*This Work*	*ASHRAE*	*Expt.*
220	21.38	20.81	
239.67	20.20	19.97	20.12
273.15	18.03	18.18	
298.15	16.16	16.44	
320	14.23	14.55	
340	11.99	12.32	
370	6.43	7.06	

Acknowledgments

This work was supported by the Texas Engineering Experiment Station.

Literature Cited

1. Ashley, C. M. *Refrig. Eng.* **1950**, *58*(553), 603.
2. McHarness, R. C.; Chapman, D. D. *ASHRAE J.* **1962**, *4*, 49.
3. Zwolinski, B. J. et al. "Selected Values of Properties of Chemical Compounds," Thermodynamics Research Center Data Project, Thermodynamics Research Center, Texas A&M University, College Station, Texas, 1978.
4. Chen, S. S.; Kreglewski, A. *Ber. Bunsenges. Phys. Chem.* **1977**,*81*, 1048.
5. Sinka, J. V.; Murphy, K. P. *J. Chem. Eng. Data* **1967**, *12*, 315.
6. Prasad, M.; Kudchadker, A. P. *J. Chem. Eng. Data* **1978**, *23*, 190.
7. "Thermodynamic Properties of Refrigerants," *ASHRAE J.* **1969**, 102.
8. Haynes, W. M.; Hiza, M. J. *J. Chem. Thermodyn.* **1977**, *9*, 179.
9. Wagner, W.; Ewers, J.; Pentermann, W. *J. Chem. Thermodyn.* **1976**, *8*, 1049.
10. ASHRAE Guide and Data Book, Fundamentals, and Equipment," **1965**, **1966**, 288.

RECEIVED August 15, 1978.

18

Calculation of Compressed Liquid Excess Volumes and Isothermal Compressibilities for Mixtures of Simple Species

S. P. SINGH and R. C. MILLER

Department of Mineral Engineering, University of Wyoming, Laramie, WY 82071

Compressed liquid mixture excess volumes (V^E) *and isothermal compressibilities have been measured recently for binary and ternary mixtures containing the components nitrogen, argon, methane, and ethane. Selected liquid-phase equations of state (and some corresponding-states methods) have been tested using these new data. Simple two-parameter equations of state, based on the model of hard spheres in a uniform potential field, accurately represent* $V^E(P, T, x)$ *for these mixtures. Two empirical binary mixing rule deviation parameters were used, but they were highly correlated. Two methods of reducing the formulation to use only one parameter were shown to result in only small reductions in accuracy. The equations of state accurately predicted* V^E *values for ternary mixtures and changes in isothermal compressibility on mixing for all mixtures.*

Equations of state have been traditionally developed to describe the pressure–molar volume–temperature–composition ($PVTx$) behavior for fluid mixtures. If an equation of state can describe the $V(P, T, x)$ surface for a cryogenic liquid mixture to an accuracy approaching 0.1%, then it is satisfactory for all current industrial needs, including custody transfer calculations. Analytical equations of this accuracy have been developed for certain pure cryogens. It is nearly impossible to generalize these complicated equations of state to mixtures, other than by corresponding-states techniques. Accuracy approaching the required level has

0-8412-0500-0/79/33-182-323$05.25/1

been obtained for corresponding-states calculations when restricted to limited species and ranges of composition, as encountered in liquefied natural gas mixtures (*1, 2, 3*).

An alternative approach is to use an equation of state to describe behavior of the excess volume: $V^E(P, T, x)$. Excess volume is defined in terms of the mixture molar volume (V) and the component molar volumes (V_i) at the same pressure and temperature:

$$V^E = V - \sum x_i V_i. \quad (1)$$

An equation of state is used to calculate V and the V_i values, and the excess volume is then obtained from Equation 1. For practical applications, V^E from the equation of state must be combined with known component molar volumes to obtain the mixture molar volume. The hope is that, due to a cancellation of errors, much simpler equations of state can be used to accurately describe $V^E(P, T, x)$ than would be necessary to yield directly $V(P, T, x)$ to the required accuracy. The V^E method is only of practical use when the temperature of the solution is below the critical temperatures of the major components.

From previous work (*4, 5, 6*) the following conclusions can be drawn concerning the use of equations of state to represent $V^E(P, T, x)$ for simple liquid mixtures. The Longuet–Higgins and Widom (LHW) two-parameter equation of state (*7*), based on the model of hard spheres in a uniform potential field, can be used to accurately describe $V^E(T, x)$ at low pressures for mixtures of nonpolar, nearly spherical species. Where molecular size differences are not large, this relatively simple equation works as well or better than other equations tested, including those using more rigorous representations of the hard-sphere mixture compressibility factor.

A one-fluid theory is generally superior to a two-fluid theory in generalizing the LHW equation of state for mixtures. In the one-fluid theory the same form of equation is used for the mixture as for the components, but the equation of state parameters are assumed to be composition dependent. Van der Waals (VDW) forms, quadratic in mole fractions, prove satisfactory to describe this composition dependence for simple species.

Excess volume predictions are extremely sensitive to the value used for the unlike-molecule size parameter. Small deviations must be allowed from the commonly used arithmetic mean mixing rule for size parameters if quantitative results are to be obtained for V^E simultaneously with other measurable excess thermodynamic functions (G^E and H^E). It should be noted that similar conclusions have been drawn for mixtures of globular molecules (*8*).

The method can be extended to include nonpherical, nonpolar species (such as the lower molecular weight alkanes) by introduction of a third parameter in the equation of state, namely the Prigogine factor for chain-type molecules (*9*). This modified hard-sphere equation of state accurately describes $V^E(T, x)$ for liquefied natural gas mixtures at low pressures. Ternary and higher mixture V^E values are accurately predicted using only binary mixing rule deviation parameters.

Significant deviations between the model predictions and experimentation occur if the reduced temperature of a component present in an appreciable amount is higher than about 0.85. The pure liquid for this component becomes highly expanded at high reduced temperatures, making the uniform potential field model suspect.

A recent experimental study (*10, 11*) used dielectric constant measurements to produce $V^E(P, T, x)$ for some binary and ternary mixtures of the components nitrogen, argon, methane, and ethane. Mixture isothermal compressibilities $[\kappa(P, T, x)]$ and changes in isothermal compressibility on mixing $[\Delta\kappa^M(P, T, x)]$ also were derived from these results. The purpose of the present work is to test the ability of various simple equation-of-state forms to fit the $V^E(P, T, x)$ surfaces for the binary mixtures. An intercomparison also is made with the corresponding-states method (*1, 2*). All fitting and comparisons are done on a consistent basis. Comparisons are made also between calculated and experimental V^E values for ternary mixtures and $\Delta\kappa^M$ values for all mixtures. A recent corresponding states correlation for isothermal compressibilities (*12*) is tested also. Further details concerning the present investigation can be found in Reference *11*.

Equations of State for V^E

Development of Equations. All of the equation-of-state forms considered are based on the model of hard convex particles in a uniform potential field. In terms of compressibility factor the general form is

$$z = \frac{pV}{RT} = z^{HP} - \frac{a}{RTV}, \tag{2}$$

where z^{HP} is the compressibility factor for the hard-particle fluid, and a is a parameter considered here to be independent of both temperature and density.

Using the Prigogine approach for chain-type molecules (*9*),

$$z^{HP} = cz^{HS}, \tag{3}$$

where z^{HS} is the compressibility factor for the hard-sphere fluid and c is a parameter introduced to account for degrees of freedom which become restricted at high densities. For liquids which are not highly expanded, c is treated as independent of T and V. The forms considered for z^{HS} are due to VDW, Flory (FLR) (*13*), LHW (*7*), and Carnahan and Starling (CS) (*14*):

$$\text{VDW } z^{HS} = \frac{1}{1 - 4y}, \tag{4}$$

$$\text{FLR } z^{HS} = \frac{1}{1 - y^{1/3}}, \tag{5}$$

$$\text{LHW } z^{HS} = \frac{1 + y + y^2}{(1 - y)^3}, \tag{6}$$

$$\text{CS } z^{HS} = \frac{1 + y + y^2 - y^3}{(1 - y)^3}, \tag{7}$$

where $y = b/4V$, and b is a parameter considered independent of temperature and density.

The parameters a and b were determined such that each equation of state reproduced the experimental component V^s and κ^s values for the saturated liquids at 100 K. Values of c were taken directly from the work of Rodosevich (*6*). The P^s, V^s, κ^s, and c values which were chosen (*11*) are given in Table I, and values of a and b which were determined are given in Table II. The sensitivity of V^E calculations to the c values was tested by some calculations on the LHW model using $c = 1.00$ for ethane. Corresponding values for a and b are 97.63×10^4 J cm^3 mol^{-2} and 108.04 cm^3 mol^{-1}.

Table I. Component Liquid Molar Volumes (V^s) and Isothermal Compressibilities (κ^s) at 100 K and Vapor Pressure (P^s) Which Were Used along with Shape Factors (c or α) to Determine Equation-of-State Parameters a and b

Species	P^s/MPa	$V^s/cm^3\ mol^{-1}$	κ^s/GPa^{-1}	c	α
Nitrogen	0.7790	40.586	9.547	1.03	1.08
Argon	0.3240	30.410	3.265	1.00	0.98
Methane	0.0345	36.544	1.701	1.00	1.00
Ethane	0.00001	46.886	0.604	1.50	1.22

Table II. Equation-of-State Parameters for Components a/J cm^3 mol^{-2} and b/cm^3 mol^{-1}

	VDW	*FLR*	*LHW*	*CS*	*GIB*
		Nitrogen			
$10^{-4} \times a$	11.78	21.84	17.68	17.95	17.67
b	28.74	96.84	55.53	56.79	54.29
		Argon			
$10^{-4} \times a$	11.32	20.52	18.75	19.04	18.65
b	23.64	82.03	49.21	50.40	49.54
		Methane			
$10^{-4} \times a$	18.77	33.50	35.16	35.63	35.16
b	30.63	109.91	69.11	70.84	69.11
		Ethane			
$10^{-4} \times a$	52.34	92.44	113.69	114.66	104.74
b	41.65	154.16	102.07	104.61	103.05

A second method of handling nonspherical particles in the same general model is to use the scaled particle theory result of Gibbons (GIB) (*15*) for z^{HP}:

$$\mathrm{GIB}\ z^{\mathrm{HP}} = \frac{1 + Ay + By^2}{(1-y)^3}, \tag{8}$$

where $A = 3\alpha - 2$, and $B = 1 + 3\alpha^2 - 3\alpha$. In this formulation α is a shape factor for hard convex particles. Graboski (*16*) has developed a correlation for α in terms of the Pitzer accentric factor. Values listed in Table I were taken from this correlation. The a and b values in Table II were determined by fit of V^{s} and κ^{s} at 100 K, described above for the other equations.

The above equations of state were applied to mixtures by assuming a one-fluid theory using VDW relations:

$$a = \sum \sum x_i x_j a_{ij}, \tag{9}$$

$$b = \sum \sum x_i x_j b_{ij}, \tag{10}$$

$$c = \sum \sum x_i x_j c_{ij}, \tag{11}$$

and

$$\alpha = \sum \sum x_i x_j \alpha_{ij}. \tag{12}$$

The cross parameters were specified by (*6*)

$$a_{ij} = (a_{ii}a_{jj})^{1/2}(1 - k_{ij})\left(\frac{b_{ij}^2}{b_{ii}b_{jj}}\right)^{1/2}, \tag{13}$$

$$b_{ij} = [\tfrac{1}{2}(b_{ii}^{1/3} + b_{jj}^{1/3})(1 + j_{ij})]^3, \qquad (14)$$

and

$$c_{ij} = \tfrac{1}{2}(c_{ii} + c_{jj}) \qquad (15)$$

or

$$\alpha_{ij} = \tfrac{1}{2}(\alpha_{ii} + \alpha_{jj}) \qquad (16)$$

Some calculations also were performed on the GIB model using the mixing rule suggested by Graboski (*16*):

$$\alpha_{ij} = \frac{2\alpha_{ii}\alpha_{jj}}{\alpha_{ii} + \alpha_{jj}} \qquad (17)$$

Equations 13 and 14 follow from the Lorentz–Berthelot rules, an arithmetic mean for unlike-molecule size parameters and a geometric mean for unlike-molecule energy parameters, with deviations allowed for either rule.

There are two mixing rule deviation parameters (k_{ij} and j_{ij}) which must be evaluated for each pair of species in a mixture. In the present investigation, only binary mixture V^E data were used in the evaluation of these parameters.

Discussion of Results. For each binary system the two mixing rule deviation parameters (k_{ij} and j_{ij}) were first determined by least squares fits of the $V^E(P, T, x)$ data tabulated as "high-pressure data" in Appendix B of Reference *11*. These data are at temperatures from 91 to 115 K, with pressures from near saturation to 50 MPa. The only data not used were for binaries containing nitrogen at 115 K and pressures less than 5.5 MPa. Nitrogen at 115 K (reduced temperature of 0.91) is a highly expanded liquid at lower pressures, and the assumption of a uniform potential field is not valid (*5, 17*). These data were omitted from all fitting work reported in this chapter. Pure-fluid parameters were taken from Tables I and II, and mixing rules 9–16 were used. The resulting binary deviation parameters and standard deviations for V^E data fits are listed in Table III.

The results of the data fitting show little difference between the parameters or predictions of the LHW and CS equations. Equations 6 and 7 yield nearly the same z^{HS} for the range of y values encountered in this study.

The LHW, CS, and GIB models consistently have the lowest standard deviations; however, the VDW model is not greatly inferior. The FLR model yields significantly higher standard deviations only for N_2 + CH_4 and Ar + C_2H_6. There is little from which to choose between the LHW and GIB equation fits. In fact, calculations using LHW with c =

Table III. Binary Interaction Parameters (k_{ij} and j_{ij}) Obtained by Fitting V^E Data [Reference *11*, pp. 166–82] to Equations of State[a]

	VDW	*FLR*	*LHW*	*CS*	*GIB*
		Nitrogen + Argon			
k	−0.0175	0.0124	0.0072	0.0077	0.0067
j	0.0022	−0.0008	0.0010	0.0009	0.0015
s	0.029	0.032	0.028	0.028	0.029
		Nitrogen + Methane			
k	−0.0018	0.0597	0.0597	0.0591	0.0565
j	0.0044	0.0004	0.0036	0.0038	0.0036
s	0.050	0.074	0.044	0.044	0.045
		Argon + Methane			
k	0.0273	0.0349	0.0379	0.0373	0.0389
j	0.0052	0.0037	0.0041	0.0042	0.0043
s	0.014	0.013	0.007	0.007	0.007
		Argon + Ethane			
k	0.0279	0.1054	0.1178	0.1176	0.1261
j	0.0098	0.0113	0.0164	0.0165	0.0177
s	0.030	0.062	0.024	0.025	0.029
		Methane + Ethane			
k	−0.0425	0.0069	0.0181	0.0184	0.0292
j	0.0014	0.0023	0.0044	0.0044	0.0055
s	0.027	0.027	0.024	0.024	0.024

[a] The standard deviations (s) are in $cm^3\ mol^{-1}$.

1.00 for C_2H_6 (equivalent to GIB with $\alpha = 1.00$) gave fits of the Ar + C_2H_6 and $CH_4 + C_2H_6$ data with standard deviations of 0.026 and 0.023 $cm^3\ mol^{-1}$, respectively. These are almost identical with the values in Table III, indicating that the representation of $V^E(P, T, x)$ by these equations of state is insensitive to c (or α) values. This conclusion contradicts previous results (*6*) where both V^E and G^E data were fit simultaneously with the same form of LHW equation.

Using Equation 17 instead of Equation 16 in the mixing rules for the GIB model did not greatly change the ability of the equation to fit the V^E data. There were only slight changes in deviation parameters and standard deviations for the $N_2 + CH_4$ and Ar + C_2H_6 systems.

The ability of the equations of state to fit V^E data does depend to some extent on the component a and b values. In previous studies (*5, 6*) a and b were determined for the LHW model by fitting a saturated liquid molar volume and a heat of vaporization for each component. Using these a and b values instead of the ones from Table II led to standard deviations for the binary V^E fits which were larger than those in Table III (cf. Ref. *11*). The maximum increase was from 0.024 to 0.048 $cm^3\ mol^{-1}$

for Ar + C_2H_6. The comparison indicates that pure-fluid parameters are best determined from volumetric properties of the liquid state, if the purpose is to represent only V^E data.

Binary Deviation Parameters. The problem with fitting only one type of excess property is that the two binary deviation parameters are highly correlated. Figures 1 and 2 show some confidence ellipses for the N_2 + Ar and Ar + C_2H_6 binary parameters in the LHW model. Any point inside a confidence ellipse represents a set of k_{ij} and j_{ij} values which can reproduce the experimental data within the standard deviation specified for the ellipse. The solid points represent the optimum parameters (from Table III), for which the standard deviations are 0.028 cm^3 mol^{-1} for N_2 + Ar and 0.024 cm^3 mol^{-1} for Ar + C_2H_6. There is a rather wide range of sets of j_{ij} and k_{ij} values for either system which will fit the V^E data within 0.04 cm^3 mol^{-1}, which corresponds to about 0.1% of the mixture molar volumes. This is a reasonable level for the average uncertainty in these V^E data obtained by the dielectric constant method.

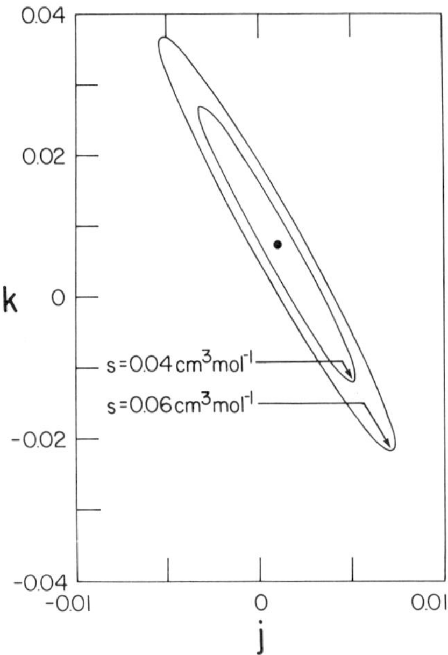

Figure 1. Optimum binary deviation parameters (k_{ij} and j_{ij}) and confidence ellipses for the fit of the LHW equation of state to the N_2 + Ar V^E data of Ref. 11, pp. 166–168

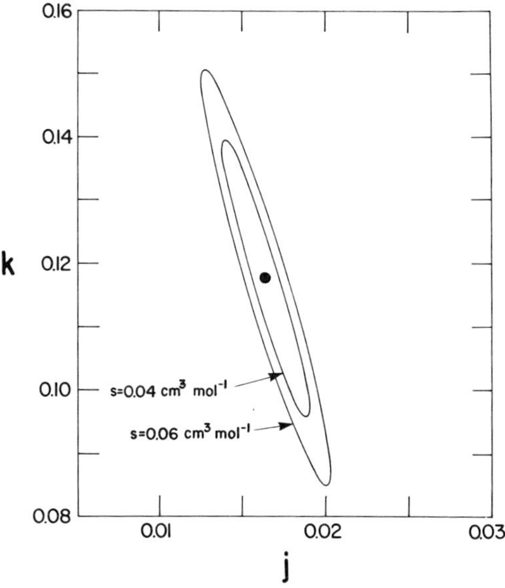

Figure 2. Optimum binary deviation parameters (k_{ij} and j_{ij}) and confidence ellipses for the fit of the LHW equation of state to the Ar + C_2H_6 V^E data of Ref. 11, pp. 175–179

If the high-pressure limiting V^E can be determined, then j_{ij} can be fixed independently. When applied to a binary mixture, any of the models except VDW yield the following relations in the limit as the pressure approaches infinity:

$$b = 4V, \tag{18}$$

$$V^{E\infty} = \frac{(b - x_1 b_{11} - x_2 b_{22})}{4}, \tag{19}$$

and

$$j_{12} = \frac{2[2V^{E\infty}/x_1 x_2 + (b_{11} + b_{22})/2]^{1/3}}{(b_{11}{}^{1/3} + b_{22}{}^{1/3})} - 1 \tag{20}$$

For systems such as Ar + C_2H_6 and CH_4 + C_2H_6 it is obvious from the V^E data that 50 MPa is not high enough pressure to give a close estimate of $V^{E\infty}$. For other systems (N_2 + Ar, N_2 + CH_4, and Ar + CH_4) it is not clear from the data alone whether or not $V^{E\infty}$ has been attained. Extrapolations of the V^E data have been made using the CS model and the parameters from Tables I, II, and III. Figures 3, 4, and 5 show the

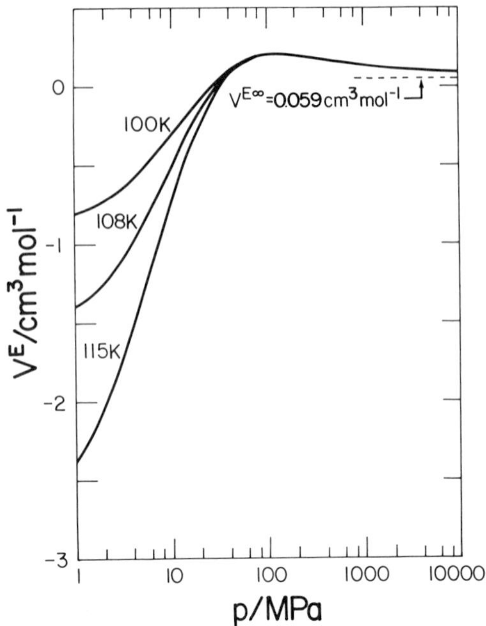

Figure 3. V^E vs. P for equimolar $N_2 + CH_4$ as calculated by the CS equation of state (parameters from Tables I, II, and III)

results for equimolar mixtures of $N_2 + CH_4$, $CH_4 + C_2H_6$, and Ar + CH_4. The $V^{E\infty}$ values were calculated from the j_{12} values from Table III using Equation 20. These figures indicate that, even for pressures as high as 10,000 MPa, the high-pressure limit ($V^{E\infty}$) is yet to be attained. A pressure of 10,000 MPa is considerably higher than the solidifying pressures of these mixtures. If the model behavior is qualitatively correct, it appears that direct determination of $V^{E\infty}$ is not possible. Thus, j_{12} can not be evaluated in this fashion.

There are a number of other methods which can be used to uncouple the j_{12} and k_{12} parameters. One of the best ways would be to simultaneously fit G^E, H^E, and V^E data. This work has not proceeded to that stage as yet.

The first method tried to uncouple j_{12} and k_{12} was to set $V^{E\infty}$ equal to zero. This is not equivalent to setting j_{12} equal to zero. The result from Equations 14 and 20 is to use an arithmetic mean rule directly for the cross b parameter,

$$b_{12} = \tfrac{1}{2}(b_{11} + b_{22}) \tag{21}$$

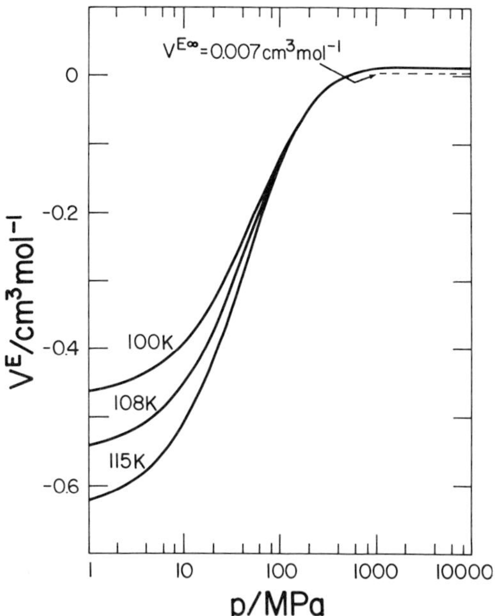

Figure 4. V^E vs. P for equimolar $CH_4 + C_2H_6$ as calculated by the CS equation of state (parameters from Tables I, II, and III)

The mixture b becomes the mole fraction average of component values:

$$b = x_1 b_{11} + x_2 b_{22} \tag{22}$$

The remaining binary deviation parameter k_{12} was determined by again fitting the 50MPa V^E data of Reference *11*. These k_{12} values, the corresponding j_{12} values from Equation 20, and the standard deviations for the V^E data fit are listed in Table IV for the LHW ($V^{E\infty} = 0$) model. By comparison with the Table III LHW values (also listed), only for $N_2 + CH_4$ and $Ar + C_2H_6$ are the standard deviations significantly increased for this method.

A second method used to uncouple j_{12} and k_{12} was to use the Robinson and Hiza (*18*) empirical observation that $k_{12} = 6j_{12}$. The binary parameters k_{12} and j_{12} were determined for the LHW model by least squares fit of the same V^E data, subject to the constraint $k_{12} = 6j_{12}$. The resulting parameters and standard deviations for the fits are listed in Table IV as the LHW ($k_{12} = 6j_{12}$) model. The $Ar + C_2H_6$ data are

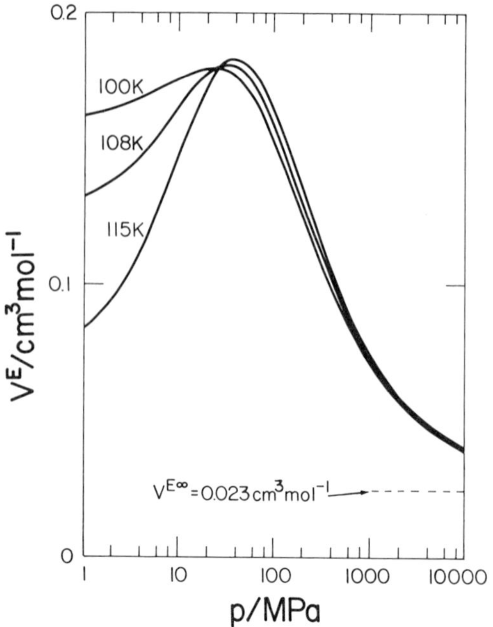

Figure 5. V^E vs. P for equimolar Ar + CH_4 as calculated by the CS equation of state (parameters from Tables I, II, and III)

more closely fit than by the $V^{E\infty} = 0$ method, but the $N_2 + CH_4$ fit is slightly worse. For the other systems the deviations are again nearly the same as the Table III results.

Either of the above methods is useful in reducing the number of binary parameters which must be fit to experimental V^E data from two to one. If other excess property data are considered, a re-evaluation of these methods will be necessary.

Predictions for Ternary Mixtures. Excess volumes were calculated from the equations of state for the nearly equimolar ternary mixtures of N_2 + Ar + CH_4 and Ar + CH_4 + C_2H_6 for which data are given in Appendix B of Reference *11*. The root mean square deviations between the experimental and calculated V^E values are given in Table V. Only component and binary parameters from Tables I, II, and III were used in these calculations.

The deviations for the ternary mixtures are on the same order as those for the constituent binaries. These simple equations of state can predict ternary V^E values for mixtures of N_2, Ar, CH_4, and C_2H_6 without requiring any ternary parameters in the formulation.

Table IV. Binary Interaction Parameters (k_{ij} and j_{ij}) and Standard Deviations (s/cm^3 mol^{-1}) between Model Calculations and the Experimental V^E Data from Reference *11*, pp. 166–82

	LHW (*Table III*)	*LHW* ($v^{E\infty} = 0$)	*LHW* ($k_{ij} = 6j_{ij}$)	*MOL 1*	*MOL 2*
		Nitrogen + Argon			
k	0.0072	0.0099	0.0067	0.0007	—
j	0.0010	0.0004	0.0011	0.0043	—
s	0.028	0.028	0.028	0.022	—
		Nitrogen + Methane			
k	0.0597	0.0717	0.0414	0.0340	0.032
j	0.0036	0.0013	0.0069	0.0117	0.015
s	0.044	0.053	0.062	0.027	0.044
		Argon + Methane			
k	0.0379	0.0435	0.0313	0.0250	—
j	0.0041	0.0032	0.0052	0.0165	—
s	0.007	0.011	0.011	0.015	—
		Argon + Ethane			
k	0.1178	0.1348	0.1062	0.0542	—
j	0.0164	0.0144	0.0177	0.0319	—
s	0.024	0.042	0.032	0.020	—
		Methane + Ethane			
k	0.0181	0.0198	0.0231	−0.0092	−0.007
j	0.0044	0.0042	0.0038	0.0048	0.004
s	0.024	0.025	0.026	0.025	0.026

Table V. Root Mean Square Deviations (cm^3 mol^{-1}) between Model Predictions and the Experimental V^E Data for Nearly Equimolar Ternary Mixtures from Rerefence *11*, pp. 183–184

System	*VDW*	*FLR*	*LHW*	*CS*
$N_2 + Ar + CH_4$	0.039	0.054	0.029	0.029
$Ar + CH_4 + C_2H_6$	0.034	0.044	0.022	0.022

Predictions of Change in Isothermal Compressibility on Mixing. A quantity closely related to the change in excess volume with pressure at constant temperature and composition is the change in isothermal compressibility on mixing:

$$\Delta\kappa^{\mathrm{M}} = \kappa - \sum x_i\kappa_i \tag{23}$$

In this equation κ is the mixture isothermal compressibility, and the x_i and κ_i are the component mole fractions and isothermal compressibilities at the same temperature and pressure as the mixture. Values of $\Delta\kappa^{\mathrm{M}}$ were derived from $V^{\mathrm{E}}(P, T, x)$ in Reference *11*. This quantity was quite large for all the systems studied (*10, 11*). Whereas the maximum V^{E} observed was on the order of 10% of the mixture V, the maximum $\Delta\kappa^{\mathrm{M}}$

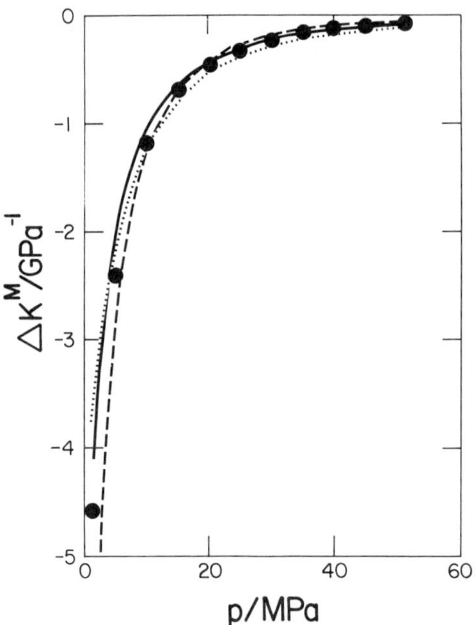

Figure 6. Change in isothermal compressibility on mixing ($\Delta\kappa^M$) vs. P for a nearly equimolar mixture of $N_2 + CH_4$ at 108 K: (●), Ref. 11, p. 170; (—), VDW; (— —), FLR; (- - -), LHW.

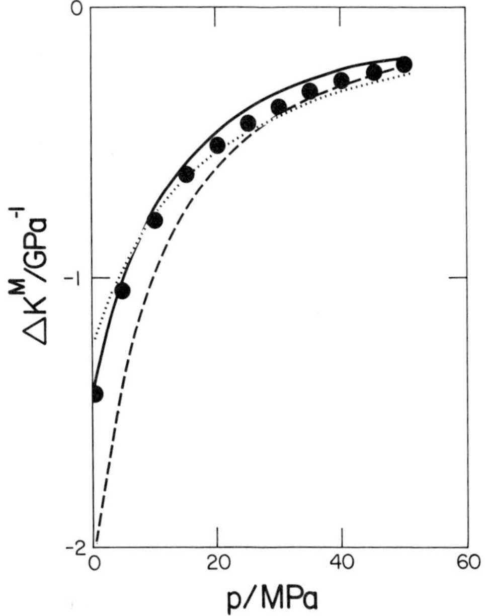

Figure 7. Change in isothermal compressibility on mixing ($\Delta\kappa^M$) vs. P for a nearly equimolar mixture of Ar + C_2H_6 at 115 K: (●), Ref. 11, p. 176; (—), VDW; (— —), FLR; (- - -), LHW.

was about 150% of the mixture κ. The mixtures were always less compressible than a mole-fraction average of component κ_i values would indicate.

Comparisons have been made between $\Delta\kappa^M$ values derived from the experimental V^E data and predictions of the equations of state. Figures 6, 7, and 8 are plots of $\Delta\kappa^M$ vs. P for some nearly equimolar mixtures of N_2 + CH_4, Ar + C_2H_6 and Ar + CH_4 + C_2H_6. The data are from Appendix B of Reference *11*. The curves were drawn by using the VDW, FLR, and LHW equations with parameters from Tables I, II, and III.

Predictions of $\Delta\kappa^M$ by all models appear to be in fairly good agreement with the experimental data. Maximum discrepancies are observed for the FLR equation at the lowest pressures. The VDW predictions are in slightly better overall agreement with experiment than the LHW model, but both are quite satisfactory with average deviations on the order of 0.1 GPa^{-1}. The CS and GIB models predict $\Delta\kappa^M$ values very close to those of the LHW model.

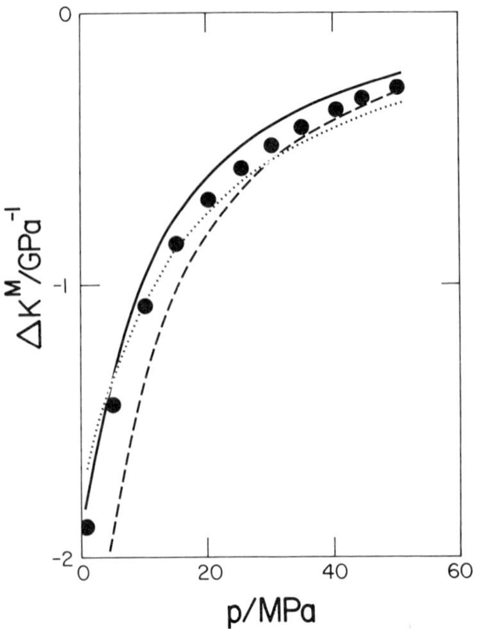

Figure 8. Change in isothermal compressibility on mixing ($\Delta\kappa^M$) vs. P *for a nearly equimolar mixture of* $Ar + CH_4 + C_2H_6$ *at 115 K: (●), Ref.* 11, *p. 184; (—), VDW; (— —), FLR; (- - -), LHW.*

Corresponding-States Calculations

Extended Corresponding States. An extended corresponding states method has recently been developed by Mollerup and Rowlinson (*1, 2, 3*) for application to liquefied natural gas and other simple liquid mixtures. Methane is used as a reference substance with properties for this fluid taken from Goodwin (*19*). Size and energy-reducing parameters are taken to be temperature dependent by use of the shape factors developed by Leach et al. (*20*). Mixture properties are calculated by using VDW one-fluid equations, similar in form to Equations 9, 10, 11, and 12. Two binary interaction parameters are used which are equivalent to j_{ij} and k_{ij} in Equations 13 and 14. All calculations reported here were made using a computer program obtained directly from Mollerup (*21*).

The two binary parameters were determined by least squares fit of the binary V^E data (0–50 MPa) from Appendix B of Reference *11,* exactly as was done for the equations of state discussed above. Binary parameters and standard deviations for the fits are compared with those from Table III for the LHW model in Table IV. MOL1 is the extended corresponding states with optimized binary parameters from the V^E fits, and MOL2 uses binary parameters as determined by Mollerup (*21*) from simultaneously fitting liquefied natural gas and petroleum gas densities and phase equilibria data. Mollerup did not report binary parameters for systems containing argon. From the values in Table IV it is obvious that the extended corresponding states are capable of describing the $V^E(P, T, x)$ surfaces as accurately as the best of the simple equation-of-state methods. In fact, there is a slight improvement for the nitrogen-containing mixtures. It is encouraging that the parameters optimized to the V^E data are not greatly different than those determined by simultaneous fit of densities and phase equilibria for $N_2 + CH_4$ and $CH_4 + C_2H_6$.

Figures 9 and 10 show comparisons of LHW and MOL1 calculations with experimental V^E values for nearly equimolar mixtures of $N_2 + CH_4$ and $Ar + C_2H_6$. The two calculational methods yield curves slightly different in shape, but they both fit the data about equally well.

Values of $\Delta\kappa^M$ have not been calculated using extended corresponding states. Since good V^E vs. P curves at constant T and x are obtained, values of $\Delta\kappa^M$ should be in reasonable agreement with experimentation.

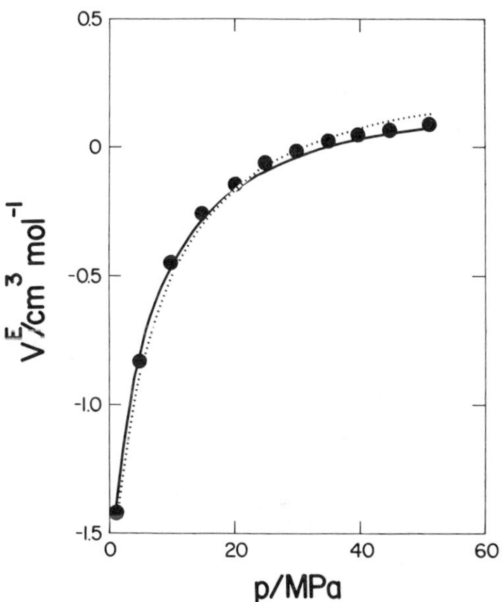

Figure 9. V^E vs. P for a nearly equimolar mixture of $N_2 + CH_4$ at 108 K: (●), Ref. 11, p. 170; (—), MOL1; (- - -), LHW.

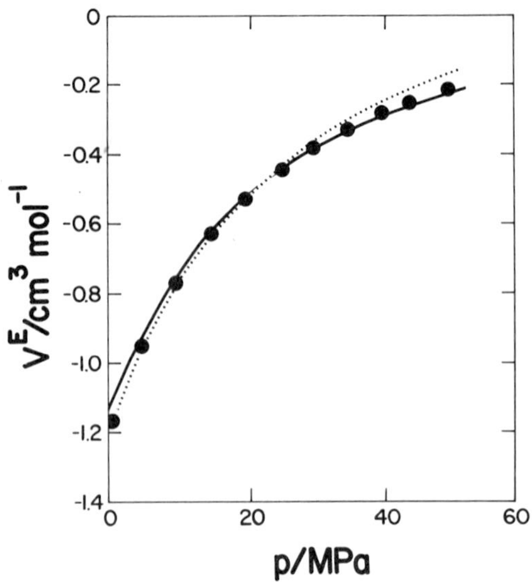

Figure 10. V^E *vs.* P *for a nearly equimolar mixture of* $Ar + C_2H_6$ *at 115 K:* (●), *Ref.* 11, *p. 176;* (—), *MOL1;* (- - -), *LHW.*

Corresponding-States Correlation for Isothermal Compressibility. Brelvi and O'Connell (*12*) developed a corresponding-states correlation for liquid isothermal compressibilities based on a form suggested by statistical thermodynamics. A knowledge of the temperature, molar volume, and critical volume are required to calculate the isothermal compressibility at any given state for simple liquids. These authors applied the correlation to mixtures (*22*); however, there were no data available for simple liquid mixtures. Based on data for more complex mixtures near room temperature, they concluded that mole-fraction or volume-fraction averages of component values yield close approximations to mixture isothermal compressibilities. Recent data (*10, 11*) for simple liquid mixtures at low temperatures do not substantiate this conclusion. Extremely large deviations from simple component averages were observed for binary and ternary mixtures of N_2, Ar, CH_4, and C_2H_6.

In this work component critical volumes were taken from Reid et al. (*23*). They are given in Table VI, along with root mean square deviations between the correlation predictions and the isothermal compressibilities tabulated in Appendix A of Reference *11* (pressures up to 50 MPa). For Ar, CH_4, and C_2H_6 the correlation is quite accurate, giving isothermal compressibilities with average deviations within 5%. Deviations for N_2 are on this order for 91 and 100 K, but they become

Table VI. Component Critical Volumes (V_c) from Reference *23* and Root Mean Square Deviations (s) between the Brelvi–O'Connell Correlation (*12*) and Component Isothermal Compressibilities from Reference *11*, pp. 124–128

Species	$V_c/cm^3\ mol^{-1}$	s/GPa^{-1}
N_2	89.5	1.00
Ar	74.9	0.09
CH_4	99.0	0.08
C_2H_6	148.0	0.02

progressively larger at 108 and 115 K. Nitrogen is highly compressible at 115 K. The average deviation could be reduced somewhat by using a slightly higher V_c value for nitrogen.

Three methods were tested for calculating mixture isothermal compressibilities. The first used a mole fraction average of component values:

$$\kappa = \sum x_i \kappa_i \tag{24}$$

Experimental component values were used rather than correlation predictions. The second and third methods were based on using a one-fluid mixture theory to apply the Brelvi–O'Connell corresponding-states correlation to mixtures:

$$V_c = \sum \sum x_i x_j V_{cij} \tag{25}$$

Cross parameters were calculated from

$$V_{cij} = [\tfrac{1}{2}(V_{ci}^{1/3} + V_{cj}^{1/3})(1 + j_{ij})]^3, \tag{26}$$

with the deviation parameter (j_{ij}) either taken as zero or optimized to fit binary mixture data.

In Table VII a comparison is made between root mean square deviations from experimental binary data for the three methods. Experimental mixture isothermal compressibilities were taken from Appendix B of Reference *11*. Component values for use in Equation 24 came from Appendix A of the same reference. Use of κ_i values from the correlation would have yielded slightly higher deviations for the systems containing N_2.

It is generally much better to use the Equation 25 approach than the mole-fraction average of the component values to obtain mixture isothermal compressibilities. Very little is gained by trying to optimize the deviation parameters to binary data. It would appear that $j_{ij} = 0$ is a satisfactory approximation in this method.

Table VII. Root Mean Square Deviations (GPa^{-1}) between Calculated Mixture Isothermal Compressibilities and Experimental Values from Reference *11*, pp. 166–184, and Optimum Deviation Parameters (*j*) for Equation 26

System	*RMS Deviations in κ Using Equations 25 and 26 with* $\sum x_i\kappa_i$	$(j_{ij} = 0)$	*(Optimum* j_{ij}*)*	*Optimum* j_{ij} *Values*
$N_2 + Ar$	1.20	0.46	0.39	0.0059
$N_2 + CH_4$	2.27	0.44	0.43	0.0035
$Ar + CH_4$	0.13	0.13	0.07	0.0052
$Ar + C_2H_6$	0.61	0.03	0.01	0.0030
$CH_4 + C_2H_6$	0.19	0.06	0.03	−0.0050
$N_2 + Ar + CH_4$	1.68	0.41	—	—
$Ar + CH_4 + C_2H_6$	0.53	0.05	—	—

Acknowledgment

The authors are grateful to the U.S. National Science Foundation for financial support of this investigation.

Literature Cited

1. Mollerup, J.; Rowlinson, J. S. *Chem. Eng. Sci.* **1974,** *29,* 1373.
2. Mollerup, J. *Adv. Cryog. Eng.* **1975,** *20,* 172.
3. Haynes, W. M.; Hiza, M. J.; McCarty, R. D. *Pap.—Int. Conf. Liquified Nat. Gas* **1977,** *2,* paper 11.
4. Liu, Y.-P.; Miller, R. C. *J. Chem. Thermodyn.* **1972,** *4,* 85.
5. Massengill, D. R.; Miller, R. C. *J. Chem. Thermodyn.* **1973,** *5,* 207.
6. Rodosevich, J. B.; Miller, R. C. *AIChE J.* **1973,** *19,* 729.
7. Longuet-Higgins, H. C.; Widom, B. *Mol. Phys.* **1964,** *8,* 549.
8. Herring, W. A.; Winnick, J. *J. Chem. Thermodyn.* **1974,** *6,* 957.
9. Prigogine, I.; Trappeniers, N.; Mathot, V. *Faraday Discuss. Chem. Soc.* **1953,** *15,* 93.
10. Singh, S. P.; Miller, R. C. *J. Chem. Thermodyn.* **1978,** *10,* 747.
11. Singh, S. P. "Dielectric Constants, Excess Volumes and Compressibilities for Liquid Mixtures of Nitrogen, Argon, Methane and Ethane from 91 to 115 K at Pressures to 50 MPa," Ph.D. Thesis, University of Wyoming, Laramie, 1978.
12. Brelvi, S. W.; O'Connell, J. P. *AIChE J.* **1972,** *18,* 1239.
13. Flory, P. J. *J. Am. Chem. Soc.* **1965,** *87,* 1833.
14. Carnahan, N. F.; Starling, K. E. *AIChE J.* **1972,** *18,* 1184.
15. Gibbons, R. M. *Mol. Phys.* **1970,** *18,* 809.
16. Graboski, M. "A Generalized Hard Sphere Equation of State for Vapor–Liquid Equilibria Prediction," Ph.D. Thesis, Pennsylvania State University, University Park, 1977.
17. Miller, R. C. *J. Chem. Phys.* **1971,** *55,* 1613.
18. Robinson, R. L.; Hiza, M. J. *Adv. Cryog. Eng.* **1975,** *20,* 218.
19. Goodwin, R. D. *Natl. Bur. Stand. (U. S.), Tech. Note* **1974,** No. 653.
20. Leach, J. W.; Chappelear, P. S.; Leland, T. W. *Proc. Am. Pet. Inst.* **1966,** *46,* 223.

21. Mollerup, J. "The Computer Program LNG PROPERTY for the Calculation of the Thermodynamic Properties of Natural Gas and Petroleum Gas Mixtures," Instituttet for Kemiteknik, Danmarks Tekniske Hojskole, Denmark, 1977.
22. Brelvi, S. W.; O'Connell, J. P. *AIChE J.* **1975**, *21*, 1024.
23. Reid, R. C.; Prausnitz, J. M.; Sherwood, T. K. "The Properties of Gases and Liquids," 3rd ed.; McGraw-Hill: New York, 1977.

RECEIVED August 15, 1978.

19

The Nonanalytic Equation of State for Pure Fluids Applied to Propane

ROBERT D. GOODWIN

Thermophysical Properties Division, Center for Mechanical Engineering and Process Technology, National Bureau of Standards, Boulder, CO 80303

An isochoric equation is designed for computing thermodynamic functions of fluids. It has its origin on the liquid–vapor coexistence boundary, and it yields a maximum in isochoric specific heats at the critical point. Its basic structure is similar to that of the Beattie–Bridgeman equation. With only five least-squares coefficients, it describes a $P(\rho,T)$ *surface free of irregularities. A modified function in the equation is presented, for the problem of behavior in the limit of low densities, especially as required for integration of the thermodynamic equation of state, to obtain the change of internal energy along isotherms. Recently derived vapor pressures for propane at low temperatures also have been introduced. Constants are reported for all equations, as needed for computations on propane.*

An isochoric equation has been developed for computing thermodynamic functions of pure fluids. It has its origin on a given liquid–vapor coexistence boundary, and it is structured to be consistent with the known behavior of specific heats, especially about the critical point. The number of adjustable, least-squares coefficients has been minimized to avoid irregularities in the calculated $P(\rho,T)$ surface by using selected, temperature-dependent functions which are qualitatively consistent with isochores and specific heats over the entire surface. Several nonlinear parameters appear in these functions. Approximately fourteen additional constants appear in auxiliary equations, namely the vapor-pressure and orthobaric-densities equations, which provide the boundary for the $P(\rho,T)$ equation-of-state surface.

The range of validity of the present equation of state is for all fluid states at densities to the triple-point liquid density, temperatures from the triple-point to infinity, and pressures to at least 700 bar. Vapor pressures of the solid at temperatures below the triple point are replaced by an extrapolation of the vapor-pressure curve to absolute zero temperature, and a completely new type of formulation is presented for densities of saturated vapor from absolute zero to the critical-point temperature.

Some comments on accuracy are appropriate in this introduction. The single motivation for the present work is the fact that experimental $P\rho T$ data almost invariably are of much lower absolute accuracy than required for derived thermodynamic properties, such as specific heats. With an excessive number of adjustable, least-squares coefficients, one obtains a $P(\rho,T)$ surface with irregularities in the derivatives. At the critical point, e.g., the derivatives $\partial\rho/\partial T$ and $\partial\rho/\partial P$ become infinite, such that the slightest inconsistencies in temperature and/or pressure scales between different laboratories must yield gross deviations of densities from any calculated surface. For compressed liquid at low temperatures the derivative $\partial P/\partial\rho$ becomes extremely large, such that very small irregularities in the experimental densities yield deviations of pressure greatly exceeding the accuracy of pressure measurements. In this domain, any equation of state should be used only to find $\rho(P,T)$.

We have attempted to depart from the fruitless procedure of using more least-squares coefficients to obtain a best possible "fit" of a mass of $P\rho T$ data, which do not have the accuracy needed to define an equation of state consistent with the known behavior of specific heats. The present type of equation of state is more highly constrained than any previously known, and therefore serves as a reliable smoothing and interpolation function by which means the inaccuracies and inconsistencies of various experimental data may be observed and intercompared. We prefer to replace the question "How accurate is the equation of state?" by the question "How accurate and consistent are the experimental data used here?" In present work, the answer is that, generally, densities are within a few tenths of one percent over the entire fluid domain (except for the critical region where they are much larger, but are within experimental uncertainties) (*see* Tables II and III).

An advantage of this equation is that specific heat data need not be included in the least-squares determination of coefficients in order to obtain acceptable agreement with those data. The present equation thus may be used to estimate specific heats for some domains of the $P(\rho,T)$ surface in the absence of data other than those for ideal gas states.

Disadvantages of this equation include: (a) lengthy time of development for each substance; (b) it cannot be integrated analytically (numerical integrations are performed for each thermodynamic value); and (c) it is not defined inside the liquid–vapor coexistence envelope.

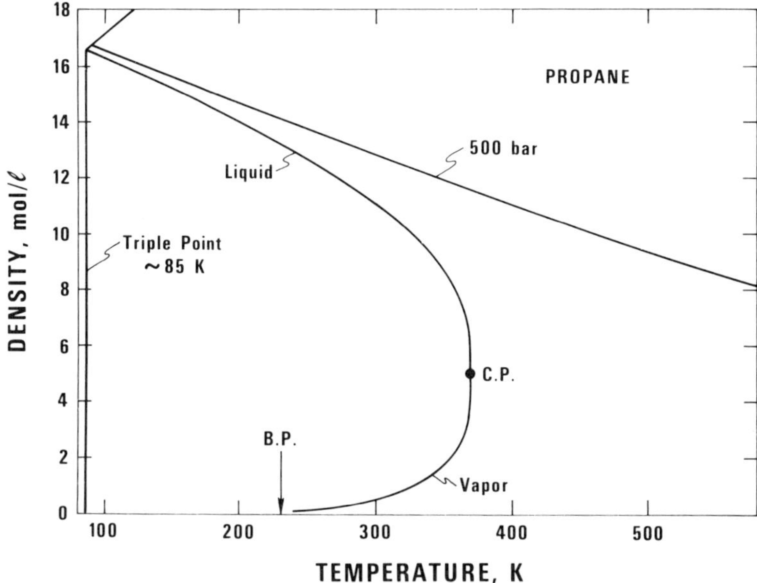

National Bureau of Standards (U.S.), Interagency Report

Figure 1. Density-temperature diagram of propane (5)

In this chapter, we shall at first give a full description of the logical development of the equation of state, necessarily repeating some published material (*1, 2, 3, 4, 5*). Parameters and coefficients for propane are given in the text, but the auxiliary equations (vapor pressures and orthobaric densities) are presented at the end of this chapter.

In the second part of this report we describe alternative functions for use in the equation of state, and comment on their merits.

Finally, we resolve the long-standing problem of behavior of this equation of state in the limit of low densities, which arises in the integration of the thermodynamic equation of state, to obtain the change of internal energy as a function of density along isotherms.

The density–temperature diagram for propane is given by Figure 1. The upper, left-hand corner of Figure 1 gives the freezing liquid line. Symbols and units are given in Appendix A, and fixed-point values used for propane are in Table I.

Table I. Fixed Points Used for Propane

	Triple Point	*Boiling Point*	*Critical Point*
Temperature (K)	85.47	231.0679	369.80
Pressure (bar)	$1.6609 \cdot 10^{-9}$	1.01325	42.3974
Density (mol/L)			
vapor	$2.3373 \cdot 10^{-10}$	0.05479	4.96
liquid	16.620	13.1687	4.96

Developing the Equation of State

Experimental specific heats, $C_v(\rho,T)$, are known to increase apparently without limit on the close approach to the critical point. This nonanalytic behavior influences a far greater portion of the $P(\rho,T)$ surface than generally is appreciated. The thermodynamic relation between specific heats and the equation of state along isotherms is

$$\Delta C_v(\rho,T) = -T \cdot \int_0^\rho (\partial^2 P/\partial T^2) \cdot d\rho/\rho^2 \qquad (1)$$

Figure 2 shows behavior of the curvatures of isochores as indicated by Rowlinson (*6*) for consistency with Equation 1.

Figure 3 shows the zero slope and curvature of the critical isotherm at the critical point, which are needed for thermodynamic consistency with the relations (*6*),

$$C_p(\rho,T) = C_v(\rho,T) + T \cdot (\partial P/\partial T)^2/(\partial P/\partial \rho)/\rho^2 \qquad (2)$$

$$W(\rho,T) = [C_p \cdot (\partial P/\partial \rho)/C_v]^{1/2} \qquad (3)$$

In our experience, a necessary but insufficient condition for a well-behaved critical isotherm is that, at the critical point, the slope of the critical isochore from the equation of state be equal to the slope of the vapor–pressure equation, $\partial P/\partial T = dP_\sigma/dT$. This constraint always is applied in the following work via the least-squares program (*7*).

To obtain isochores whose curvatures become very large, approaching the critical density along the critical isotherm, as required by Equation 1, we have designed an infinite curvature for isochores at an origin, $\theta(\rho)$,

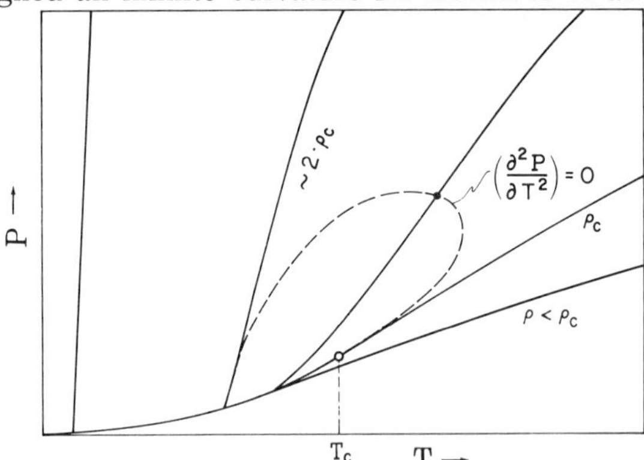

Journal of Research of the National Bureau of Standards

Figure 2. The locus of isochore inflection points (1)

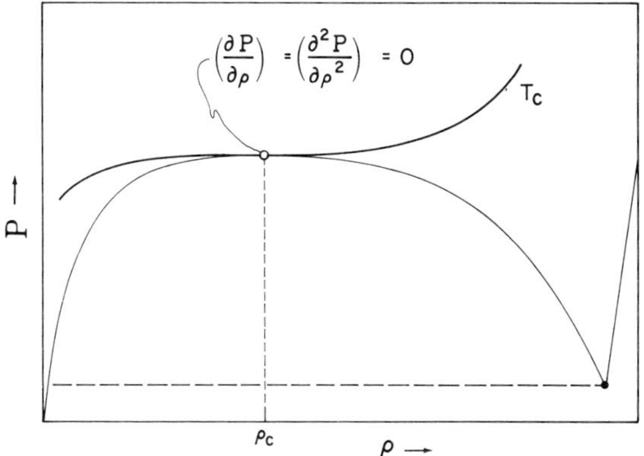

Journal of Research of the National Bureau of Standards

Figure 3. Behavior of the critical isotherm (1)

inside the coexistence envelope, (*see* Figure 4). The function giving this curvature shall have a density-dependent coefficient with a root at the critical density, such that the very large curvatures will change sign when integrating Equation 1 through the critical density, and the critical isochore will be characterized by $\partial^2 P/\partial T^2 = 0$ at the critical point.

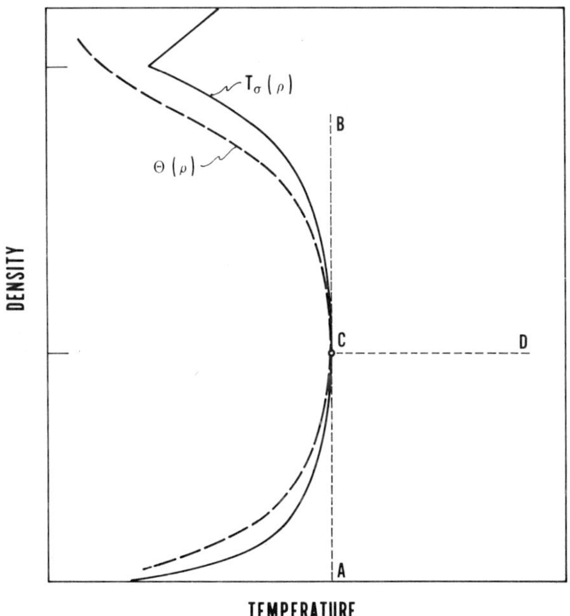

Journal of Research of the National Bureau of Standards

Figure 4. Behavior of the locus, $\Theta(\rho)$ (1)

With the above objectives in mind, we constrain the equation to the liquid–vapor coexistence boundary. For any density, the coexistence temperature, $T_\sigma(\rho)$, is obtained by iteration from equations for the orthobaric densities. Thus the vapor pressure, $P_\sigma[T_\sigma(\rho)]$, is a function of density. By subtraction

$$P - P_\sigma(\rho) = \rho R^* \cdot [T - T_\sigma(\rho)] + \rho^2 R^* T_c \cdot f(\rho,T) \tag{4}$$

Only two temperature-dependent functions (in addition to $\rho R^* T$) are needed to describe the sigmoid shape of isochores at $\rho_c < \rho < 2 \cdot \rho_c$ (*see* Figure 2)

$$f(\rho,T) \equiv B(\rho) \cdot \Phi(\rho,T) + C(\rho) \cdot \Psi(\rho,T) \tag{5}$$

where $B(\rho)$ and $C(\rho)$ are polynomial coefficients to be developed from $P\rho T$ data by least squares.

The function

$$\Phi(\rho,T) \equiv x^{1/2} \cdot \ln [T/T_\sigma(\rho)] \tag{6}$$

(*see* Figure 5) has the valuable property that $\partial^2\Phi/\partial T^2 = 0$ everywhere on the coexistence boundary at $T = T_\sigma(\rho)$. Its weak, negative curvature at very high temperatures corresponds to the decline of virial coefficients toward zero at these temperatures.

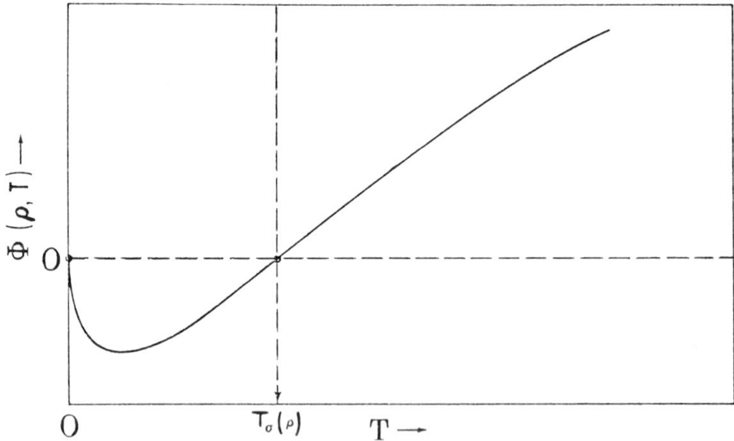

Journal of Research of the National Bureau of Standards

Figure 5. Behavior of the function, Φ(ρ,T) (1)

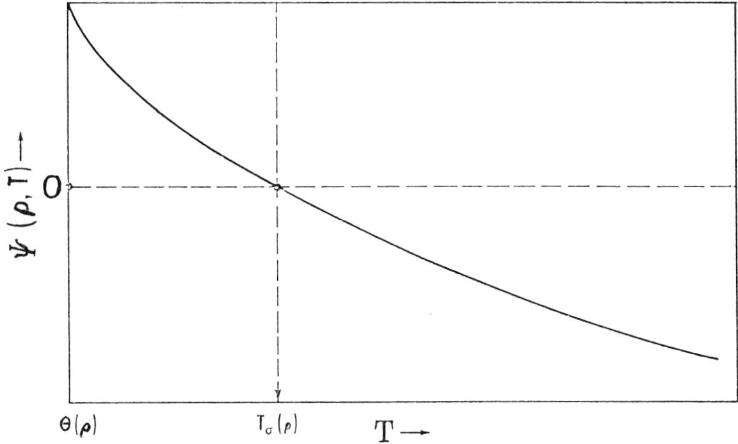

Journal of Research of the National Bureau of Standards

Figure 6. Behavior of the function, Ψ(ρ,T) (1)

The function $\Psi(\rho,T)$ (*see* Figure 6) yields a maximum in $C_v(\rho,T)$ at the critical point via Equations 1 and 4. Its origin is

$$\theta(\rho) \equiv T_\sigma(\rho) \cdot \exp\left[-\alpha \cdot \mathrm{f}(\rho)\right]$$
$$\mathrm{f}(\rho) \equiv |\rho - 1|^3/(\rho_t - 1)^3 \quad (7)$$

where ρ_t is the reduced density at the liquid triple point. The function $\omega(\rho,T) \equiv (1 - \theta(\rho)/T)$ is an argument for Equation 9 below. As Ψ must be zero at coexistence, it is defined as the difference

$$\Psi(\rho,T) \equiv \psi(\rho,T) - \psi_\sigma(\rho) \quad (8)$$

where $\psi_\sigma(\rho)$ is obtained from $\psi(\rho; T)$ merely by replacing T with $T_\sigma(\rho)$,

$$\psi(\rho,T) \equiv \delta \cdot \exp\left[\epsilon \cdot (1 - x)\right] + \left[1 - \omega + \omega \cdot \ln(\omega)\right] \quad (9)$$

The parameter, $0 \leq \delta \leq 1$, in Equation 9 is for relative weighting of the analytic and nonanalytic parts.

Table II summarizes the $P\rho T$ data used here, as detailed in Ref. 5. Earlier compilations on propane are found in Refs. 8 and 9. Parameters and coefficients for Equation 1, developed as described in Refs. 3 and 5 are

$$B(\rho) \equiv B_1 + B_2 \cdot \rho + B_3 \cdot \rho^2 + B_4 \cdot \rho^3 \quad (10)$$

$$C(\rho) \equiv C_1 \cdot (\rho - 1) \cdot (\rho - 2) \cdot \exp(-\gamma \cdot \rho^4) \quad (11)$$

$$\alpha = 1, \quad \gamma = 0.07, \quad \delta = 2/3, \quad \epsilon = 2$$

$B_1 =$ 0.2555 5013 087 $\qquad B_4 =$ 0.0961 8894 561

$B_2 =$ 0.8275 9684 245 $\qquad C_1 =$ −0.3681 8676 338

$B_3 =$ −0.2818 2856 341

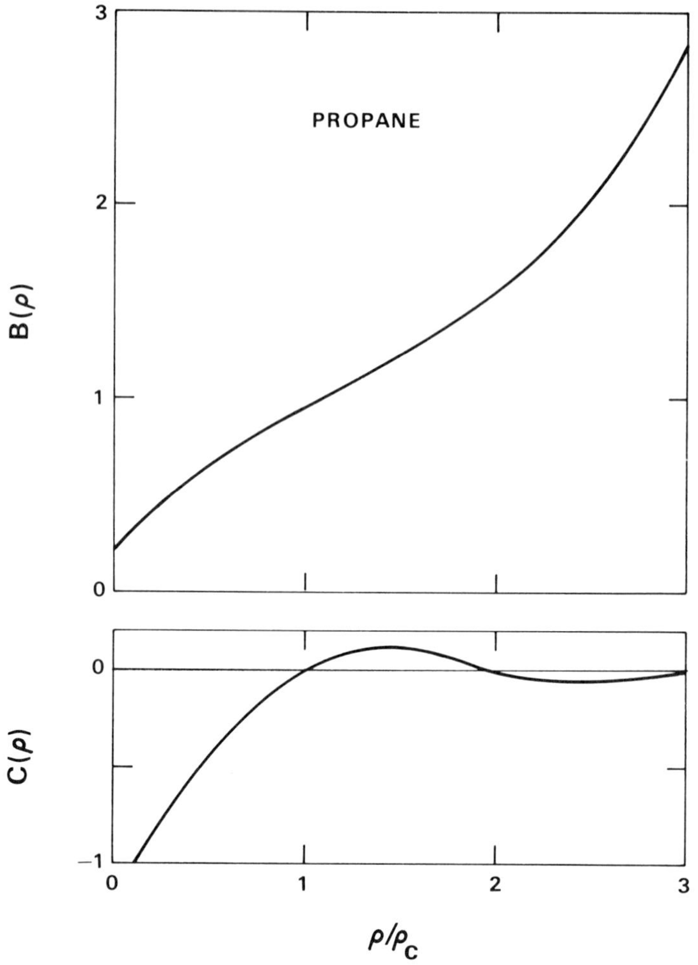

National Bureau of Standards (U.S.), Interagency Report

Figure 7. Behavior of coefficients B(ρ), *C*(ρ) *for propane (5)*

The behavior of $B(\rho)$ and $C(\rho)$ for propane is shown in Figure 7. The number of $P\rho T$ data used here for adjusting the equation of state is 843, with different least-squares weightings than in Refs. *3* and *5*. Overall deviations, with equal weighting for all points, are 2.07 bar for the mean of absolute pressure deviations and 0.34% for the rms of relative density deviations.

We always examine the behavior of the calculated critical isotherm in minute detail near the critical density ($\pm$ 10%). Small adjustments in values of parameters (including the critical density) may be made to eliminate any negative slopes $(\partial P/\partial \rho)_{T_c}$, as described in Ref. *3*.

Table II. Summary of $P\rho T$ Data for Propane

Authors	Range of the Data: d *(mol/L)*	T *(K)*	P *(bar)*	Deviations[a]: *NP*	$\Delta d/d$ *(rms %)*	*Mean* ΔP *(bar)*
Goodwin *(5)*	0.30	290–700	6 – 17	42	0.04	0.004
Beattie *(12)*	1.0 –10.0	370–548	23 – 310	110	1.39	1.00
Cherney *(13)*	0.04– 2.6	323–398	11 – 50	25	0.37	0.038
Dawson *(14)*	0.02– 0.07	243–348	0.9– 1.8	18	0.14	0.002
Deschner *(15)*	0.03– 9.5	303–609	1.0– 142	236	1.93	1.04
Dittmar *(16)*	7.3 –13.4	273–413	10 –1035	336	0.41	3.02
Ely *(17)*	11.5 –14.8	166–322	2.5– 428	222	0.06	2.54
Reamer *(18)*	0.02–13.0	311–511	1.0– 690	306	0.42	1.42
Tomlinson *(19)*	10.3 –12.0	277–327	20 – 137	40	0.07	0.94

[a] Using Equations 7 and 8 for $B(\rho)$, $C(\rho)$.

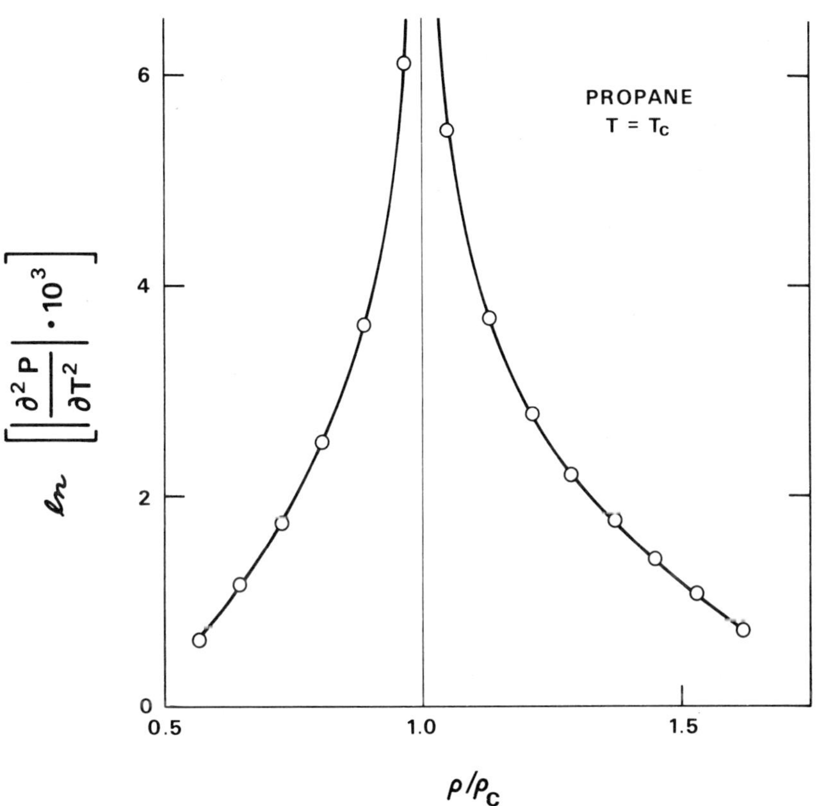

National Bureau of Standards (U.S.), Interagency Report

Figure 8. Behavior of $\partial^2P/\partial T^2$ *on the critical isotherm* (5)

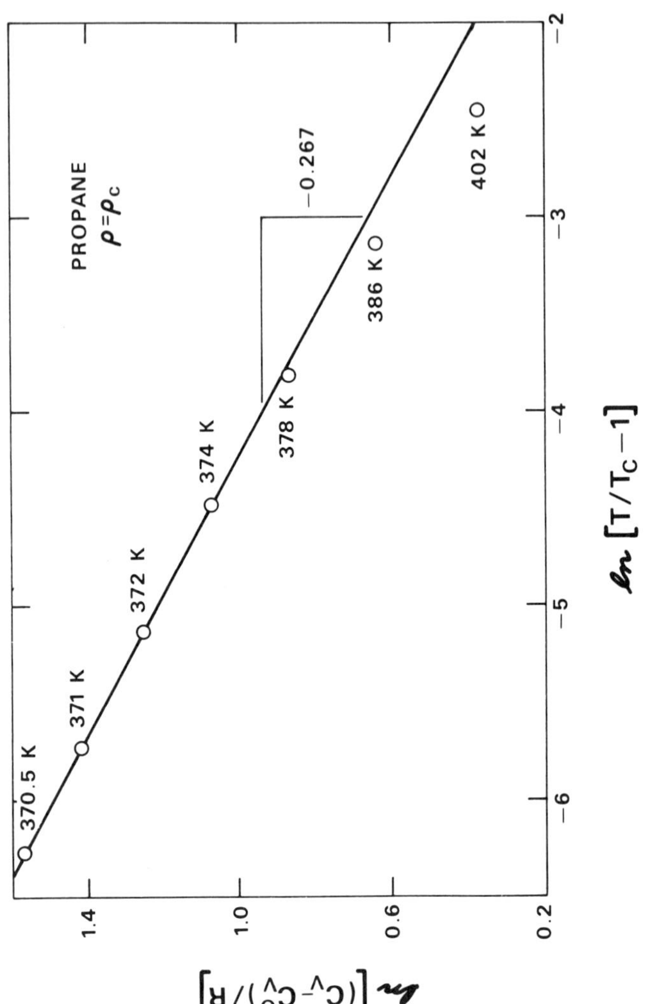

National Bureau of Standards (U.S.), Interagency Report

Figure 9. Behavior of $C_v(T)$ *on the critical isochore* (5)

The nonanalytic character of Equation 1 is seen clearly in Figure 8, which is a logarithmic plot of absolute values of isochore curvatures $|\partial^2 P/\partial T^2|$ along the critical isotherm. The values of $\partial^2 P/\partial T^2$ are negative at $\rho < \rho_c$. The linear behavior of calculated specific heats along the critical isochore, in coordinates of Figure 9, and close to the critical temperature, is in agreement with experimental behavior, as discussed in Ref. 3.

Modified Functions

The following modifications are presented in this separate section because they may be less generally applicable for a range of different substances than equations given above. It is valuable to recognize that different functions can be used to accomplish essentially the same constraints on the $P\rho T$ surface, and we wish to give these options for investigation in future work. Some modifications may give a better overall fit of $P\rho T$ data than do others, while at the same time giving a less satisfactory representation in a particular domain of the surface.

The function $\Phi(\rho,T)$ of Equation 6 approaches infinity as $\rho \to 0$, $(T_\sigma(\rho) \to 0)$. This brings in the question of its validity for use in the thermodynamic equation of state. At very low densities, however, the coexistence temperature diminishes roughly only as the logarithm of density, $1/T_\sigma(\rho) \sim \ln(1/\rho)$, and hence this behavior must yield a finite integral from the thermodynamic equation of state.

An alternative function for $\Phi(\rho,T)$ has not been used for propane because it gave a slightly less satisfactory representation of PρT data than Equation 6. It has been used successfully however for ethylene (*10*). This modified form is finite at $\rho = 0$,

$$\Phi(\rho,T) \equiv x^\beta \cdot \exp\left[b \cdot (1 - T_\sigma/T)\right] - x_\sigma{}^\beta \tag{12}$$

The value $\beta = 3/4$ yields weak, negative isochore curvatures at high temperatures. For an inflection at $T = T_\sigma(\rho)$ it is necessary that

$$b \equiv (1 - \beta) + (1 - \beta)^{1/2}$$

The following two more highly constrained modifications of the equation state may be useful in future work. We at first removed the second root from $C(\rho)$, allowing it, instead, to diminish strongly toward zero at high densities,

$$C(\rho) \equiv C_1 \cdot (\rho - 1) \cdot \exp(-\gamma \cdot \rho^4) \tag{13}$$

$$\alpha = 1, \qquad \gamma = 0.30, \qquad \delta = 2/3, \qquad \epsilon = 2$$

$B_1 =$	0.3153 3388 360	$B_4 =$	0.0752 9552 029
$B_2 =$	0.6834 4896 212	$C_1 =$	0.6652 5273 915
$B_3 =$	−0.2765 7101 050		

Overall deviations are $\overline{\Delta P} = 2.20$ bar, and the rms of $\Delta d/d = 0.39\%$. Deviations for individual authors appear in the first part of Table III (which is a continuation of Table II).

We next removed the possibility for an inflection in $B(\rho)$, demanding ethane-like behavior (*2, 4*),

$$B(\rho) \equiv B_1 + B_2 \cdot \exp(\eta \cdot \rho) \tag{14}$$

With the new coefficients, Equations 13 and 14,

$$\alpha = 1, \quad \gamma = 0.30, \quad \delta = 2/3, \quad \epsilon = 2, \quad \eta = 0.65$$

$$B_1 = 0.1679\ 3192\ 483 \qquad C_1 = 0.4921\ 7046\ 074$$

$$B_2 = 0.3808\ 7177\ 242$$

Overall deviations are $\overline{\Delta P} = 2.02$ bar and the rms of $\Delta d/d = 0.33\%$. Deviations of the data of individual authors appear in the second part of Table III.

Some comments are necessary concerning the different formulations investigated here for $B(\rho)$ and $C(\rho)$, namely Equations 10, 11, 13, and 14.

Prior to the investigation of propane, the molecularly simpler substances investigated apparently have required only a quadratic for $B(\rho)$, (compare with Equation 10). For the reader planning similar work on simple substances, therefore, Equation 14 might well be used initially.

With the exception of hydrogen (*1*), the second root in $C(\rho)$, Equation 11, not only seems unnatural, but it leads both to redundancy at $\rho > 2$, and in some cases to isochores having derivatives $\partial^3 P/\partial T^3$ which are irregular. Starting any new work with Equation 13, therefore, could be advantageous.

Table III. Deviations with Alternate Coefficients

	Equations for B(ρ), C(ρ)			
	Equations 7 and 9		*Equations 10 and 9*	
Authors	$\Delta d/d$ (*rms %*)	*Mean* ΔP (*bar*)	$\Delta d/d$ (*rms %*)	*Mean* ΔP (*bar*)
Goodwin (*5*)	0.05	0.003	0.19	0.017
Beattie (*12*)	1.07	1.13	1.14	0.94
Cherney (*13*)	0.27	0.03	0.55	0.075
Dawson (*14*)	0.17	0.002	0.22	0.002
Deschner (*15*)	1.70	0.82	1.80	0.92
Dittmar (*16*)	0.40	3.03	0.35	2.86
Ely (*17*)	0.06	2.60	0.06	2.26
Reamer (*18*)	0.54	1.76	0.43	1.60
Tomlinson (*19*)	0.07	1.04	0.07	1.00

Behavior at Very Low Densities

Placing Equation 4 in the thermodynamic equation of state yields a leading term as follows

$$\Delta E = \int_0^\rho [(Z_\sigma(\rho) - 1) \cdot RT_\sigma(\rho)/\rho + \ldots] \cdot d\rho \qquad (15)$$

where ρ is the reduced density. At $\rho < 1$, Z_σ is the compressibility factor for saturated vapor,

$$Z_\sigma(\rho) \equiv P_\sigma(\rho)/[\rho \cdot R^* \cdot T_\sigma(\rho)]$$

An initial problem with Equation 15 is loss of significant figures as $Z_\sigma \rightarrow 1$. The more serious problem is that if independent equations are used for the vapor pressures, $P_\sigma[T_\sigma(\rho)]$, and for the saturated vapor densities, $\rho_\sigma(T)$, we find that in the limit $\rho \rightarrow 0$ (and hence $T_\sigma \rightarrow 0$), Z_σ may approach values from zero to infinity, depending on the formulations for P_σ and ρ_σ.

This difficulty arises only at extremely low densities, because under these conditions the temperature diminishes merely as a logarithmic function of density, $1/T \sim \ln(1/\rho)$. Our numerical integrations for thermodynamic properties were performed such that the equation of state was not called at these extremely low densities (*4,5*).

A solution for the problems mentioned above is to replace conventional formulations of the saturated vapor densities by a formulation of the compressibility factor, $Z_\sigma(T)$, for saturated vapor (*see* Saturated Vapor Densities). By using this new formulation for substitutions, Equation 15 can be transformed to

$$\Delta E = \int_0^\rho [(Z_c - 1) \cdot (Z_\sigma/Z_c) \cdot RT_c \cdot f(x_\sigma) + \ldots] \cdot d\rho \qquad (16)$$

where Z_c is the value of the compressibility factor at the critical point, $Z_\sigma \equiv Z_\sigma[T_\sigma(\rho)]$ is the compressibility factor for saturated vapor, $x_\sigma \equiv T_\sigma/T_c$, and $f(x_\sigma)$ is finite in the range $0 \leq x_\sigma \leq 1$. For any given density, $\rho \leq 1$, the value of $T_\sigma(\rho)$ is obtained by iteration from Equation 19 for $Z_\sigma(T)$.

Auxiliary Equations

Vapor Pressures. Approaching the triple point, propane vapor pressures become immeasurably small. From Ref. 5 we derived new data from the triple point to the boiling point by thermal loops in a procedure related to that recently presented by Yarbrough and Tsai (*11*). Our new data now have been used in an equation of the same form as that for ethane (*2, 4*). The arguments are defined as

$$x(T) \equiv (T - T_t)/(T_c - T_t) \qquad u(T) \equiv (1 - T_t/T)/(1 - T_t/T_c)$$

when the equation (for P in bar) is

$$\ln (P) = a + b \cdot u + c \cdot x + d \cdot x^2 + e \cdot x^3 + f \cdot x \cdot (1 - x)^{\epsilon} \quad (17)$$

where $\epsilon = 1.30$ and

$a = -20.2158\ 8450$	$d = 7.3102\ 3303$
$b = 29.0773\ 3466$	$e = -1.8342\ 9015$
$c = -10.5903\ 0558$	$f = 0.7545\ 5673$

The rms relative deviation for 104 equally weighted pressures is 0.21%. The exponent, $\epsilon = 1.30$, was adjusted for a best fit of $P\rho T$ data, because Equation 17 is not highly sensitive to this exponent.

Saturated Liquid Densities. Precise and consistent data are available from the triple point to the critical point (*5*). The variables are defined as

$$x(T) \equiv (T_c - T)/(T_c - T_t); \qquad y(\rho) \equiv (d - d_c)/(d_t - d_c)$$

Propane-saturated liquid densities are described by

$$y = x + (x^{\epsilon} - x) \cdot (\mathrm{a} + \mathrm{b} \cdot x^2 + \mathrm{c} \cdot x^3) \quad (18)$$

$\epsilon = 0.35$	$b = -0.1695\ 1563$
$a = 0.7760\ 0858$	$c = 0.0818\ 6846$

For 83 equally weighted densities, the relative deviation is 0.044%, comparable with the accuracy of the data.

Saturated Vapor Densities. We formulate the compressibility factor for saturated vapor as a function of temperature by using the vapor-pressure equation. Subscripts are omitted because we refer always to saturated vapor and to the vapor pressure. We define the constant, $A_0 \equiv Z_c - 1$, where Z_c is the value of the compressibility factor at the critical point, and the variables

$$\Pi \equiv P/P_c, \qquad x \equiv T/T_c, \qquad u \equiv 1 - x,$$

when the equation for saturated vapor densities, $d \equiv P/[Z \cdot R \cdot T]$, is given by

$$Z = 1 + A_0 \cdot \Pi \cdot x^{-2} \cdot f(x) \tag{19}$$

$$f(x) = 1 + A_1 \cdot u^{\epsilon} + \sum_{i=2}^{5} A_i \cdot u^{i-1}$$

$\epsilon = 0.38$	$A_3 = 1.466\ 673$
$A_1 = -0.962\ 9549$	$A_4 = -6.459\ 775$
$A_2 = 1.090\ 712$	$A_5 = 11.837\ 60$

Data at low pressures were estimated from the present vapor-pressure equation and the virial equation from Ref. 5. The rms relative deviation for 30 equally weighted, saturated-vapor densities is 0.17%. For most substances, the vapor pressures of the solid are extremely small; hence we assume that differences at $T < T_t$ will be negligible for present purposes.

Iteration for Coexisting Densities. Orthobaric densities near the critical point generally cannot be obtained accurately from isochoric $P\rho T$ data by extrapolation to the vapor-pressure curve because the isochore curvatures become extremely large near the critical point. The present, nonanalytic equation of state, however, can be used to estimate these densities by a simple, iterative procedure. Assume that nonlinear parameters in the equation of state have been estimated in preliminary work. For data along a given experimental isochore (density), it is necessary merely to find the coexistence temperature, $T_\sigma(\rho)$, by trial (iteration) for a best, least-squares fit of these data.

Conclusion

In previous reports we have shown graphically the behavior of functions $\Phi(\rho,T)$, $\Psi(\rho,T)$, $B(\rho)$, and $C(\rho)$ for Equation 4. In Ref. 3 we illustrated nonanalytic behavior in relation to the maximum in specific heats at the critical point. In this chapter we have given a solution for the long-standing problem of behavior in the limit of low densities, namely, a completely new type of formulation for the saturated-vapor densities, which extrapolates to $Z_\sigma = 1$ at $\rho = 0$, $T_\sigma = 0$.

Utility of the present type of equation of state for tabulating thermodynamic properties has been demonstrated in major NBS publications on methane, ethane, and propane. For readers accustomed to BWR-type equations, with their attendant difficulties, the programming of the present equation, including numerical integrations, probably is no more complicated, and may be logically much simpler.

Glossary of Symbols

c, t = critical and liquid triple points
σ = liquid–vapor coexistence
o = ideal gas states
$\alpha, \beta, \gamma, \delta, \epsilon, \eta$ = nonlinear parameters in the equation of state
$B(\rho), C(\rho)$ = density-dependent coefficients in Equation 1
$C_v(\rho,T)$ = isochoric specific heat, J/mol/K
$C_p(\rho,T)$ = isobaric specific heat, J/mol/K
d = density, mol/L
$E(\rho,T)$ = internal energy, J/mol
J = joule, 1 N · m
L = liter, $10^{-3}m^3$
mol = 44.09721 grams of propane (C^{12} scale)
P = pressure in bars, 1 bar $\equiv 10^5 N/m^2$, (1 atm = 1.01325 bar)
$P_\sigma(T)$ = vapor pressure, bar
$P_\sigma(\rho) = P_\sigma[T_\sigma(\rho)]$ for Equation 4
$\Phi(\rho,T)$ = defined function for Equation 4
$\Psi(\rho,T)$ = defined function for Equation 4
R = gas constant, 0.0831434 (bar–L/mol)/K
R^* = $(0.0831434) \cdot d_c$ bar/K for Equation 4
$\rho = d/d_c$, reduced density
$\rho_t = d_t/d_c$, reduced density at the liquid triple point
T = temperature, K
$T_\sigma(\rho)$ = liquid–vapor coexistence temperature
$\theta(\rho)$ = defined locus of temperatures, Equation 7
$\omega(\rho,T) = [1 - \theta(\rho)/T]$, for Equation 9
$W(\rho,T)$ = speed of sound
$x(T) = T/T_c$, reduced temperature for Equation 4
$x_\sigma(\rho) = T_\sigma(\rho)/T_c$, reduced coexistence temperature
$Z(P,\rho,T) = P/[d \cdot R \cdot T]$, the "compressibility factor"

Acknowledgment

We are indebted to the American Gas Association, Inc., for their generous support, without which this work might not have been accomplished. R. D. McCarty contributed the essential least-squares program, providing for constraints. Discussions, suggestions, references, and experimental data from other members of this research group have contributed substantially to make possible the developments reported here.

Literature Cited

1. Goodwin, R. D. "Equation of State for Thermodynamic Properties of Fluids," *J. Res. Natl. Bur. Stand., Sect. A* **1975,** *79,* 71.
2. Goodwin, R. D. "An Equation of State for Thermodynamic Properties of Pure Fluids," In "The Proceedings of the Fifth Biennial International CODATA Conference"; Dreyfus, B., Ed.; Pergamon: Oxford, **1976;** pp. 441–444.
3. Goodwin, R. D. "On the Nonanalytic Equation of State for Propane," *Adv. Cryog. Eng.* **1977,** *23,* Paper M-8, 611–618.
4. Goodwin, R. D.; Roder, H. M.; Straty, G. C. "Thermophysical Properties of Ethane, from 90 to 600 K at Pressures to 700 bar," *Natl. Bur. Stand. (U.S.), Tech. Note* **1976,** *684.*
5. Goodwin, R. D. "Provisional Thermodynamic Functions of Propane, from 85 to 700 K at Pressures to 700 bar," *Natl. Bur. Stand. (U.S.), Interagency Rept.,* **1977,** NBSIR 77-860.
6. Rowlinson, J. S. *"Liquids and Liquid Mixtures,"* Butterworths Scientific Publications: London, 1959; p. 98.
7. McCarty, R. D. "Least-Squares Computer Subroutine," unpublished data.
8. Eubank, P. T.; Das, T. R.; Reed, Jr., C. O. "Thermodynamic Properties of Propane, Part II: PVT Surface and Corresponding Thermodynamic Properties," *Adv. Cryog. Eng.* **1973,** *18,* 220.
9. Kuloor, N. R.; Newitt, D. M.; Bateman, J. S. In "Thermodynamic Functions of Gases"; Din, F., Ed.; Butterworths Scientific Publications: London, 1956; Vol. 2.
10. Goodwin, R. D. "On the Nonanalytic Equation of State for Ethylene," unpublished data.
11. Yarbrough, D. W.; Tsai, C.-H. "Vapor Pressures and Heats of Vaporization for Propane and Propene from 50 K to the Normal Boiling Point," submitted for publication in *Adv. Cryog. Eng.*
12. Beattie, J. A.; Kay, W. C.; Kaminsky, J. "The Compressibility of, and an Equation of State for Gaseous Propane," *J. Am. Chem. Soc.* **1937,** *59,* 1589.
13. Cherney, B. J.; Marchman, H.; York, R. J. "Equipment for Compressibility Measurements," *Ind. Eng. Chem.* **1949,** *41,* 2653.
14. Dawson, P. P.; McKetta, J. J. "Zs for Propane and Methylacetylene," *Pet. Refiner* **1960,** *39*(4), 151.
15. Deschner, W. W.; Brown, G. G. "P-V-T Relations for Propane," *Ind. Eng. Chem.* **1940,** *32*(6), 836.
16. Dittmar, P.; Schulz, F.; Strese, G. "Druck/Dichte/Temperatur-Werte für Propan und Propylen," *Chem.–Ing.–Tech.* **1962,** *34*(6), 437.
17. Ely, J. F.; Kobayashi, R. "Isochoric PVT Measurements for Compressed Liquid Propane," personal communication.
18. Reamer, H. H.; Sage, B. H.; Lacey, W. M. "Phase Equilibria in Hydrocarbon Systems: Volumetric Behavior of Propane," *Ind. Eng. Chem.* **1964,** *14*(3), 3662.
19. Tomlinson, J. R. "Liquid Densities of Ethane, Propane, and Ethane–Propane Mixtures"; Technical Publication TP-1, Natural Gas Processors Assoc.: Tulsa, OK, 1971.

RECEIVED August 18, 1978.

Table IV. Numerical

Density	T_{sat}	$f(x_\sigma)$	Z_{sat}
.1000E − 02	164.868	1.30236	.99848
.3000E − 02	178.349	1.15794	.99625
.5000E − 02	185.554	1.09453	.99434
.7000E − 02	190.688	1.05445	.99258
.9000E − 02	194.746	1.02551	.99093
.1100E − 01	198.133	1.00306	.98936
.1300E − 01	201.060	.98486	.98785
.1500E − 01	203.648	.96963	.98639
.1700E − 01	205.975	.95659	.98498
.1900E − 01	208.095	.94523	.98360
.2100E − 01	210.046	.93520	.98226
.2300E − 01	211.856	.92625	.98095
.2500E − 01	213.546	.91817	.97966
.2700E − 01	215.133	.91084	.97840
.2900E − 01	216.631	.90413	.97715
.3100E − 01	218.050	.89796	.97593
.3300E − 01	219.400	.89225	.97473
.3500E − 01	220.687	.88695	.97355
.3700E − 01	221.918	.88200	.97238
.3900E − 01	223.099	.87738	.97123
.4100E − 01	224.233	.87303	.97009
.4300E − 01	225.325	.86894	.96896
.4500E − 01	226.379	.86508	.96785
.4700E − 01	227.397	.86143	.96675
.4900E − 01	228.382	.85796	.96566

[a] Numerical integration for ΔE of propane at 231/K and up to a density of .050 mol/L.

Addendum

The behavior of Equation 4 at low densities is important. This equation may be written as

$$Z(\rho,T) = 1 + [Z_\sigma(\rho) - 1] \cdot T_\sigma(\rho)/T + (\rho/x) \cdot f(\rho,T) \qquad (20)$$

where $Z_\sigma(\rho) \rightarrow 1$ as $\rho \rightarrow 0$ (*see* Equation 19) and the last term vanishes; following are two examples of computations which start at zero density.

Table IV presents a step-by-step numerical integration of Equation 16 for propane at $T = 231$ K up to a density of 0.05 mol/L. The first column gives density at the midpoint of the 0.002 mol/L interval.

Table V presents fugacity coefficients, f/P, at $T = 260$ K for gaseous and liquid propane up to $P = 700$ bar, computed via

$$\ln (f/P) = \int_0^P (Z - 1) \cdot dP/P \qquad (21)$$

by using Equation 19 at $\rho < \rho_c$.

Integration for ΔE[a]

Z	$f(\rho,T)$	$T \cdot \partial f(\rho,T)/\partial T$	ΔE
.99917	.7882E + 00	.2116E + 01	−.5823E + 01
.99769	.5998E + 00	.2127E + 01	−.1142E + 02
.99627	.5060E + 00	.2136E + 01	−.1694E + 02
.99487	.4417E + 00	.2144E + 01	−.2241E + 02
.99349	.3924E + 00	.2150E + 01	−.2785E + 02
.99212	.3522E + 00	.2157E + 01	−.3328E + 02
.99076	.3181E + 00	.2163E + 01	−.3869E + 02
.98940	.2885E + 00	.2169E + 01	−.4410E + 02
.98804	.2623E + 00	.2174E + 01	−.4949E + 02
.98669	.2387E + 00	.2180E + 01	−.5489E + 02
.98534	.2172E + 00	.2185E + 01	−.6082E + 02
.98399	.1975E + 00	.2191E + 01	−.6567E + 02
.98264	.1793E + 00	.2196E + 01	−.7106E + 02
.98129	.1623E + 00	.2202E + 01	−.7645E + 02
.97995	.1464E + 00	.2207E + 01	−.8184E + 02
.97860	.1315E + 00	.2212E + 01	−.8723E + 02
.97725	.1174E + 00	.2218E + 01	−.9263E + 02
.97590	.1041E + 00	.2223E + 01	−.9803E + 02
.97456	.9138E − 01	.2229E + 01	−.1034E + 03
.97321	.7929E − 01	.2234E + 01	−.1088E + 03
.97186	.6773E − 01	.2240E + 01	−.1143E + 03
.97051	.5667E − 01	.2245E + 01	−.1197E + 03
.96916	.4604E − 01	.2251E + 01	−.1251E + 03
.96781	.3582E − 01	.2257E + 01	−.1305E + 03
.96646	.2598E − 01	.2263E + 01	−.1360E + 03

Table V. Fugacity Coefficients[a]

Pressure, bar	*Density, mol/L*	Z	*ln (f/P)*	f/P
.10000E + 00	.46388E − 02	.99723	−.00298	.997021
.50000E + 00	.23430E − 01	.98718	−.01348	.986606
.10133E + 01	.48113E − 01	.97420	−.02657	.973778
.15000E + 01	.72161E − 01	.96158	−.03900	.961755
.20000E + 01	.97568E − 01	.94824	−.05187	.949457
.30000E + 01	.15079E + 00	.92032	−.07807	.924899
.31082E + 01	.15677E + 00	.91719	−.08095	.922240
.31082E + 01	.12375E + 02	.01162	−.08095	.922240
.40000E + 01	.12378E + 02	.01495	−.32984	.719036
.50000E + 01	.12382E + 02	.01868	−.54924	.577388
.60000E + 01	.12386E + 02	.02241	−.72782	.482959
.70000E + 01	.12390E + 02	.02614	−.87824	.415515
.80000E + 01	.12394E + 02	.02986	−1.00803	.364936
.10000E + 02	.12402E + 02	.03730	−1.22370	.294139
.12000E + 02	.12409E + 02	.04473	−1.39856	.246952
.14000E + 02	.12417E + 02	.05216	−1.54526	.213257

Table V. Continued

Pressure, bar	*Density, mol/L*	Z	*ln (f/P)*	*f/P*
.16000E + 02	.12425E + 02	.05957	−1.67134	.187995
.18000E + 02	.12432E + 02	.06698	−1.78168	.168356
.20000E + 02	.12440E + 02	.07437	−1.87959	.152652
.24000E + 02	.12454E + 02	.08914	−2.04705	.129116
.28000E + 02	.12469E + 02	.10388	−2.18634	.112327
.32000E + 02	.12484E + 02	.11858	−2.30504	.099754
.36000E + 02	.12498E + 02	.13325	−2.40801	.089994
.40000E + 02	.12512E + 02	.14788	−2.49857	.082202
.41000E + 02	.12516E + 02	.15154	−2.51957	.080494
.42000E + 02	.12520E + 02	.15519	−2.53997	.078869
.43000E + 02	.12523E + 02	.15884	−2.55981	.077320
.44000E + 02	.12527E + 02	.16249	−2.57910	.075842
.46000E + 02	.12534E + 02	.16978	−2.61617	.073082
.48000E + 02	.12541E + 02	.17706	−2.65135	.070556
.50000E + 02	.12547E + 02	.18434	−2.68480	.068235
.52000E + 02	.12554E + 02	.19161	−2.71665	.066096
.55000E + 02	.12565E + 02	.20249	−2.76169	.063185
.60000E + 02	.12582E + 02	.22060	−2.83030	.058995
.70000E + 02	.12615E + 02	.25669	−2.94773	.052459
.80000E + 02	.12647E + 02	.29261	−3.04464	.047614
.90000E + 02	.12679E + 02	.32836	−3.12589	.043898
.10000E + 03	.12710E + 02	.36396	−3.19481	.040974
.11000E + 03	.12740E + 02	.39940	−3.25377	.038628
.12000E + 03	.12770E + 02	.43469	−3.30451	.036717
.13000E + 03	.12799E + 02	.46985	−3.34837	.035142
.14000E + 03	.12828E + 02	.50486	−3.38638	.033831
.16000E + 03	.12883E + 02	.57451	−3.44794	.031811
.18000E + 03	.12937E + 02	.64366	−3.49406	.030377
.20000E + 03	.12988E + 02	.71234	−3.52804	.029362
.22000E + 03	.13038E + 02	.78059	−3.55225	.028660
.25000E + 03	.13109E + 02	.88219	−3.57393	.028045
.30000E + 03	.13221E + 02	1.04965	−3.58057	.027860
.35000E + 03	.13326E + 02	1.21500	−3.56047	.028425
.40000E + 03	.13423E + 02	1.37846	−3.52107	.029568
.45000E + 03	.13515E + 02	1.54021	−3.46714	.031206
.50000E + 03	.13602E + 02	1.70042	−3.40191	.033309
.55000E + 03	.13685E + 02	1.85920	−3.32770	.035876
.60000E + 03	.13763E + 02	2.01666	−3.24618	.038923
.65000E + 03	.13838E + 02	2.17292	−3.15862	.042484
.70000E +03	.13909E + 02	2.32805	−3.06602	.046606

[a] Propane fugacity coefficients at 260 K.

20

An Equation for Liquid–Vapor Saturation Densities as a Function of Pressure

PHILIP A. THOMPSON[1]

Max–Planck-Institut für Strömungsforschung, D 34 Göttingen, Federal Republic of Germany

The explicit formula $|\rho^{\lambda}_{r} - 1| = (1 - P_r)^{\beta}$ *for reduced saturation density as a function of reduced pressure is proposed for the entire liquid–vapor saturation boundary. The expression* $\lambda \sim 1$ *depends on* P_r*;* $\beta \sim 0.35$ *depends weakly on* P_r*, corresponding at* $P_r = 1$ *to the critical exponent* β_c*. The parameters* λ *and* β *can be related to the Pitzer factor* ω*. Special cases include the power law* $|\rho_r - 1| = C(1 - T_r)^{\beta_c}$ *. . . and the low-pressure vapor equation* $\rho_r{}^{\lambda}{}_0 = \beta_0 P_r$*. The function* $\lambda - \lambda_c = g(P_r)$ *is found from data to be a universal function for nonpolar substances. If* λ_c *is correlated with* ω*, the formula takes on the corresponding-states form* $\rho_r = \rho_r(P_r, \omega)$*. This form predicted the density of saturated liquid and vapor with 0.4% and 0.9% accuracy, respectively, for 38 substances.*

The smooth curve passing through the critical point and bounding the two-phase liquid–vapor region in a pressure–volume diagram is familiar to every student of thermodynamics. The mathematical description $\rho(P)$ of this coexistence curve or saturation boundary is the subject of this chapter.

Descriptions in terms of temperature rather than pressure are well known. The notable equations of Guggenheim (*1*) are, for liquid and vapor, respectively,

$$\rho_{r(1)} = 1 + \frac{3}{4}(1 - T_r) + \frac{7}{4}(1 - T_r)^{1/3} \qquad (1)$$

[1] Present address: Department of Mechanical Engineering, Rensselaer Polytechnic Institute, Troy, NY 12181.

0-8412-0500-0/79/33-182-365$05.00/1

and

$$\rho_{r(v)} = 1 + \frac{3}{4}(1 - T_r) - \frac{7}{4}(1 - T_r)^{1/3} \quad (2)$$

where the subscript r denotes a reduced property, e.g., $T_r = T/T_c$. These equations incorporate the law of the rectilinear diameter $\rho_{(l)} + \rho_{(v)} \propto T$ and the near-critical power law $\rho_{(l)} - \rho_{(v)} \propto (T_c - T)^{1/3}$. In the latter result, the exponent 1/3 now is referred to as the critical exponent β_c. In the interest of just attribution, it may be noted that the latter result, described by Guggenheim as new, was given much earlier by others (*2, 3, 4, 5*).

Examples of modern substance–specific density–temperature relations are found in Goodwin (*6*) and in Pentermann and Wagner (*7*).

The equation $\rho(P)$ for the saturation boundary presented in this chapter has the form

$$|\rho_r^{\lambda} - 1| = (1 - P_r)^{\beta} \quad (3)$$

where both $\lambda \sim 1$ and $\beta \sim 0.35$ are weak functions of pressure. This form includes as special cases the near-critical temperature-form power law,

$$|\rho_r - 1| = A(1 - T_r)^{\beta_c} + \ldots \quad (4)$$

the low-pressure expression for the vapor,

$$\rho^{\lambda}{}_{r(v)} = \beta_0 P_r + \ldots \quad (5)$$

and the low-pressure expression for the liquid

$$\rho^{\lambda}{}_{r(l)} = 2 - \beta_0 P_r + \ldots \quad (6)$$

(These special forms will be discussed in the following section.) The basic Equation 3 is a fortuitous result, rather than the product of an orderly development program. Accordingly, the appearance of P (rather than T) as independent variable does not represent a deliberate choice. However, it would be a defensible choice on three grounds: (a) by application of a vapor-pressure equation, normally accurately known, pressure and temperature are interchangeable; (b) pressure is a natural physical constraint in both static and dynamic problems—the idea of a boiling point is a simple example; and (c) there appears to be an inherent advantage, manifested in Equation 5, in that $\rho_r \sim P_r$ on the vapor side (on the liquid side $\rho_r \sim 1$ and there is little advantage either way).

Guggenheim Equations 1 and 2 are in corresponding-states form: that is, the reduced-form equations $\rho_r(T_r)$ explicitly contain no material constants and are formally applicable to all substances (in practice, to substances having small nearly spherical molecules, with modest accuracy). The desirable corresponding-states formulation will be retained in the equation presented here in a modified form, by including (as is now usual) the Pitzer–Curl acentric factor ω as an explicit material constant:

$$\omega \equiv -\log_{10} P_{r(s)}\,(0.7) - 1 \tag{7}$$

Then Equation 3 takes the form $\rho_r = \rho_r(\omega, P_r)$ and will be conveniently applicable to nonpolar or weakly polar substances.

The aim of the equation development is the accurate and reasonably simple description of the entire saturation boundary, with a clear representation of the corresponding-states principle.

Development of the Equation

The basic Equation 3 becomes explicit as soon as the parameters $\beta(P_r)$ and $\lambda(P_r)$ are specified. This specification is the result of a lengthy trial procedure, based on the following factors: (a) consistency with known near-critical power laws, (b) approximate consistency with the law of the rectilinear diameter; (c) the tendency of the low-pressure vapor curve to form a straight line in logarithmic coordinates, as predicted by Equation 5; (d) imposition of a definite form of the corresponding-states principle; and (e) consistency with a large collection of experimental data.

The numerical value of the exponent β proves to be quite close to that of the critical exponent $\beta_c \sim 0.35$ over the entire range of pressures. Generally the exponent λ takes on different values on the vapor and liquid branches of the saturation boundary. The numerical range of λ easily can be estimated, using the critical point, the low-pressure vapor, and the low-pressure liquid as reference points. Near the critical point, the vapor-pressure equation is

$$\ln P_r = -w_1\,(1 - T_r) + \ldots \tag{8}$$

where w_1 is a positive constant sometimes called the Riedel parameter. In conjunction with the power law (Equation 4) and the basic Equation 3 this yields the critical value for λ:

$$\lambda_c = w_1^{\beta_c}/A \tag{9}$$

Using the somewhat representative value $w_1 = 6.33$ from ethylene data (*8*) and the Guggenheim values $A = 7/4$ and $\beta_c = 1/3$ yields $\lambda_c \sim 1.06$. For vapor in the limit of zero pressure, application of a vapor-pressure equation such as $\ln P_r = w_1 (T_r - 1)/T_r$ and assuming that the compressibility factor $Z = 1$ yields, after calculation with Equation 5, the limiting value $\lambda = 1$. Similarly, assuming a limiting low-pressure liquid density such as the Van der Waals (VDW) value $\rho_r = 3$ yields with Equation 6 the limiting value $\lambda \sim 0.63$. (It turns out that the low-pressure vapor and liquid estimates are independent of the value of β.) Thus the value of λ is estimated to be in the range 0.6 to 1.1.

Fixing the Value of β. The exponent β is assumed to vary linearly between a high-pressure limit β_c (the critical exponent) and a low-pressure limit β_o according to

$$\beta = \beta_c + (\beta_o - \beta_c)(1 - P_r) \tag{10}$$

The material constants β_c and β_o will be related to ω. The value of $\beta_c(\omega)$ is taken from numerous experimental determinations (as shown in Figure 1) which are represented by the empirical equation

$$\beta_c = .340 + \frac{.023}{1 + e^{12(\omega - .1)}} \tag{11}$$

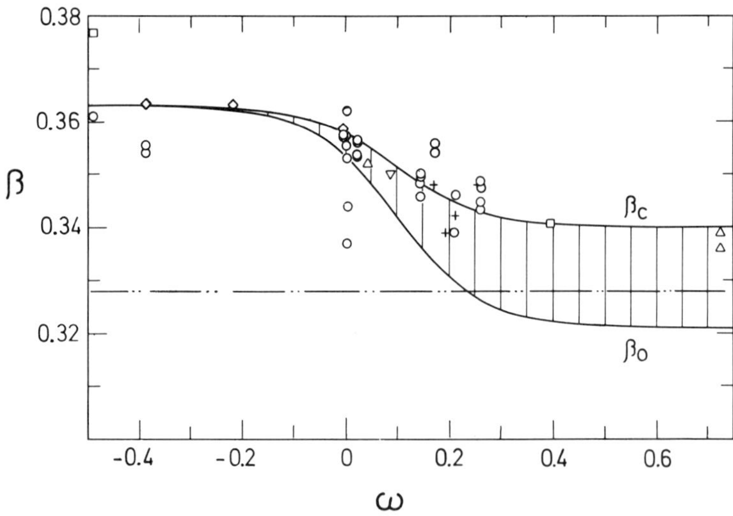

Figure 1. Material constants β_c and β_o correlated with ω. Data from various sources are shown for the scaling exponent β_c. (—), Equations 11 and 12; (— · · —), trend of β_c data from Refs. 9, 10, and 11. (□), 12; (▽), 13; (△), 14; (○), 15; (+), 16; and (◇), A.

(This is not an easy choice; several recent and somewhat controversial determinations put the value of β_c close to .328 (*9, 10, 11*).)

The value of $\beta_o(\omega)$ was determined initially from Equation 5, by fitting a straight line, in logarithmic coordinates, to low-pressure vapor data (this procedure also established corresponding low-pressure values for $\lambda_{(v)}$). Data conveniently were generated for this purpose by using accurate substance-specific vapor-pressure equations in conjunction with a two-term virial equation, using the second virial coefficient of Tsonopoulos (*17*). The initial results then were refined by comparing actual saturation-density data (without assuming any particular form for the density curve) to yield

$$\beta_o = .321 + \frac{.042}{1 + e^{12(\omega - .1)}} \tag{12}$$

The straight-line form (Equation 5) of the low-pressure pressure–density was found by Young (*18, 19*) and independently by Thompson and Sullivan (*15*) some 70 years later. It is interesting to compare Young's results for β_o with those from Equation 12, and his results for $\lambda_{(v)}$ with those found below. In Table I, Young's results are indicated by subscript Y and the results of this work by subscript T. Young's values of λ were taken directly from his work; his values of β were computed from his data, using modern values for P_c and ρ_c. The values $\lambda_T^{(v)}$ tabulated here will be defined in the following section.

Table I. Comparison with Young's Results for the Low-Pressure Vapor

Substance	β_Y^0	β_T^0	$\lambda_Y^{(v)}$	$\lambda_T^{(v)}$
Benzene	.299	.330	1.079	1.068
n-Pentane	.316	.327	1.070	1.064
n-Hexane	.307	.325	1.068	1.061
n-Heptane	.309	.323	1.058	1.056
n-Octane	.308	.322	1.058	1.053

Fixing the Value of λ. The function $\lambda(P_r)$ has two branches corresponding to saturated vapor and saturated liquid: these branches meet at the critical point where the common value of λ is designated as λ_c. The two branches can be calculated from experimental $\rho_r(P_r)$ data from Equation 3 rewritten in the form

$$\lambda = \frac{\ln\,(1 \mp (1 - P_r)^{\beta})}{\ln \rho_r} \tag{13}$$

where the upper and lower signs correspond to the vapor and liquid branches, respectively. From Equation 3, the vapor value $\lambda_{(v)}$ and liquid value $\lambda_{(l)}$ are related by

$$\rho_{r(v)}\,(P_r)^{\lambda_{(v)}(P_r)} + \rho_{r(l)}\,(P_r)^{\lambda_{(l)}(P_r)} = 2 \tag{14}$$

For convenience, the pressure scale is expanded by the substitute independent variable x,

$$x \equiv -\ln P_r \tag{15}$$

(Note that $x \approx 1 - P_r$ for $x << 1$.)

It was observed early in this study that the experimental curves $\lambda(x)$ for various substances tended to coincide if the curve for each substance were shifted in amplitude. This corresponds to the form

$$\lambda(x) = \lambda_c + g(x) \tag{16}$$

where λ_c is a material property and $g(x)$ is a universal function with two branches $g_{(v)}$ and $g_{(l)}$, where $g_{(v)}(0) = g_{(l)}(0) = 0$. Equation 16 is an explicit statement of the corresponding-states principle.

The quality of the approximation corresponding to Equation 16 could be improved by adjustments in the function $\beta_0(\omega)$, leading to Equation 12. The degree of coincidence obtained can be seen in Figure 2, which includes data for 34 substances shifted according to an optimum choice for λ_c. Thus the plot is an experimental representation of the function $g(x)$. The data investigated covered the range $0 < x < 18.9$ (larger values of x are not shown in the figure because of format limitations).

Considerable scatter in the neighborhood of the critical point is a result of the difficulty of measurement and the deviation amplifying effect of the function $\lambda(P_r)$ for P_r near unity. In the case of smoothed data, a smoothing β_c which departs from Equation 11 will lead to a singularity in λ at the critical point. Similarly, deviant critical properties P_c and V_c will lead to a singularity in λ at the critical point (for a detailed discussion, *see* Thompson (*20*)).

The analytical forms for the functions $g(x)$ are based on fits to the oxygen data of Weber (*21*). The form

$$g(x) = \frac{x^m}{A + Bx^n} \tag{17}$$

was well suited to both vapor and liquid sides. The values of the exponents m and n were found by trial optimization, as illustrated for

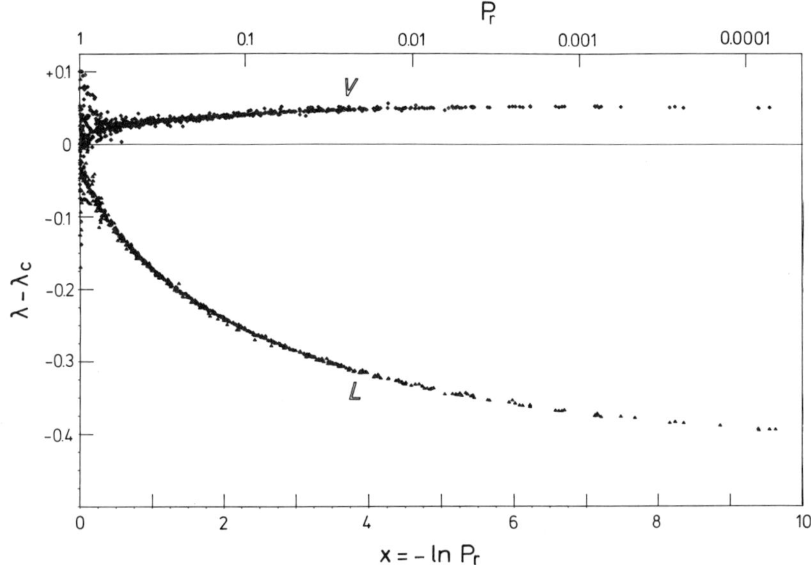

Figure 2. Universal plot of all $\lambda - \lambda_c$ data (x < 10) for the substances investigated. Labels V and L designate vapor and liquid branches, respectively. For quantum substances ($\omega < -0.1$), only vapor points are shown.

vapor in Figure 3. The vapor values $m = 0.35$ and $n = 2.00$, and the liquid values $m = 0.60$ and $n = 0.95$ were determined in this way. With coefficients and λ_c determined by least squares, the final results for $g(x)$ were

$$g_{(v)}(x) = \frac{x^{.35}}{31 + .14x^2} \tag{18}$$

and

$$g_{(l)}(x) = \frac{-x^{.6}}{5.25 + .54x^{.95}} - .0009\,(x-9)^{1/3} \tag{19}$$

where

$$(x-9) = \begin{cases} x-9 & x \geq 9 \\ 0 & x < 9 \end{cases}$$

The last term in Equation 19, 0 for $x < 9$, is a small ad hoc correction to improve the fit to liquid data extending to large values of x, i.e., for the substances ethane and propane (*22, 23*), and certain others which have extremely low (reduced) triple-point pressures.

The values of $\lambda_T^{(v)}$ shown in Table I were calculated from the maximum amplitude ($\approx .052$) in Equation 18.

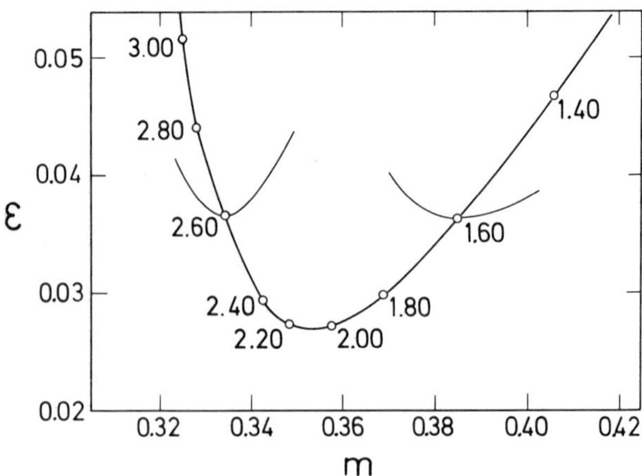

Figure 3. Average absolute-value error ε (in percent) in the computed λ for oxygen vapor (21) as a function of the exponent m, *for optimum exponent* n. *The numbers along the curve show the value of the optimum* n; $\lambda_c = 1.0313$.

A Near-Critical Series Development for λ(x). It is interesting to compare Equations 18 and 19 with the forms which would be required to satisfy a Guggenheim-type power law,

$$\rho_r = 1 \mp \frac{w_1^{\beta_c}}{\lambda_c} \tau^{\beta_c} + R^{\mp} \tau \tag{20}$$

where $\tau = 1 - T_r$ and the upper and lower signs consistently will designate vapor and liquid states, respectively. In conjunction with Equation 13 and a scaling form vapor-pressure equation,

$$x = w_1 \tau + w_2 \tau^{2-\Theta} \ldots \tag{21}$$

which includes the scaling exponent $\theta \sim 0.1$, this defines a series expansion for $\lambda(x)$ near $x = 0$. The result is, with β considered identical to β_c,

$$\lambda = \lambda_c \pm \frac{\lambda_c - 1}{2} x^{\beta} \pm \frac{(\lambda_c - 1)(4\lambda_c + 1)}{12\lambda_c} x^{2\beta}$$
$$\pm \frac{R^{\mp} \lambda_c^2}{w_1} x^{1-\beta} + \frac{\beta \lambda_c w_2}{w_1^{2-\Theta}} x^{1-\Theta}$$
$$\pm \frac{\beta \lambda_c w_2}{2w_1^{2-\Theta}} x^{1+\beta-\Theta} + \frac{\lambda_c}{2}\left(\frac{R^{\mp}\lambda_c}{w_1} - \beta\right) x \ldots \tag{22}$$

It is apparent that many of the above powers of x are not represented in Equations 18 and 19. In particular, the leading term x^{β} is not represented on the liquid side, although it manifests itself very distinctly on the vapor side, as shown by the exponent optimization (Figure 3). The rectilinear diameter term $x^{1-\beta}$, on the other hand, is represented explicitly only on the liquid side. Remarkably, still more powers of x would appear in Equation 22 if a modern form of the density expansion, consistent with the renormalization theory (*24*) had been used instead of the Guggenheim form.

An experimental test of the Guggenheim form (Equation 20) is shown in Figure 4, where the numerical value of $\lambda(x)$ required to exactly reproduce the power law (omitting the linear, rectilinear diameter term) is compared with experimental data for ethylene.

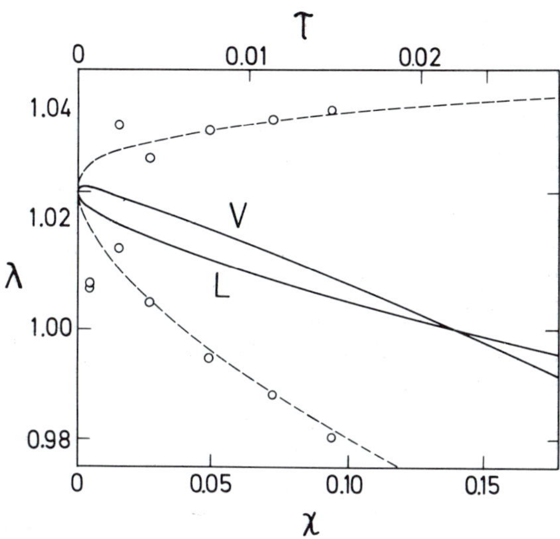

Figure 4. The λ(x) required to reproduce the power law 20 for ethylene (—). Experimental data are from Ref. 13. (- - -), Equations 18 and 19. Vapor pressure data are from Ref. 8; the assumed constants are $\beta_c = 0.35$ and $\lambda_c = 1.025$.

Computed Results

With λ defined by Equations 16, 18, and 19, and β defined by Equations 10, 11, and 12, the optimum values for λ_c were found for the 38 substances (46 data sets) listed in Table II by minimizing the average error in the predicted density for each substance.

The overall average error for all data sets was 0.50% on the vapor side and 0.23% on the liquid side. These calculations do not include liquid-side data for the quantum substances ($\omega < -0.1$), because they did not conform to the corresponding-states relation (Equation 19); however, vapor data seem to conform quite well.

Table II. Substance Data and Optimum λ_c[a]

Substance	N x_m	$\%_{(v)}$ $\%_{(l)}$	λ_c ω	β_c β_o	$P_c(bar)$ $T_c(K)$	$V_c(cm^3/g)$ Z_c
He^3	18	.40	1.2085	.3630	1.147	24.125
(25, A, 25)	.3	.40	—.490	.3630	3.310	.4025
He^4	8	.33	1.1328	.3629	2.274	14.360
(26, 27, 27)	3.8	.65	—.387	.3629	5.189	.3032
n-H_2	8	.32	1.0754	.3625	12.930	31.850
(28, 29, 29)	5.2	1.17	—.217	.3621	32.980	.3028
p-H_2	7	.55	1.0760	.3625	12.850	31.857
(28, 27, 27)	5.1	.14	—.218	.3621	32.935	.3014
Ar	35	.41	1.0334	.3579	48.650	1.8692
(30, 27, 27)	4.3	.10	—.005	.3537	150.725	.2899
Kr	8	.42	1.0322	.3578	54.931	1.0810
(31, 27, A)	4.0	.19	—.002	.3535	209.286	.2860
Xe	12	.09	1.0322	.3575	57.800	.9009
(32, A, 27)	4.1	.21	.003	.3530	289.734	.2838
O_2	26	.12	1.0313	.3565	50.430	2.2925
(21, 27, 27)	10.4	.03	.022	.3512	154.580	.2878
O_2	21	.07	1.0314	.3565	50.430	2.2925
(33, 27, 27)	6.0	.04	.022	.3512	154.580	.2878
O_2 (v)	71	.27	1.0316	.3565	50.430	2.2925
(7, 27, 27, 34)	7.1	—	.022	.3512	154.580	.2878
O_2 (l)	99	—	1.0306	.3565	50.430	2.2925
(7, 27, 27, 34)	8.5	.07	.022	.3512	154.580	.2878
N_2	13	.34	1.0315	.3556	33.980	3.1857
(35, 27, 27)	5.3	.23	.038	.3495	126.240	.2889
N_2O	13	1.40	1.0165	.3483	72.060	2.2084
(36, 37, A)	4.3	.19	.147	.3362	309.540	.2722
CO_2	16	.97	1.0164	.3429	73.753	2.1377
(38, 27 27)	2.7	.38	.260	.3264	304.127	.2744
$CClF_3$	33	.41	1.0211	.3468	38.700	1.7254
(39, 39, A)	7.2	.14	.172	.3335	302.100	.2777
CCl_2F_2	20	.38	1.0195	.3465	41.240	1.7750
(40, 41, A)	9.4	.31	.177	.3329	384.950	.2765
CCl_4	21	.44	1.0192	.3456	46.200	1.7920
(36, A, 36)	3.7	.22	.194	.3313	556.300	.2932
C_3F_8	13	.86	1.0224	.3414	26.800	1.5975
(42, 42, A)	5.0	.37	.325	.3236	345.050	.2806
$C_3F_5H_3$	13	.22	1.0099	.3418	31.500	2.0137
(43, A, A)	4.5	.13	.304	.3243	380.120	.2690

Table II. Continued

Substance	N x_m	$\%_{(v)}$ $\%_{(l)}$	λ_c ω	β_c β_o	P_c(bar) T_c(K)	V_c(cm^3/g) Z_c
C_6F_5H (44, 45, 44, 45)	14 6.2	— .26	1.0002 .374	.3408 .3225	35.310 530.960	1.9320 .2597
C_6F_6 (44, 46, 44, 47)	14 5.9	— .18	.9982 .396	.3406 .3222	32.732 516.670	1.8169 .2576
C_6F_6 (46, 46, 44)	6 .3	.30 1.08	.9982 .396	.3406 .3222	32.732 516.670	1.8169 .2576
$C_7F_5H_3$ (44, 45, 44, 45)	14 7.8	— .32	.9966 .416	.3405 .3219	31.260 566.520	2.1119 .2552
CH_4 (48, 27, 27)	18 6.0	.68 .37	1.0305 .007	.3573 .3526	45.950 190.555	6.1463 .2860
CH_4 (l) (23, 27, 27)	11 4.4	— .03	1.0326 .007	.3573 .3526	45.950 190.555	6.1463 .2860
C_2H_4 (13, 13, 13)	14 1.1	.10 .14	1.0214 .086	.3525 .3438	50.420 282.350	4.6690 .2813
C_2H_4 (49, 13, 13)	14 5.2	.47 .40	1.0234 .086	.3525 .3438	50.420 282.350	4.6690 .2813
C_2H_6 (50, 22, 51)	12 1.4	.11 .19	1.0196 .091	.3521 .3431	48.714 305.330	4.8477 .2797
C_2H_6 (22, 22, 51)	22 15.3	.70 .07	1.0235 .091	.3521 .3432	48.714 305.330	4.8477 .2797
C_2H_6 (l) (23, 22, 51, 22)	23 13.0	— .06	1.0240 .091	.3521 .3432	48.714 305.330	4.8477 .2797
C_3H_6 (49, 49, A)	17 5.1	.25 .21	1.0197 .143	.3486 .3367	46.130 364.900	4.3430 .2779
C_3H_6 (52, 52, A)	22 2.2	.37 .31	1.0222 .214	.3447 .3295	55.790 398.300	3.8312 .2716
C_3H_8 (53, 53, 53)	26 3.7	.40 .18	1.0194 .153	.3480 .3355	42.497 369.800	4.5356 .2764
C_3H_8 (l) (23, 53, 53, 54)	16 18.9	— .08	1.0193 .145	.3484 .3364	42.497 369.820	4.5356 .2764
iso-C_4H_8 (55, A, A)	29 4.1	.62 .21	1.0131 .197	.3455 .3310	39.800 418.000	4.2392 .2724
n-C_4H_{10} (53, 56, A)	11 3.6	.62 .21	1.0143 .199	.3454 .3308	37.970 425.160	4.3952 .2744
iso-C_4H_{10} (56, A, A)	16 3.6	1.60 .44	1.0166 .182	.3463 .3324	36.300 408.130	4.4221 .2750
n-C_5H_{12} (36, 58, A)	17 3.4	.84 .23	1.0119 .243	.3435 .3274	33.690 469.650	4.3144 .2686

Table II. Continued

Substance	N / X_m	$\%_{(v)}$ / $\%_{(l)}$	λ_c / ω	β_c / β_o	$P_c(bar)$ / $T_c(K)$	$V_c(cm^3/g)$ / Z_c
n-C_5H_{12}	28	.41	1.0122	.3432	33.690	4.3144
(58, 58, A)	3.5	.68	.252	.3268	469.650	.2686
iso-C_5H_{12}	18	.50	1.0177	.3441	34.110	4.2704
(59, 59, A)	5.4	.18	.227	.3285	461.000	.2742
C_5H_{12}	19	.08	1.0196	.3455	31.530	4.3280
(60, A, A)	1.7	.15	.196	.3311	433.750	.2730
C_6H_6	20	1.02	1.0085	.3447	48.980	3.3112
(36, 41, 41)	3.6	.11	.213	.3296	562.090	.2711
n-C_6H_{14}	16	.72	1.0081	.3420	30.150	4.2940
(36, A, A)	3.4	.07	.296	.3247	507.400	.2491
n-C_7H_{16}	17	.50	1.0014	.3411	27.360	4.3097
(36, 41, 41)	3.3	.17	.350	.3230	540.200	.2631
n-C_8H_{18}	29	.61	1.0011	.3407	25.000	4.2886
(61, A, A)	7.2	.12	.394	.3222	568.760	.2590
n-$C_{10}H_{22}$	7	.76	.9976	.3402	21.600	4.3493
(62, A, A)	6.7	.20	.490	.3214	617.400	.2604
Average Values:		.50				
		.23				

[a] The given citations listed beneath the compound name refer respectively to: density data, P_c, V_c, and vapor pressure (where conversion from temperature to pressure is required). The citation A refers to a value by the author.

In some cases (*see* Table II), the critical volume was adjusted by the author such that an independent fit of vapor and liquid data yielded the same value of λ_c. In a few cases, also noted in the table, the critical pressure was adjusted so as to avoid a sharp singularity in λ_c (*20*). A representative case of adjusted critical properties is n-octane:

	$P_c(bar)$	$V_c(cm^3/g)$
Data Source	24.97	4.2540
Author	25.00	4.2886

In every case where several experimental determinations of P_c and v_c were available, the adjusted values fell well within the field of the measured values. Of course, one can take the viewpoint that critical properties adjusted in this way are actually pseudocritical properties which are relevant only to the model given here. In any case, it is quite easy to calculate critical properties from the model, e.g., from low-pressure liquid data (*2*); *see also* Ref. 63.

The optimum values λ_c from Table II are plotted in Figure 5 as a function of ω (values for $\omega < 0.02$ are not shown). The fitted functions $\lambda_c(\omega)$ are

$$\lambda_c = 1.033 - .082\,\omega \qquad (\omega \geq 0) \tag{23}$$

and

$$\lambda_c = 1.033 - .192\,\omega - 2.9\,\omega^5 \qquad (\omega < 0) \tag{24}$$

With these equations, the corresponding-states form $\rho_r(\omega, P_r)$ is complete. This correlation form yields an average error for the predicted density of 0.85% on the vapor side and 0.37% on the liquid side for the substances shown in Table II. Much of this increased error stems from the data for perfluoropropane and *n*-decane.

A plot of the saturation boundary in pressure–volume coordinates is shown in Figure 6 for three different values of ω. On the scale of this plot, no deviation of the curves (calculated from the corresponding-states model) from the data can be detected.

A reasonably accurate representation of the entire vapor boundary is obtained with a constant β, constant λ model with

$$\beta = \beta_o(\omega), \qquad \lambda = \lambda_c(\omega) + .05 \tag{25}$$

This value of λ corresponds approximately to the slope of the low-pressure, nearly straight-line curve in the P, V diagram. Tested on oxygen data in

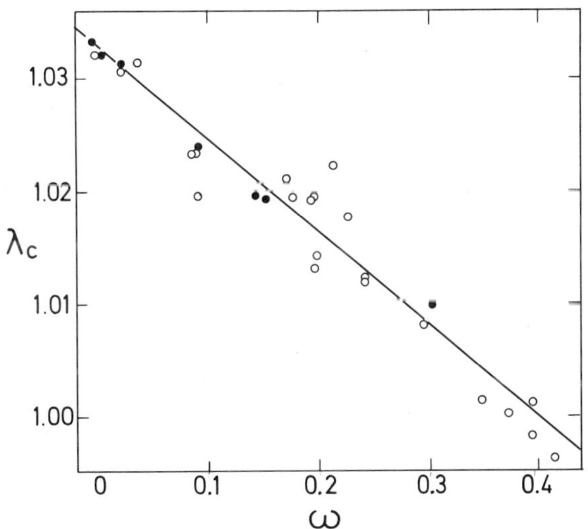

Figure 5. Optimum values of λ_c as a function of ω, with the fit given by Equations 23 and 24. Only values with $\omega > -.02$ are shown.

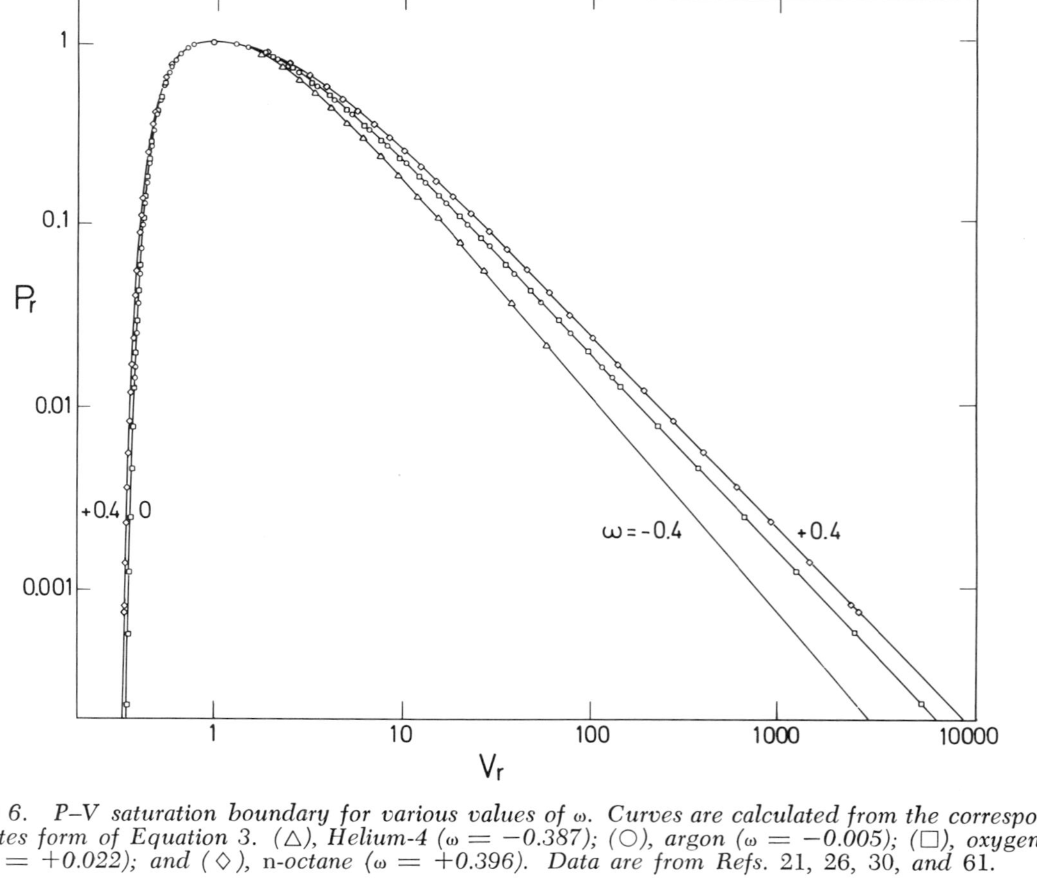

Figure 6. P–V saturation boundary for various values of ω. Curves are calculated from the corresponding-states form of Equation 3. (△), *Helium-4* ($\omega = -0.387$); (○), *argon* ($\omega = -0.005$); (□), *oxygen* ($\omega = +0.022$); *and* (◇), n-*octane* ($\omega = +0.396$). *Data are from Refs.* 21, 26, 30, *and* 61.

Equation 3, this simple model yielded the following average percent errors (right-hand column):

Data Source	*Optimum*	*Correlation*	*Constant* β, λ
Weber (*21*)	.12	.13	1.52
Wagner (*7*)	.27	.29	1.25

Discussion

The basic Equation 3 seems to be successful in predicting the saturation densities and providing a clear statement of the corresponding-states principle. However, vexatious questions about the model remain, particularly for the near-critical region. The existence of a unique critical exponent β_c, which approaches a stationary value in the neighborhood of the critical point, is called into question by recent experiments (*9, 10, 11*). The complexity of modern critical-region theoretical models (*24*) might even suggest that no fully consistent closed-form algebraic description of the type attempted here is possible. However, it is well to keep in mind that current theoretical descriptions are themselves approximate models of the physical world.

The available data have a direct influence on the numerical constants derived here. The data used were of highly variable quality (*20*) and a more critical data selection process (*7*) would be desirable. The critical properties P_c, and especially V_c, have great influence and small accuracy. The accuracy problem is in part caused by the difficulty of measurement.

Glossary of Symbols

Notation

A = coefficient in Equations 1 and 2
C = coefficient
g = universal function in Equation 16
m, n = exponents in Equation 17
N = number of data points
P = pressure
R^+, R^- = coefficients in Equation 20
T = absolute temperature
V = ρ^{-1}, specific volume
w_1, w_2 = coefficients in vapor–pressure equation
$x \equiv -\ln P_r$
$Z \equiv PV/RT$, compressibility factor
e = base of natural log

Greek

β = exponent in Equation 3
ϵ = absolute percent error
θ = scaling exponent
λ = exponent in Equation 3
$\rho = V^{-1}$, density
$\tau \equiv 1 - T_r$
ω = Pitzer–Curl acentric factor

Subscripts and Superscripts

c = critical state
l = liquid (saturation) state
r = reduced
rl = reduced saturated liquid state
rv = reduced saturated vapor state
s = saturation state
T = Thompson
v = vapor (saturation) state
Y = Young
o = zero-pressure condition

Acknowledgment

The author is grateful to D. A. Sullivan, H. Vogel, G. E. A. Meier, W. Wagner, J. Straub, and D. Rathmann for helpful discussions and assistance of various kinds. This work would not have been possible without the support of E.–A. Müller and the Max–Planck–Gesellschaft.

Literature Cited

1. Guggenheim, E. A. "The Principle of Corresponding States," *J. Chem. Phys.* **1945**, *13*, 253–261.
2. Ambrose, D., personal communication, 1977.
3. Verschaffelt, J. "Measurements on Capillary Ascension of Liquefied Carbonic Acid near the Critical Temperature," *Commun. Phys. Lab. Univ. Leiden* **1896**, *28*, 3–16.
4. Goldhammer, D. A. "Studien über die Theorie der übereinstimmenden Zustände," *Z. Phys. Chem.* **1910**, *71*, 577–623.
5. von Jüptner, H. "Verdampfungsstudien VII," *Z. Phys. Chem.* **1913**, *85*, 1–61.
6. Goodwin, R. D. "Estimation of Critical Constants T_c, ρ_c from $\rho(T)$ and $T(\rho)$ Relations at Coexistence," *J. Res. Natl. Bur. Stand.* **1970**, *74A2*, 221–227.
7. Pentermann, W.; Wagner, W. "New Pressure-Density-Temperature Measurements and New Rational Equations for the Saturated Liquid and Vapour Densities of Oxygen," submitted for publication in *J. Chem. Thermodyn.*

8. Hastings, J. R.; Levelt Sengers, J. M. H. "Vapor Pressure, Critical Pressure and Critical Isochore of Ethylene," Seventh Symposium on Thermophysical Properties, 1977.
9. Levelt Sengers, A.; Hocken, R.; Sengers, Jan V. "Critical-Point Universality and Fluids," *Phys. Today* **1977**, *30*(12), 42–51.
10. Ley-Koo, M.; Green, M. S. "Revised and Extended Scaling for Coexisting Densities of SF_6," *Phys. Rev. A* **1977**, *16*, 2483–2487.
11. Hayes, C. E.; Carr, H. Y. "NMR Measurement of the Liquid–Vapor Critical Exponents β and β_1," *Phys. Rev. Lett.* **1977**, *39*, 1558–1561.
12. Thompson, Philip A.; Sullivan, Daniel A. "A Simple Formula for Saturated Vapor Volume," submitted for publication in *Ind. Eng. Chem. Fundam.*
13. Douslin, D. R.; Harrison, R. H. "Pressure, Volume, Temperature Relations of Ethylene," *J. Chem. Thermodyn.* **1976**, *8*, 301–330.
14. Hall, K. R.; Eubank, P. T. "Empirical Description of the Liquid–Vapor Critical Region Based upon Coexistence Data," *Ind. Eng. Chem. Fundam.* **1976**, *15*, 323–330.
15. Levelt Sengers, J. M. H. "From Van der Waals' Equation to the Scaling Laws," *Physica* **1974**, *73*, 73–106.
16. Rathjen, W.; Straub, J. In "Heat Transfer in Boiling," Hahne, E.; Grigull, U., Eds.; Academic: New York, 1977.
17. Tsonopoulos, Constantine. "An Empirical Correlation of Second Virial Coefficients," *AIChE J.* **1974**, *20*, 263–272.
18. Young, Sydney. "Densités des Vapeurs Saturantes et Valeurs du Coefficient de Variation par Millimètre de Mercure sous la Pression Normale," *J. Phys. (Paris)* **1909**, *8*, 5–16.
19. Young, S. "The Specific Volumes of the Saturated Vapors of Pure Substances," *Z. Phys. Chem.* **1910**, *70*, 620–626.
20. Thompson, P. A. "An explicit Formula for the Liquid–Vapor Saturation Densities as a Function of Pressure," Max-Planck-Institut für Strömungsforschung, 1977, Internal Report 111.
21. Weber, L. A. "PVT, Thermodynamic and Related Properties of Oxygen from the Triple Point to 300 K at Pressures to 33 MN/m^2," *J. Res. Natl. Bur. Stand., Sect. A* **1970**, *74*, 93–129.
22. Goodwin, R. D.; Roder, H. M.; Straty, G. C. "Thermophysical Properties of Ethane, from 90 to 600 K at Pressures to 700 bar," *Nat. Bur. Stand. (U.S.), Tech Note* **1976**, *684*.
23. Haynes, M. W.; Hiza, M. J. "Measurements of the Orthobaric Liquid Densities of Methane, Ethane, Propane, Isobutane and Normal Butane," *J. Chem. Thermodyn.* **1977**, *9*, 179–187.
24. Wegener, F. J. In "Phase Transitions and Critical Phenomena," Domb, C.; Green, M. S., Eds.; Academic: New York, 1977; Vol. 6.
25. Wallace, B. J.; Meyer, H. "Equation of State of He-3 Close to the Critical Point," *Phys. Rev. A* **1970**, *2*, 1563–1575.
26. McCarty, R. D. "Thermodynamic Properties of Helium-4 from 2 to 1500 K at Pressures to 10^8 Pa.," *J. Phys. Chem. Ref. Data* **1973**, *2*, 923–1041.
27. Levelt Sengers, J. M. H.; Geer, W. L.; Sengers, J. V. "Scaled Equation of State Parameters for Gases in the Critical Region," *J. Phys. Chem. Ref. Data* **1976**, *5*, 46.
28. Vargaftik, N. B. "Tables of the Thermophysical Properties of Liquids and Gases," 2nd ed.; John Wiley and Sons, Inc.: New York, 1975.
29. Mathews, J. F. "The Critical Constants of Inorganic Substances," *Chem. Rev.* **1972**, *72*, 71–100.
30. Gosman, A. L.; McCarty, R. D.; Hust, J. G. "Thermodynamic Properties of Argon from the Triple Point to 300 K at Pressures to 100 Atmospheres," *Natl. Stand. Ref. Data Ser., Natl. Bur. Stand.* **1969**, *27*.
31. Streett, W. B.; Staveley, L. A. K. "Experimental Study of the Equation of State of Liquid Krypton," *J. Chem. Phys.* **1971**, *55*, 2495–2506.

32. Streett, W. B.; Sagan, L. S.; Staveley, L. A. K. "An Experimental Study of the Equation of State of Liquid Xenon," *J. Chem. Thermodyn.* **1973,** *5,* 633–650.
33. Roder, H. M.; Weber, L. A. "Thermophysical Properties," ASRDI Oxygen Technology Survey Vol. 1, NASA SP-3071, 1972.
34. Wagner, W.; Ewers, J.; Pentermann, W. "New Vapour-Pressure Measurements and a New Rational Vapour-Pressure Equation for Oxygen," *J. Chem. Thermodyn.* **1976,** *8,* 1049–1060.
35. Wagner, W. "A Method to Establish Equations of State Exactly Representing All Saturated State Variables Applied to Nitrogen," *Cryogenics* **1972,** *12,* 214–221.
36. Washburn, E. W., Ed. "International Critical Tables of Numerical Data, Physics, Chemistry and Technology"; McGraw-Hill: New York, 1928.
37. Cook, D. "The Vapour Pressure and Orthobaric Density of Nitrous Oxide," *Trans. Faraday Soc.* **1953,** *49,* 716–723.
38. Canjar, L. N.; Manning, F. S. "Thermodynamic Properties and Reduced Correlations for Gases," Gulf Publishing Co.: Houston, 1967.
39. Albright, L. F.; Martin, J. J. "Thermodynamic Properties of Chlorotrifluoromethane," *Ind. Eng. Chem.* **1952,** *44,* 188–198.
40. Perel'shtein, I. I. "Thermodynamic Properties of Freon-12 and Freon-13" In "Thermophys. Prop. of Matter and Substances"; Rabinovich, V. A., Ed.; National Bureau of Standards: Washington, D.C., 1975; Vol. 4.
41. Kudchadker, A. P.; Alani, G. H.; Zwolinski, B. J. "The Critical Constants of Organic Substances," *Chem. Rev.* **1968,** *68,* 659–735.
42. Fang, F.; Joffe, J. "Thermodynamic Properties of Perfluoropropane," *J. Chem. Eng. Data* **1966,** *11,* 376–379.
43. Shank, R. L. "Thermodynamic Properties of 1,1,1,2,2-Pentafluoropropane (Refrigerant 245)," *J. Chem. Eng. Data* **1967,** *12,* 474–480.
44. Hales, J. L.; Townsend, B. "Liquid Densities from 293 to 490 K of Eight Fluorinated Aromatic Compounds," *J. Chem. Thermodyn.* **1974,** *6,* 111–116.
45. Ambrose, D.; Sprake, C. H. S. "Thermodynamic Properties of Fluorine Compounds. Part X. Critical Properties and Vapour Pressure of Pentafluorobenzene, Chloropentafluorobenzene, 2,3,4,5,6-Pentafluorotoluene, and Pentafluorophenol," *J. Chem. Soc., Ser. A* **1971,** 1263–1266.
46. Douslin, D. R.; Harrison, R. H.; Moore, R. T. "Pressure-Volume-Temperature Relations of Hexafluorobenzene," *J. Chem. Thermodyn.* **1961,** *1,* 305–319.
47. Counsell, J. F.; Green, J. H. S.; Hales, J. L.; Martin, J. F. "Thermodynamic Properties of Fluorine Compounds. Part 2: Physical and Thermodynamic Properties of Hexafluorobenzene," *Trans. Faraday Soc.* **1965,** *61,* 212–218.
48. Goodwin, R. D. "The Thermophysical Properties of Methane from 90 to 500 K at Pressures to 700 bar," *Natl. Bur. Stand. (U.S.), Tech. Note* **1974,** *653.*
49. Bender, E. "Equations of State for Ethylene and Propylene," *Cryogenics* **1975,** *15,* 667–673.
50. Douslin, D. R.; Harrison, R. H. "Pressure-Volume-Temperature Relations of Ethane," *J. Chem. Thermodyn.* **1973,** *5,* 491–512.
51. Sliwinski, P. "The Lorenz-Lorentz Function of Gaseous and Liquid Ethane, Propane and Butane," *Z. Phys. Chemie,* Neue Folge **1969,** *63,* 263.
52. Lin, D. C.-K.; Silberberg, I. H.; McKetta, J. J. "Volumetric Behavior, Vapour Pressure, and Critical Properties of Cyclopropane," *J. Chem. Eng. Data* **1970,** *15,* 483.
53. Das, T. R.; Eubank, P. T. "Thermodynamic Properties of Propane: Vapour-Liquid Coexistence Curve," *Adv. Cryog. Eng.* **1973,** *18,* 208.

54. Goodwin, R. D., personal communication, 1977.
55. Barron, C. H.; Chaty, J. C.; Gonzales, R. A.; Eldridge, J. W. "Thermodynamic Properties of *n*-Butane," *J. Chem. Eng. Data* **1962,** *7,* 394–397.
56. Das, T. R.; Reed, C. O., Jr.; Eubank, P. T. "PVT Surface and Thermodynamic Properties of *n*-Butane," *J. Chem. Eng. Data* **1973,** *18,* 244–253.
57. Das, T. R.; Reed, C. O., Jr.; Eubank, P. T. "PVT Surface and Thermodynamic Properties of Isobutane," *J. Chem. Eng. Data* **1973,** *18,* 253.
58. Das, T. R.; Reed, C. O., Jr.; Eubank, P. T. "PVT Surface and Thermodynamic Properties of *n*-Pentane," *J. Chem. Eng. Data* **1977,** *22,* 3–8.
59. Arnold, E. W.; Liou, D. W.; Eldridge, J. W. "Thermodynamic Properties of Isopentane," *J. Chem. Eng. Data* **1965,** *10,* 88–92.
60. Dawson, P. P.; Silberberg, I. H.; McKetta, J. J. "Volumic Behaviour, Vapour Pressure, and Critical Properties of Neopentane," *J. Chem. Eng. Data* **1973,** *18,* 7–15.
61. Das, T. R.; Kuloor, N. R. "Thermodynamic Properties of Hydrocarbons: Part IV. *n*-Octane," *Indian J. Technol.* **1967,** *5,* 51–57.
62. Couch, H. T.; Kozicki, W.; Sage, B. H. "Latent Heat of Vaporization of *n*-Decane," *J. Chem. Eng. Data* **1963,** *8,* 346–349.
63. Hales, J. L.; Townsend, R. "Liquid Densities from 293 to 490 K of Nine Aromatic Hydrocarbons," *J. Chem. Thermodyn.* **1972,** *4,* 763–772.

RECEIVED October 2, 1978.

21

Application of a Generalized Equation of State to Petroleum Reservoir Fluids

LYMAN YARBOROUGH

AMOCO Production Co., P.O. Box 5340 A, Mail Code 4507, Chicago IL 60680

Generalized temperature- and component-dependent parameters are developed for the Redlich–Kwong equation of state. Binary vapor–liquid equilibrium data are used to determine interaction parameters which will be used in the combining equations for mixtures. Comparisons of calculated and experimental vapor–liquid equilibrium for several ternary systems indicated that only nonhydrocarbon–hydrocarbon interaction parameters are required in the combining equations. A procedure for characterizing the effect of the complex molecular structure of the heptanes and heavier components on phase equilibrium was developed. The combination of this characterization procedure with a gas chromatographic analysis of the heptanes and heavier fractions allowed the equation of state to be used to successfully predict the phase behavior of petroleum reservoir fluids, both crude oils and gas condensates.

Vapor and liquid phases coexist in virtually all areas of petroleum production operations, including reservoirs, wellbores, surface-production units, and gas-processing plants. Knowledge of fluid properties and phase behavior is required to calculate the fluid in place, fluid recovery by primary depletion, and fluid recovery by enhanced oil recovery techniques such as gas cycling, hydrocarbon solvent injection, and CO_2 displacement. Because of the complex nature of petroleum reservoir fluids and the often complicated phase behavior observed at elevated temperature and pressure conditions, the fluid properties and phase behavior historically have been measured experimentally. The complex nature of the fluids arises because of the supercritical components which are dissolved in the mixture of paraffinic, naphthenic, and

0-8412-0500-0/79/33-182-385$13.75/1

aromatic hydrocarbons (or these heavier hydrocarbons dissolved in the supercritical gas mixture), which is often compounded by the presence of nonhydrocarbons (nitrogen, CO_2, or H_2S). The phase behavior is complicated sometimes because the elevated pressure makes the heavier hydrocarbons very soluble in the gas phase; it also makes the supercritical (or gaseous) components very soluble in the liquid phase, resulting in near critical and retrograde behavior. An accurate and reliable phase-equilibria prediction method allows the calculation of important information with a considerable savings of time and cost than if experimental measurements are required. This is particularly true for systems in which large composition variations occur, as in some of the enhanced oil recovery techniques.

Therefore, a phase-equilibrium prediction method generally applicable to petroleum reservoir fluids must be capable of being used with supercritical components including the nonhydrocarbons, nitrogen, CO_2, and H_2S, in addition to predicting the effect of the different molecular types of hydrocarbons on phase equilibria. The prediction method also must provide reasonable results in the retrograde, near critical, and critical regions. Because of these requirements, an equation of state which could be applied to both the vapor and liquid phases was necessary. The two constant Redlich–Kwong (RK) equations of state was selected for study because of its simplicity and accuracy relative to other simple equations of state. The simplicity was considered to be an asset because of the necessity to generalize the equation parameters based on properties readily obtained for undefined fractions of petroleum fluids.

Equation of State Parameters

The empirical RK equation of state (*1*)

$$P = RT/(V - b) - a/T^{1/2}\,V(V + b) \tag{1}$$

(where P = absolute pressure; T = absolute temperature; V = molar volume; and R = universal gas constant; a = equation parameter; b = equation parameter) has two parameters, a and b, which can be determined by applying the classical definition of the critical point

$$(\partial P/\partial V)_{T_c} = 0 \text{ and } (\partial^2 P/\partial V^2)_{T_c} = 0 \tag{2}$$

(where subscript c denotes the critical point) to the equation of state, or by fitting the equation of state to experimental volumetric data. Applying the former method results in generalized parameters which can be used for any component for which the critical temperature and pressure are known (*1*).

$$a = 0.4278R^2T_c^{2.5}/P_c \qquad b = 0.0867RT_c/P_c \tag{3}$$

Several investigators have used the latter method of fitting the constants to volumetric data or a combination of volumetric data and fugacity. A popular method for correlating the fitted parameters is to use relationships similar to those in Equation 3:

$$a = \Omega_a R^2T_c^{2.5}/P_c \qquad b = \Omega_b RT_c/P_c \tag{4}$$

Considerable success has been enjoyed by investigators who allowed Ω_a and Ω_b to be temperature-dependent coefficients.

The procedures for determining Ω_a and Ω_b which were suggested by Zudkevitch and Joffe (*2*), Chang and Lu (*3*), and Joffe, et al. (*4*) are used in this study except that generalized correlations of vapor pressure (*5, 6*), saturated liquid density (*7*), and saturated liquid fugacity (*8*) are used in place of pure component data. The resulting values of Ω_a and Ω_b are smooth functions of reduced temperature and the acentric factor. Values of Ω_a and Ω_b are calculated for a reduced temperature range of 0.15 to 1.00 for acentric factors of 0 to 1.50. Selected results are shown in Figures 1 and 2. The calculated values of Ω_a and Ω_b are fitted by spline equations which are used because the first derivative is continuous, and the temperature derivative of Ω_a and Ω_b is required to calculate enthalpies. Also, the calculation of Ω_a and Ω_b via the spline equations requires very little computer time. The final equations can be extrapolated to an acentric factor of 1.60 and to a reduced temperature of 0.10.

This work results in correlations which can be used to predict parameters for the RK equation of state for hydrocarbon and other nonpolar components for which the critical pressure, critical temperature, and acentric factor are known or can be estimated. However, the applicability of the correlations to large molecules is unproven because the generalized correlations of physical and thermodynamic properties used to develop Ω_a and Ω_b are based on components no heavier than *n*-decane (acentric factor = 0.4885). Although the predicted parameters are based on properties for the saturated liquid phase, the parameters are applied to both vapor and liquid phases. For components above their critical temperature (a reduced temperature greater than 1.00), the values of Ω_a and Ω_b determined at a reduced temperature of unity are used.

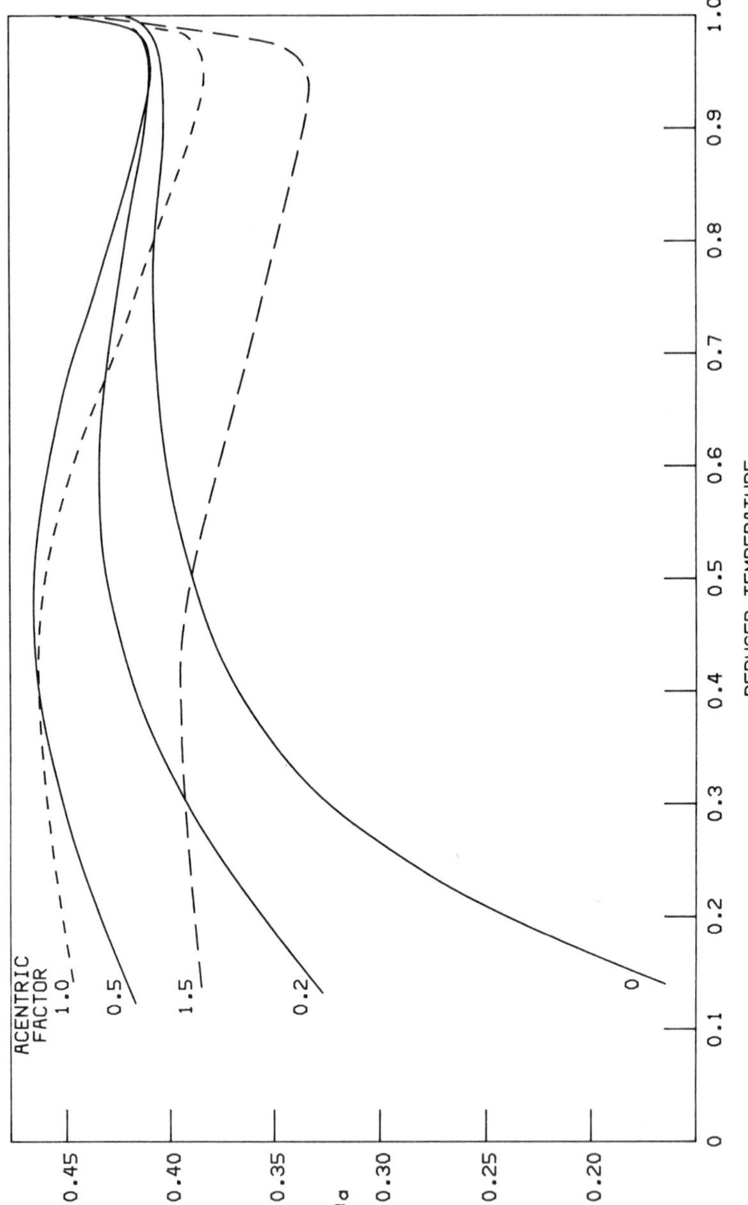

Figure 1. Derived values of Ω_a as a function of reduced temperature and the acentric factor

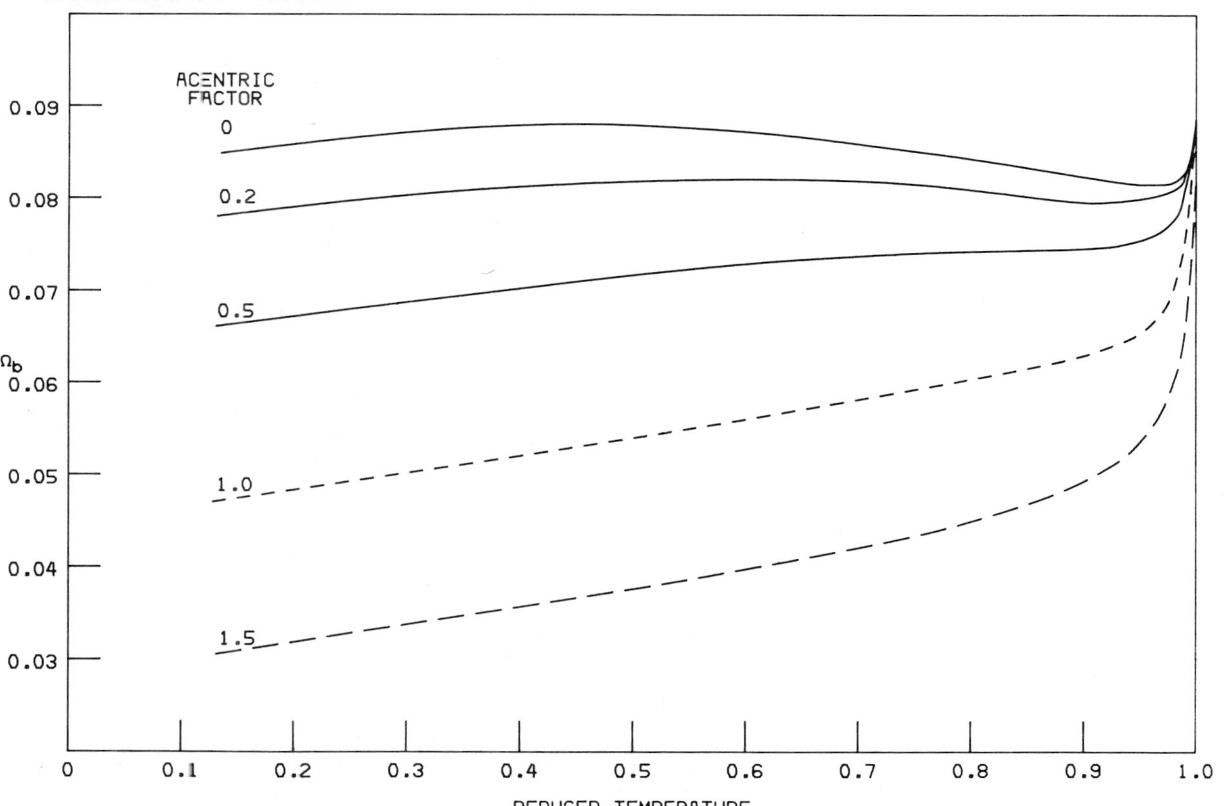

Figure 2. Derived values of Ω_b as a function of reduced temperature and the acentric factor

Application to Mixtures of Defined Components

Rules are proposed by Redlich and Kwong (*1*) for combining the pure-component parameters so the equation of state can be applied to mixtures

$$a_m = \sum_i \sum_j y_i y_j a_{ij}, \qquad \text{where } a_{ij} = (a_i a_j)^{1/2} \tag{5}$$

$$b_m = \sum_i y_i b_i \tag{6}$$

where y is the mole fraction.

Table I. Binary System Data Used

System	*No. of Points Used*	*Temperature Range, °F*	*Pressure Range, psia*	*Refs.*
Nitrogen–Methane	13	−256– −140	29–600	*9, 10*
Nitrogen–Ethane	10	−150– 50	300–900	*11*
Nitrogen–Propane	13	−148–158	600–1,400	*12*
Nitrogen–*n*-Butane	39	−184–280	236–4,020	*13, 14, 15*
Nitrogen–*n*-Hexane	24	100–340	250–5,000	*16*
Nitrogen–*n*-Heptane	18	90–360	1,020–10,025	*17*
Nitrogen–*n*-Decane	10	100–280	500–5,000	*18*
Nitrogen–Benzene	10	167–257	901–4,411	*19*
Nitrogen–H_2S	23	2–160	498–3,003	*20*
Nitrogen–CO_2	15	−100–77	764–2,015	*21, 22*
Methane–CO_2	17	−65–50	600–1,188	*23, 24*
CO_2–Ethane	15	−60–68	114–821	*25, 26*
CO_2–Propane	18	−4–180	200–900	*27. 28 29*
CO_2–*n*-Butane	14	80–280	200–1,000	*28, 30*
CO_2–*n*-Pentane	8	100–250	400–1,000	*28*
CO_2–*n*-Decane	15	40–460	300–2,000	*31*
CO_2–Benzene	12	86–140	300–1,200	*32*
CO_2–H_2S	9	40–180	400–1,200	*33, 34*
Methane–H_2S	14	−80–160	600–1,800	*35, 36*
Ethane–H_2S	6	59–150	500–1,000	*37*
H_2S–Propane	10	76–183.7	300–1,000	*38, 39*
H_2S–*n*-Butane	10	100–250	284–1,050	*40*
H_2S–*n*-Pentane	5	40–220	100–1,000	*41*
H_2S–*n*-Decane	10	40–340	100–1,400	*42*
H_2S–Benzene	3	77–158	147	*43*
Methane–Ethane	50	−225–50	35–900	*11, 44, 45*
Methane–Propane	66	−225–160	34–1,300	*44, 46 47, 48*
Methane–*iso*-Butane	7	100–220	600–1,500	*49*
Methane–*n*-Butane	19	−80–220	200–1,800	*50, 51*
Methane–*iso*-Pentane	8	160–280	400–1,000	*52*

Zudkevitch and Joffe (*2*) suggest a modification to the mixture rule for a_m for use with mixtures of unlike molecules:

$$a_m = \sum_i \sum_j y_i y_j a_{ij}, \qquad \text{where } a_{ij} = a_i, \ i = j \qquad (7)$$
$$a_{ij} = (1 - C_{ij})(a_i a_j)^{1/2}, \ i \neq j$$

and C_{ij} is the unlike pair interaction parameter.
Equations 6 and 7 are used in this work. Note that when the interaction parameter is zero, Equation 7 reduces to Equation 5.

The unlike pair interaction parameter is determined using binary vapor–liquid equilibrium data as described by Zudkevitch and Joffe (*2*). The systems used in this study are given in Table I. The interaction

to Determine Interaction Parameters

System	*No. of Points Used*	*Temperature Range, °F*	*Pressure Range, psia*	*Refs.*
Methane–*n*-Pentane	12	100–280	400–2,000	*53*
Methane–*n*-Hexane	18	77–340	500–2,500	*54, 55*
Methane–*n*-Heptane	22	−100–460	600–3,000	*56, 57*
Methane–*n*-Octane	8	77–302	588–1,029	*58*
Methane–*n*-Nonane	16	−13–302	588–4,410	*59*
Methane–*n*-Decane	19	0–460	600–4,500	*60, 61, 62*
Methane–*n*-Eicosane	3	104	588–882	*63*
Methane–Cyclohexane	15	70–340	600–4,000	*64*
Methane–Methyl-cyclohexane	8	−100–0	600–3,000	*65*
Methane–Benzene	5	150	600–4,000	*66*
Methane–Toluene	15	−100–150	500–5,000	*66, 67*
Ethane–Propane	31	−100–180	8–550	*10, 44, 68, 69*
Ethane–*iso*-Butane	24	−94–250	4.6–701	*10, 20*
Ethane–*n*-Butane	20	50–250	200–805	*70, 71*
Ethane–*n*-Pentane	11	40–280	150–900	*72*
Ethane–*n*-Hexane	10	150–350	100–950	*73*
Ethane–*n*-Heptane	6	150–350	874–1,215	*74*
Ethane–*n*-Octane	4	104–212	529–588	*75*
Ethane–*n*-Decane	12	40–460	300–1,500	*76*
Ethane–*n*-Eicosane	3	140	588–882	*63*
Ethane–Cyclohexane	14	50–450	400–1,200	*77*
Ethane–Benzene	7	230–419	800–1,400	*78*
Propane–*iso*-Butane	21	−4–200	15–489	*10, 79*
Propane–*n*-Butane	15	−4–250	10–400	*10, 80*
Propane–*iso*-Penane	8	32–167	29–368	*81*
Propane–*n*-Pentane	7	160–280	150–550	*82*
Propane–*n*-Decane	7	160–460	300–1,000	*83*
Propane–Benzene	5	280–400	400–800	*84*
n-Butane–*n*-Heptane	6	320–400	400–550	*85*
n-Butane–*n*-Decane	5	280–460	300–700	*86*
n-Pentane–*n*-Heptane	3	277–392	147–294	*87*

Table II. Interaction Parameters

Component	*Nitrogen*	CO_2	H_2S
Nitrogen			
CO_2	−0.055		
H_2S	0.1985	0.1076	
Methane	0.028	0.0762	0.0847
Ethane	0.061	0.1092	0.0905
Propane	0.124	0.1373	0.0846
iso-Butane			
n-Butane	0.169	0.1359	0.052
iso-Pentane			
n-Pentane		0.014[a]	0.075
n-Hexane	0.210		
n-Heptane	0.187		
n-Octane			
n-Nonane			
n-Decane	0.1325	0.1369	0.053
n-Eicosane			
Cyclohexane			
Methylcyclohexane			
Benzene	0.245	0.0843	0.115
Toluene			

[a] Results questionable, data suspect.

parameters are virtually independent of temperature and pressure for a binary system. Average values determined in this work are given in Table II. The interaction parameters are correlated against the acentric factor of the less volatile hydrocarbon component, as shown in Figures 3–6. These correlations are used to predict interaction parameters for missing pairs.

The interaction parameters for the hydrocarbon–hydrocarbon pairs are considerably less than those for the nonhydrocarbon–hydrocarbon pairs. To evaluate the need for the interaction parameter for hydrocarbon–hydrocarbon pairs, phase equilibria has been predicted both with and without interaction parameters for ternary systems for which experimental data are available. The ternary systems used and the results of the study are given in Table III. These results indicate that there is no significant advantage for using hydrocarbon–hydrocarbon interaction parameters in predictions for ternary systems. Therefore, for general calculations, hydrocarbon–hydrocarbon interaction parameters are set at zero, but nonzero interaction parameters are used for nonhydrocarbon–hydrocarbon pairs.

Additional evaluation of the equation of state with the generalized temperature-dependent parameters and the interaction parameters has been carried out by comparing predicted K values with experimental

Obtained from Binary Data

Methane	*Ethane*	*Propane*	n-*Butane*	n-*Pentane*
−0.003				
0.005	0.001			
0.019	0.004	0.008		
0.023	0.009	0.012		
0.023		0.014		
0.019	0.0185	0.026		
0.030	0.038			
0.013	0.034		0.030	0.020
0.027	0.006			
0.0155				
0.015	0.012	0.012	0.022	
0	0.006			
0.038	0.0276			
0.019				
0.029	0.036	0.025		
0.021				

data on a ten-component system (*109*). The ranges of composition, temperature, and pressure at which comparisons were made are given in Table IV. The overall average absolute deviations between the predicted values and the experimental data for each component are given in Table V. The nitrogen results given in Table V are calculated using the interaction parameters provided by the lower curve in Figure 4. The higher curve in Figure 4 resulted in nitrogen *K* values that are generally too high.

The information in Table V provides general comparisons but does not show details of the deviations at various temperatures and pressures. By studying the experimental and predicted *K* values on log *K* vs. log *P* plots at several temperatures, details of the deviations for each component at various conditions can be observed. Several examples of this type of comparison are shown in Figures 7–11. Knowing the percent deviation between the predicted and experimental *K* values does not provide a quantitative measure of errors in the predicted vapor-to-liquid ratio and the saturation pressure-properties which are important in reservoir-engineering calculations. The best indication of the accuracy of a *K*-value prediction method for reservoir fluids would be a detailed comparison of component *K* values as well as fractions of the fluid in the vapor and liquid phases and the saturation pressure. The latter comparisons are

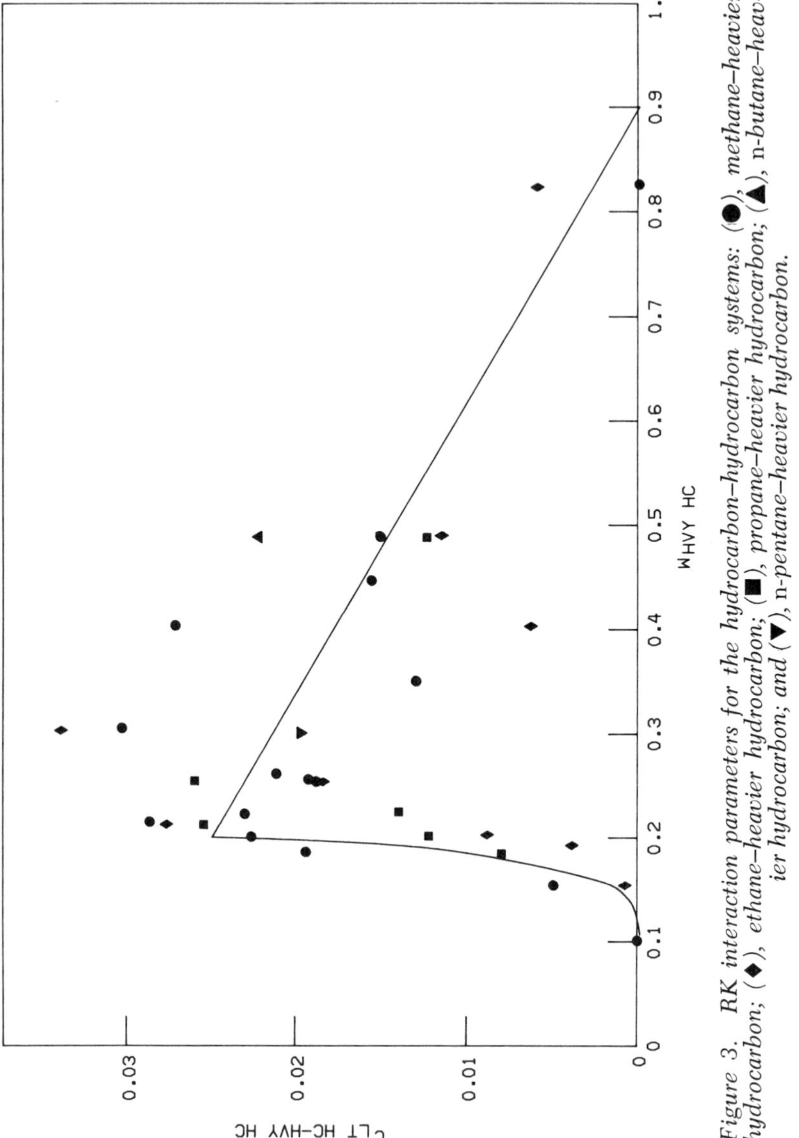

Figure 3. RK interaction parameters for the hydrocarbon–hydrocarbon systems: (●), *methane–heavier hydrocarbon;* (◆), *ethane–heavier hydrocarbon;* (■), *propane–heavier hydrocarbon;* (▲), n-*butane–heavier hydrocarbon; and* (▼), n-*pentane–heavier hydrocarbon.*

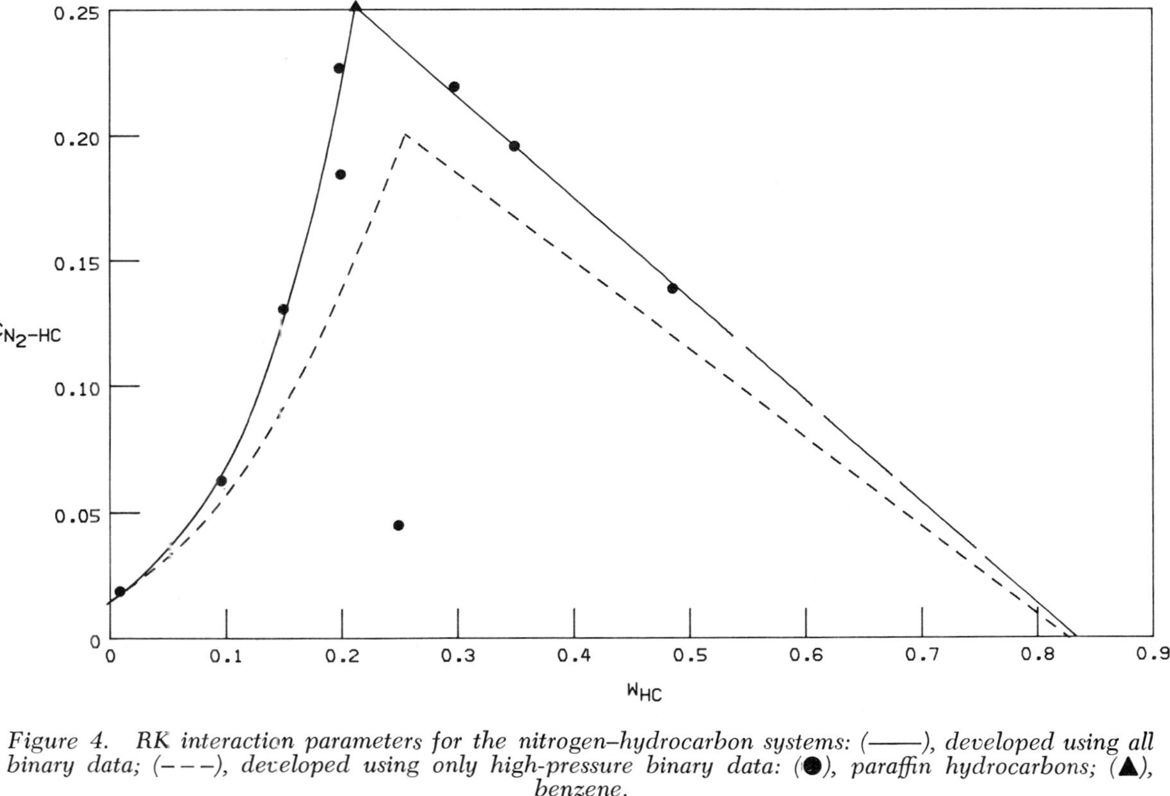

Figure 4. RK interaction parameters for the nitrogen–hydrocarbon systems: (——), developed using all binary data; (– – –), developed using only high-pressure binary data: (●), paraffin hydrocarbons; (▲), benzene.

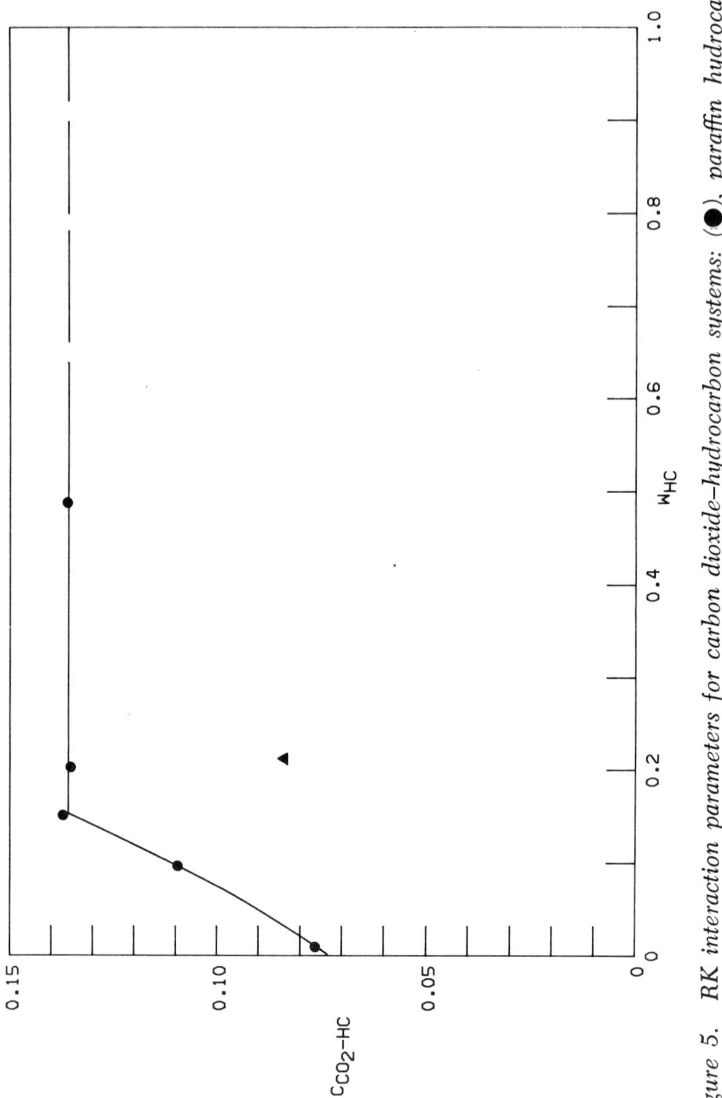

Figure 5. RK interaction parameters for carbon dioxide–hydrocarbon systems: (●), paraffin hydrocarbons; (▲), benzene.

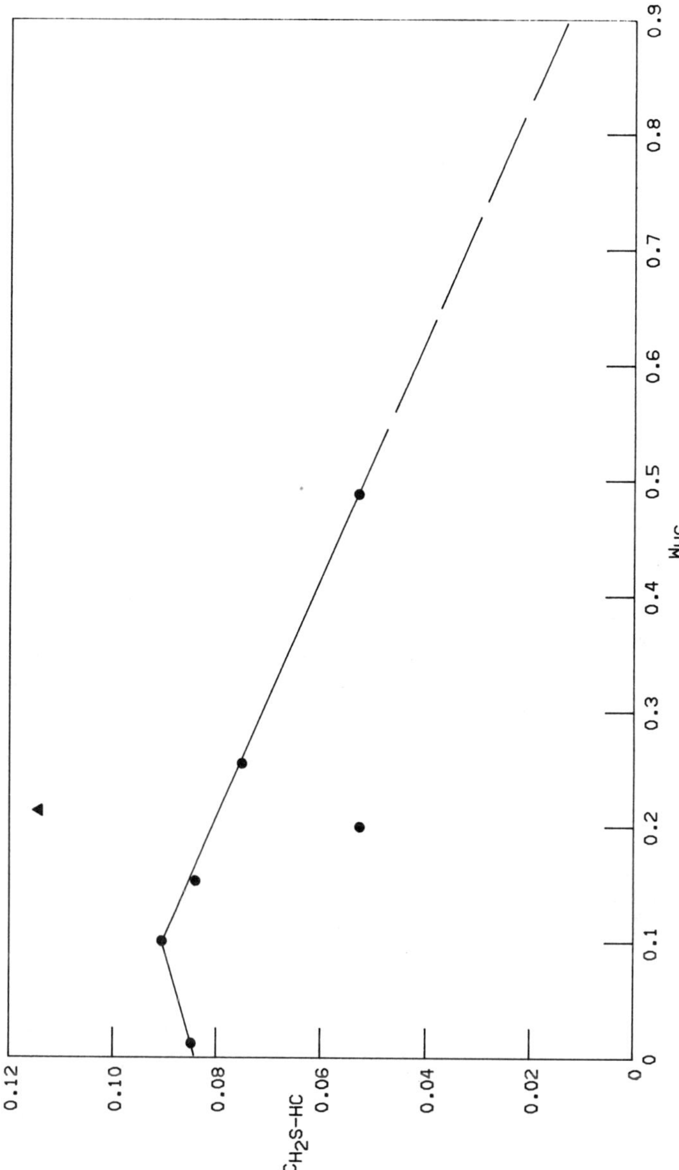

Figure 6. RK interaction parameters for hydrogen sulfide–hydrocarbon systems: (●), *paraffin hydrocarbons;* (▲), *benzene.*

Table III. Effect of Hydrocarbon–Hydrocarbon Interaction Parameters

Components in System					
1	*2*	*3*	*Refs.*	*No. of Points*	*Temperature Range, °F*
C_1	C_2	C_3	*88*	60	−150–50
C_1	C_2	C_3	*89*	45	−175–75
C_1	C_3	nC_4	*90*	47	40–160
C_1	C_2	nC_5	*91*	12	100
C_1	C_3	nC_5	*92* *93*	33	100–220
C_1	C_3	nC_{10}	*94*	55	40–400
C_1	nC_6	nC_{10}	*95* *96* *97*	59	40–280
C_1	C_2	nC_7	*98*	13	−60– −20
C_1	C_3	nC_7	*99*	6	−40– −20
C_1	C_3	nC_{10}	*99*	28	−20–70
C_1	nC_6	Cyclo-hexane	*100*	50	32–140
C_2	nC_4	nC_5	*101* *102*	54	
C_2	nC_4	nC_7	*103* *104*	58	150–350
N_2	C_1	C_2	*105*	20	−151.1
N_2	C_1	nC_4	*106*	48	100–220
N_2	C_1	nC_{10}	*107*	64	100–280
C_1	CO_2	nC_4	*108*	31	−20–100
C_1	H_2S	nC_4	*108*	28	−20–100

	Summary of Overall Absolute Average			
	C_1	C_2	C_3	nC_4
No. of points	540	262	274	325
Without HC–HC C_{ij}	4.116	6.504	8.289	11.230
With HC–HC C_{ij}	5.077	7.517	9.799	13.430

on Agreement of Predicted Values Compared to Experimental Data

Pressure Range, psia	HC–HC C_{ij}	Absolute Average Percent Deviation between Predicted & Experimental: Component 1	Component 2	Component 3
100–1,100	No	2.953	6.249	16.702
	Yes	7.346	7.468	18.757
32–875	No	3.723	8.217	13.950
	Yes	8.404	10.873	15.286
200–1,500	No	3.527	2.696	6.586
	Yes	3.698	4.106	9.268
500–2,000	No	5.266	1.771	13.467
	Yes	3.850	4.961	15.111
500–3,000	No	4.790	3.904	9.710
	Yes	3.569	4.977	9.233
400–4,000	No	3.042	3.669	31.612
	Yes	3.773	4.364	35.826
600–4,000	No	2.659	10.842	24.771
	Yes	2.379	14.190	25.239
200–1,000	No	4.734	3.347	—
	Yes	2.431	16.931	—
201–993	No	5.851	11.782	—
	Yes	8.758	7.751	—
292–1,023	No	8.011	3.853	—
	Yes	6.839	8.134	—
711–2,987	No	5.496	14.498	15.469
	Yes	7.500	23.853	28.191
451–888	No	6.383	4.911	5.202
	Yes	4.463	4.978	5.064
453–1,102	No	6.819	3.496	8.883
	Yes	6.129	7.942	10.054
197–402	No	21.127	5.538	7.717
	Yes	21.017	5.506	7.317
1,000–3,000	No	10.326	4.212	12.223
	Yes	10.607	5.551	13.833
1,000–5,000	No	7.078	4.029	45.478
	Yes	6.841	3.280	40.659
400–1,200	No	10.460	8.934	21.700
	Yes	12.605	8.964	22.716
400–1,200	No	12.613	7.815	34.752
	Yes	16.577	7.737	35.517

Percent Deviations for Components

nC_5	nC_6	*hexane*	*Cyclo-* nC_7	nC_{10}
99	50	50	58	178
7.706	14.498	15.469	8.883	34.330
7.761	23.853	28.191	10.054	34.055

Table IV. Composition, Temperature, and Pressure Ranges for the Multicomponent System Data

Component	*Composition Range in Overall Mixture, Mol %*
Methane	44.0–85.0
Ethane	2.8–7.5
Propane	2.0–3.1
n-Pentane	1.7–10.5
n-Heptane	1.3–7.4
n-Decane	0.8–5.9
Toluene	0–16.0
Nitrogen	0–15.0
Carbon Dioxide	0–15.0
Hydrogen Sulfide	0–25.0
Temperature range	−50–250°F
Pressure range	105–4496 psia

Table V. Deviations between Predicted and Experimental *K* Values for a Multicomponent System

Component	*No. of Points*	*Average Absolute Percent Deviation*[a]
Methane	434	5.57
Ethane	434	4.14
Propane	434	4.89
n-Pentane	434	7.94
n-Heptane	430	11.35
n-Decane	427	26.51
Toluene	277	14.87
Nitrogen	185	14.94
Carbon dioxide	160	6.75
Hydrogen sulfide	257	7.23

[a] AAPD = |experimental K − calculated K|/experimental K.

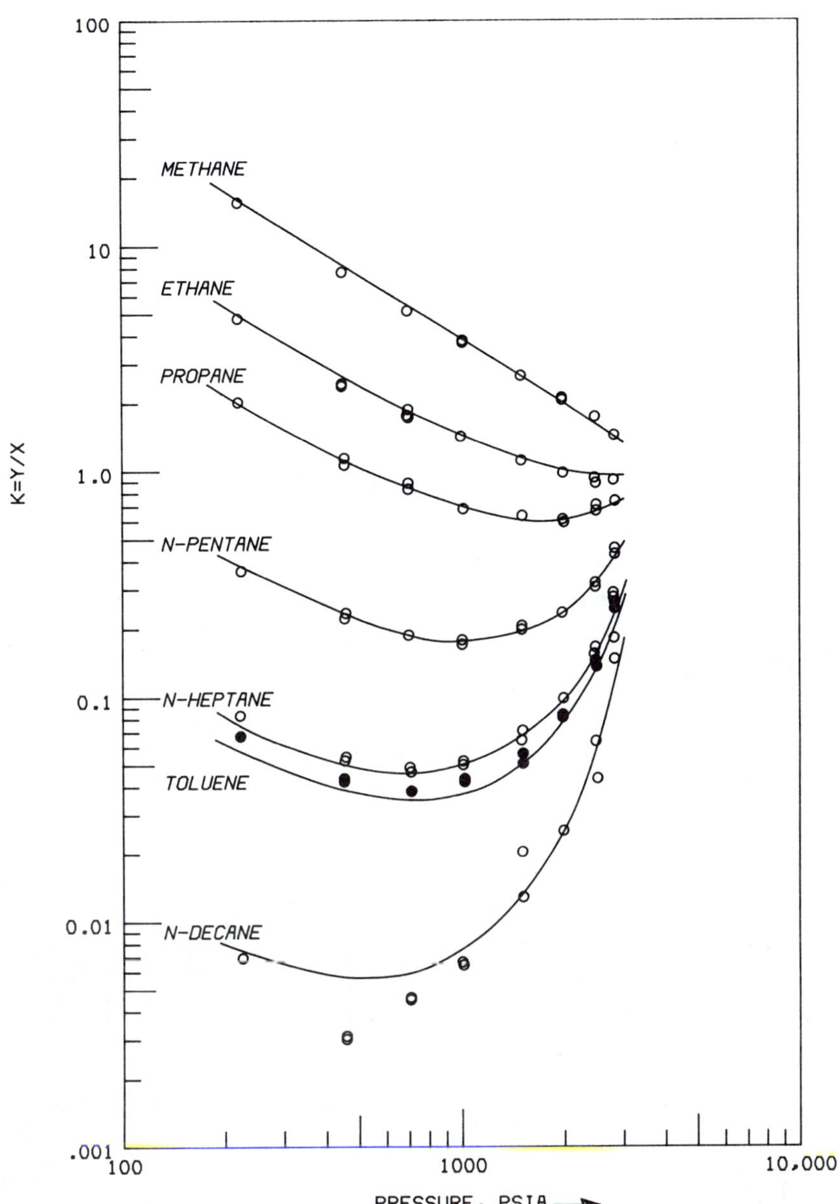

Figure 7. Comparisons between predicted and experimental K values for a multicomponent system: (——), predicted; (○), experimental data. Mixture 4, 300°F.

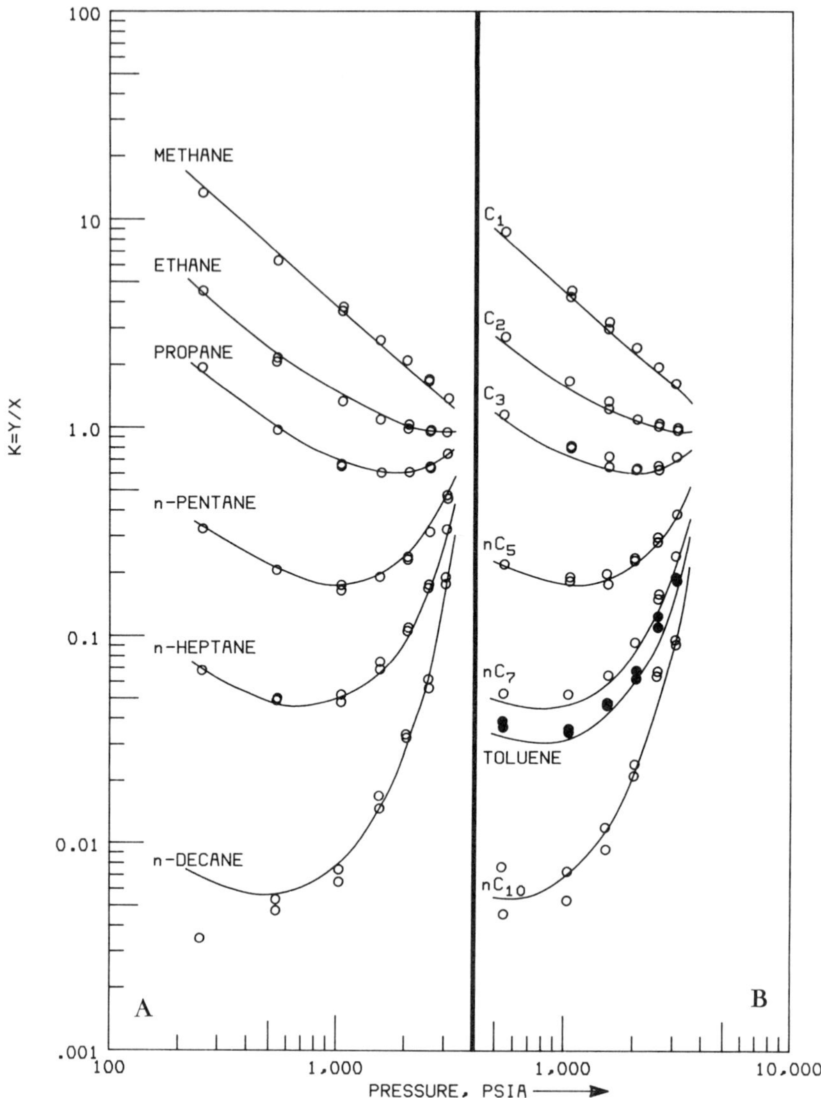

Figure 8. Comparisons between predicted and experimental K *values for a multicomponent system: (——), predicted; (○), experimental data. (A), Mixture 8, 200°F; (B), Mixture 7, 200°F.*

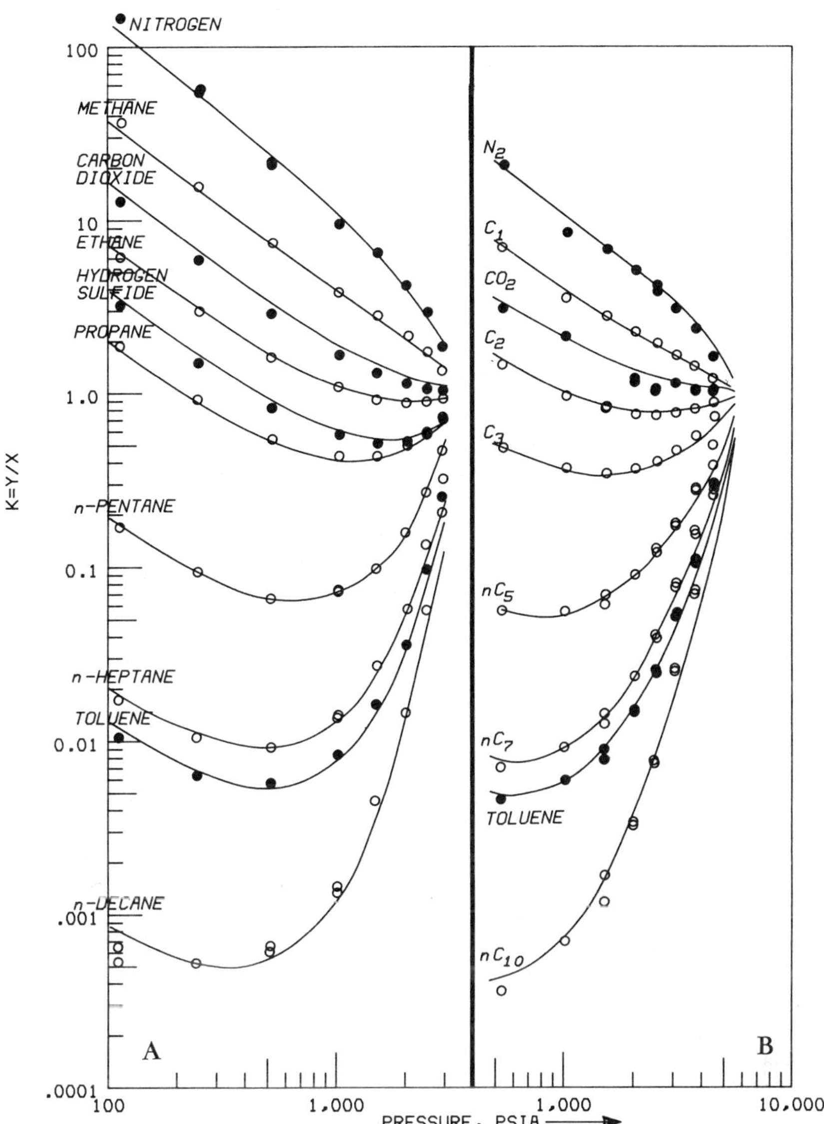

Figure 9. Comparisons between predicted and experimental K values for a multicomponent system: (——), predicted; (○), experimental data. (A), Mixture 20B, 100°F; (B), Mixture 20, 100°F.

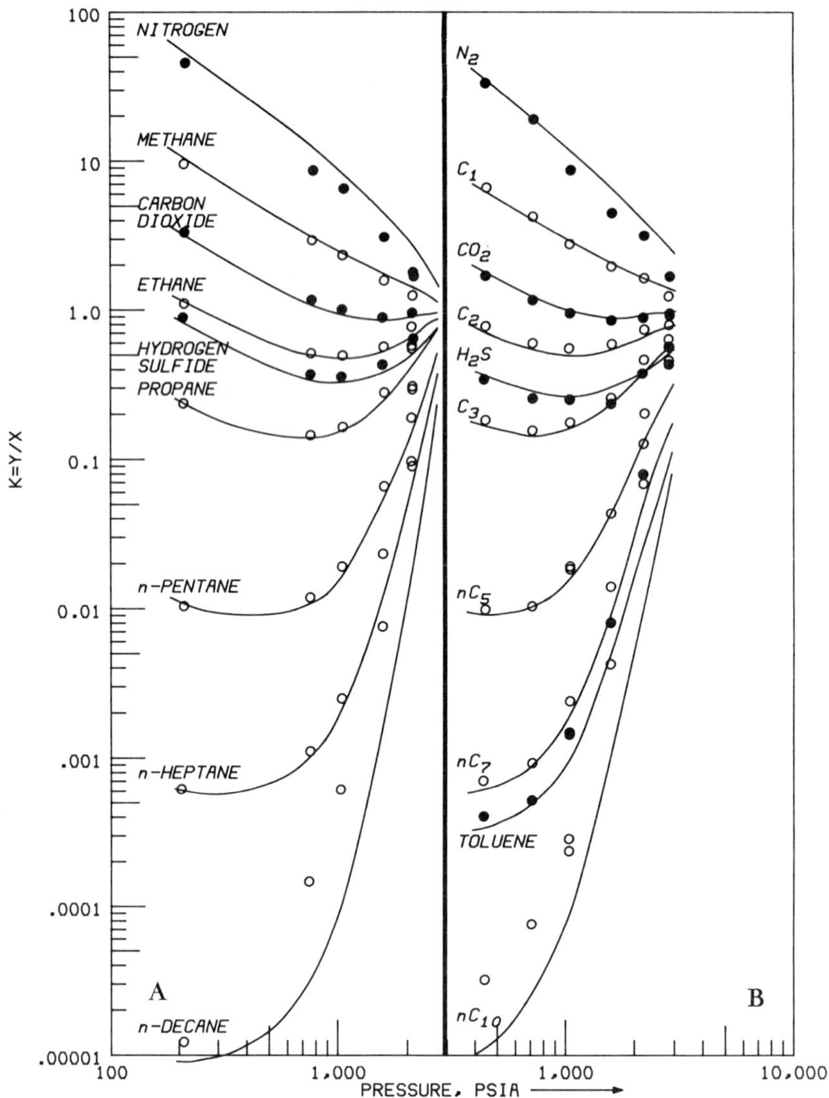

Figure 10. Comparisons between predicted and experimental K values for a multicomponent system: (——), predicted; (○), experimental data. (A), Mixture 16, 0°F; (B), Mixture 15B, 0°F.

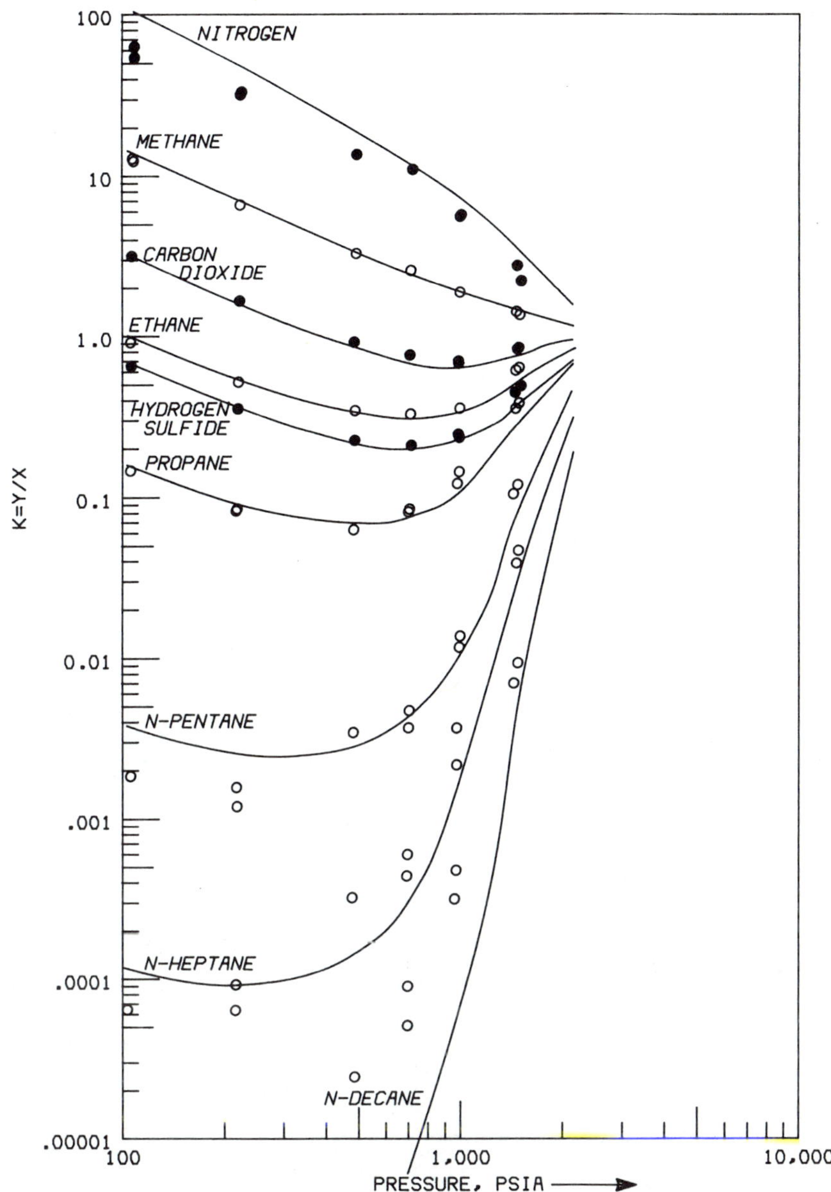

Figure 11. Comparisons between predicted and experimental K *values for a multicomponent system: (——), predicted; (○), experimental data. Mixture 16B, −45°F.*

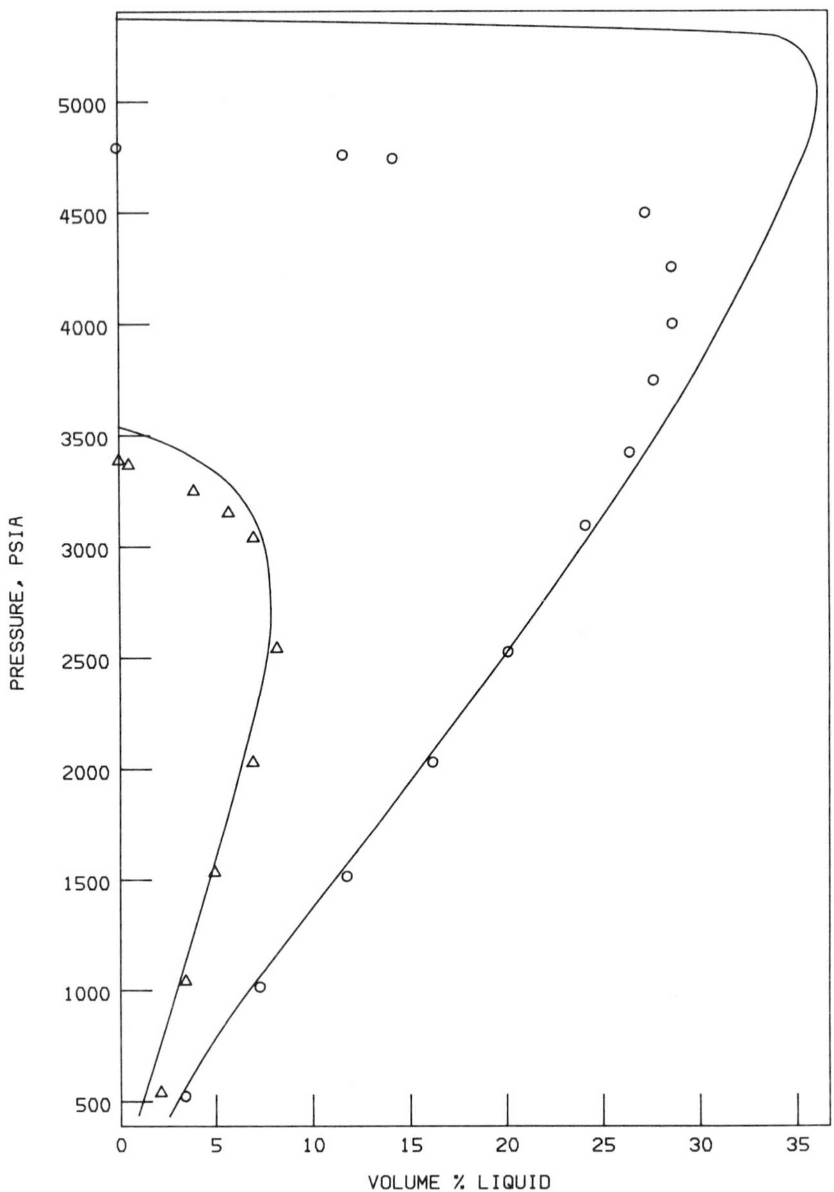

Figure 12. Comparisons between predicted and experimental phase distribution for multicomponent systems: (——), predicted. Experimental data:(△), Mixture 7, 200°F; (○), Mixture 20, 100°F.

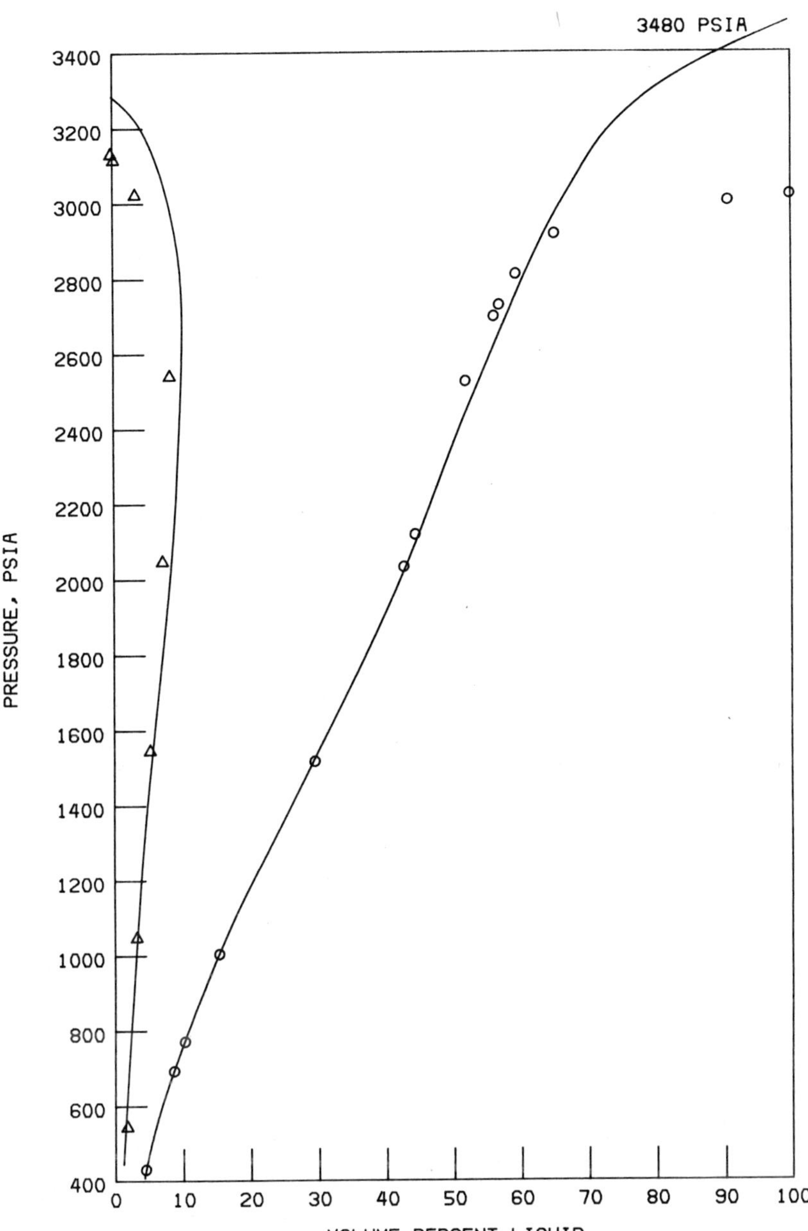

Figure 13. Comparisons between predicted and experimental phase distribution for multicomponent systems: (——), predicted. Experimental data: (○), Mixture 15B, 0°F; (△), Mixture 8, 200°F.

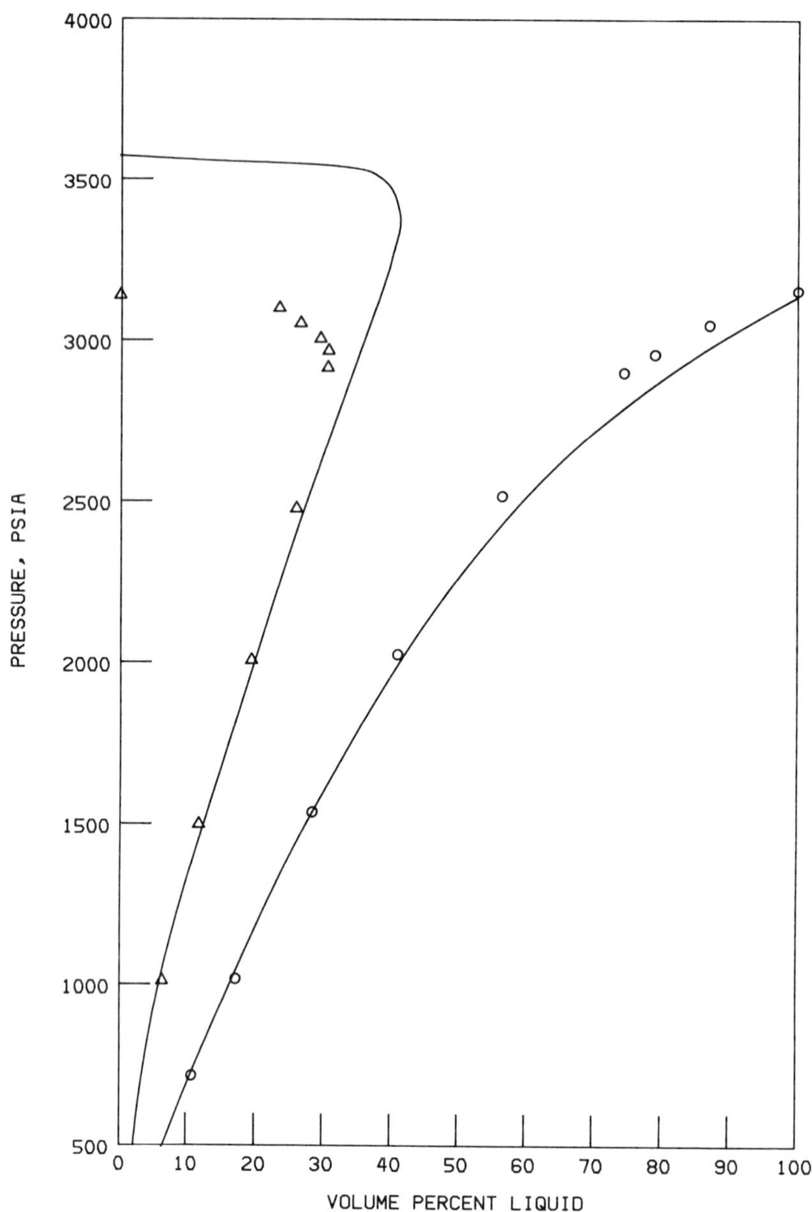

Figure 14. Comparisons between predicted and experimental phase distribution for multicomponent systems: (——), predicted. Experimental data: (○), Mixture 4, 200°F; (△), Mixture 20B, 100°F.

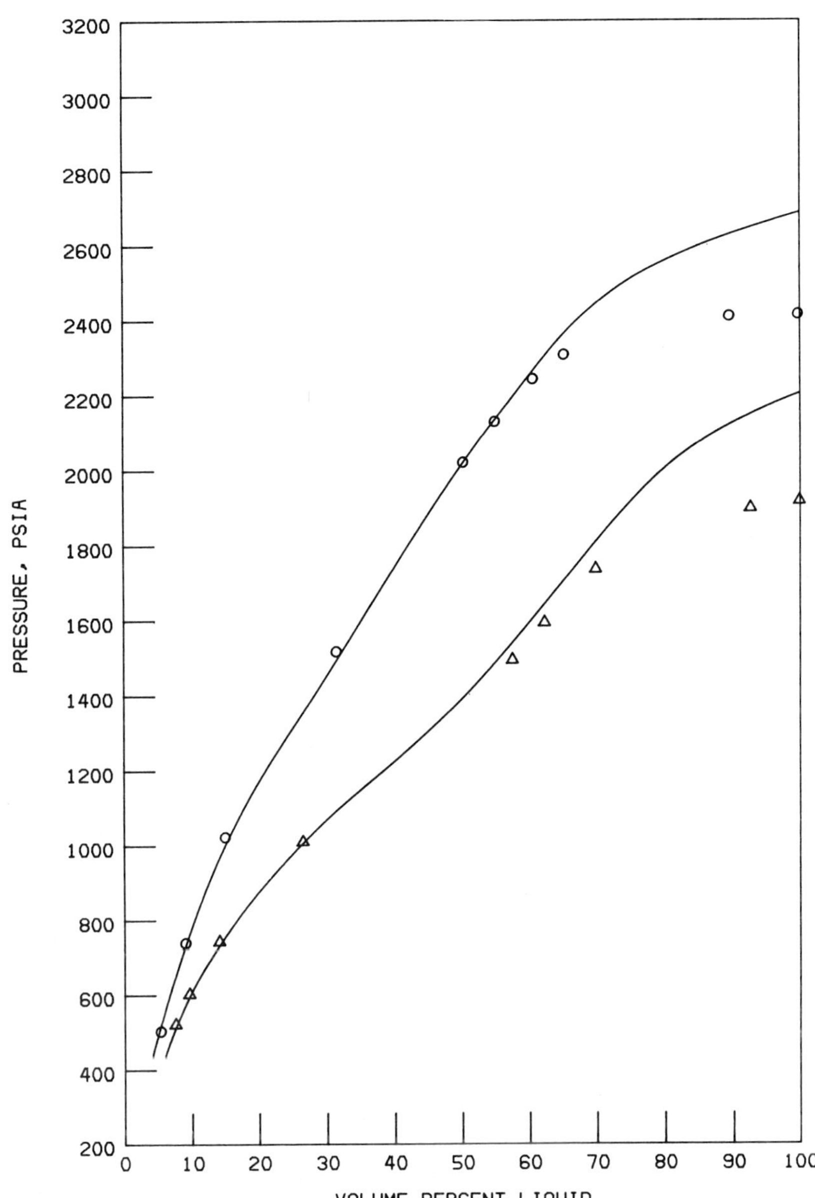

Figure 15. Comparisons between predicted and experimental phase distribution for multicomponent systems: (——), predicted. Experimental data: (○), Mixture 16. 0°F; (△), Mixture 16B, −45°F.

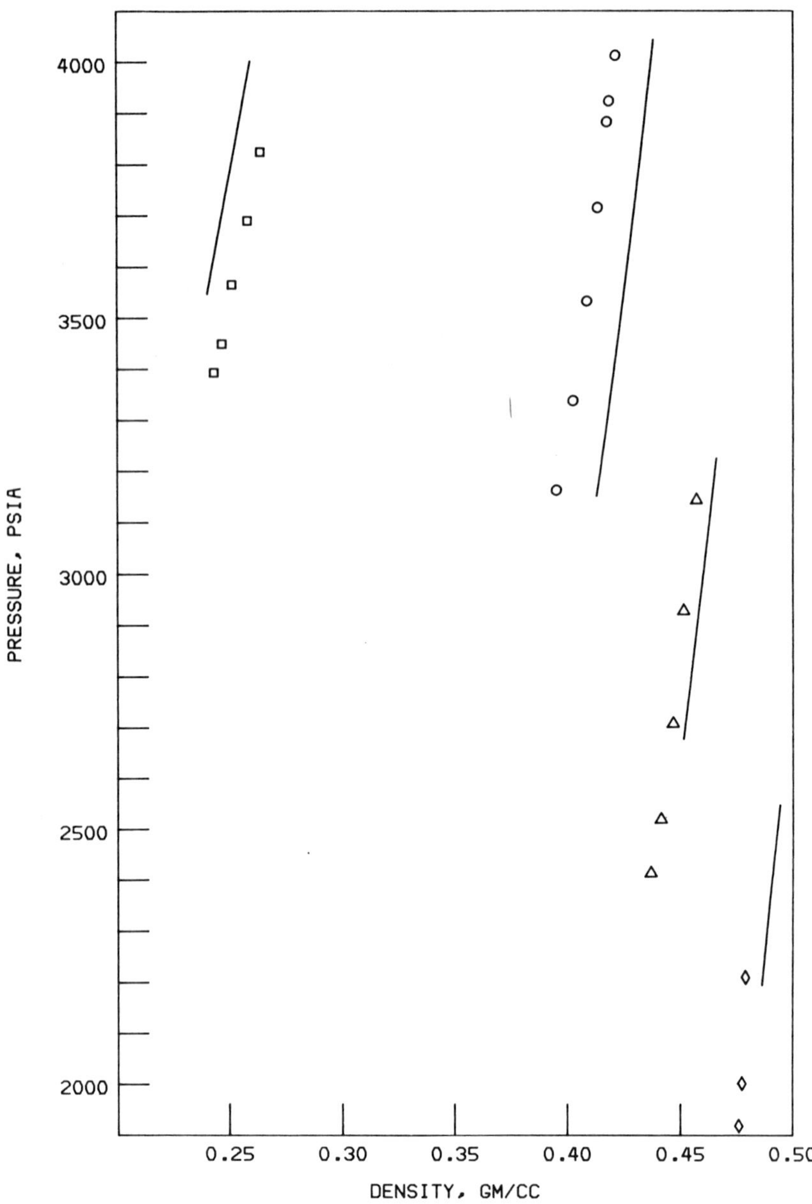

Figure 16. Comparisons between predicted and experimental densities in the single-phase region for multicomponent systems: (——), predicted. Experimental data: (○), Mixture 4, 200°F; (□), Mixture 7, 200°F; (△), Mixture 16, 0°F; (◇), Mixture 16B, −45°F.

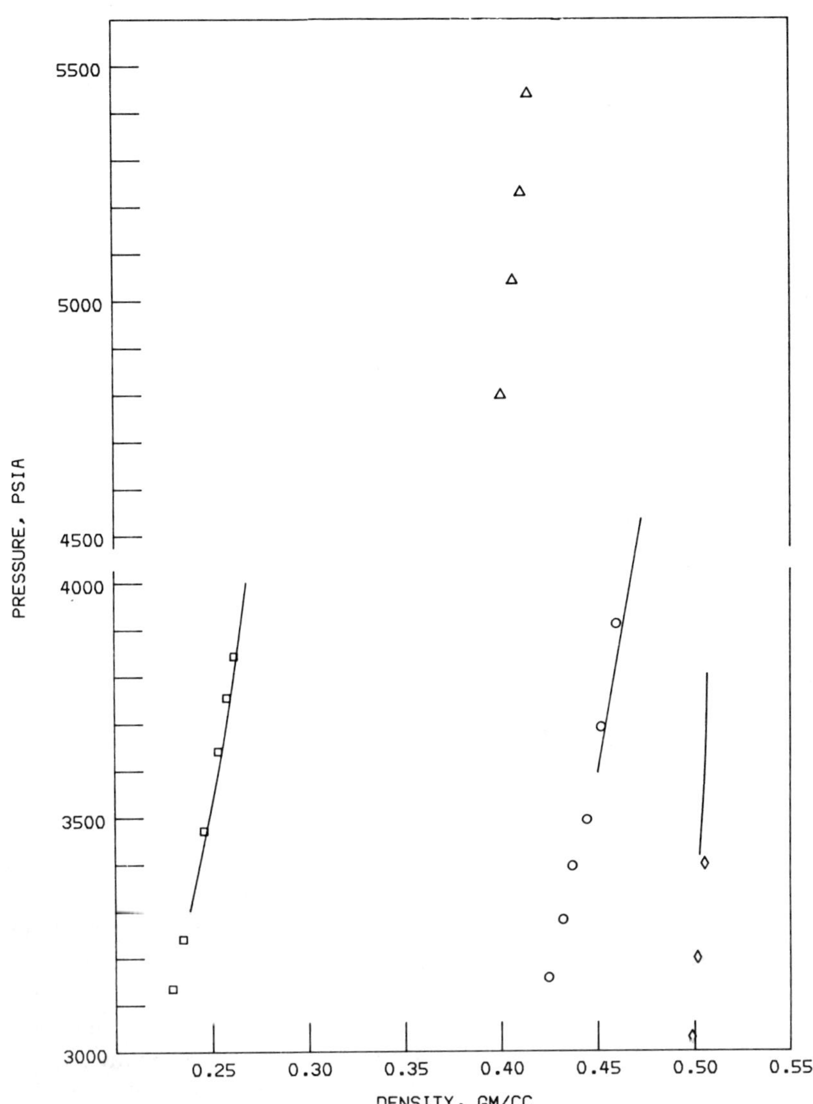

Figure 17. Comparisons between predicted and experimental densities in the single-phase for multicomponent systems (calculated 0.4332 g/cc at 5800 psia): (——), predicted. Experimental data: (□), Mixture 8, 200°F; (○), Mixture 20B, 100°F; (△), Mixture 20, 100°F; (◇), Mixture 15B, 0°F.

shown in Figures 12–15 for the same systems shown in Figures 7–11. The flash calculation provides the relative amounts of vapor and liquid on a molar basis, and molar volumes are required to calculate the relative amounts of vapor and liquid on a volume basis for comparison with the experimental data. Accurate phase densities are required, and these are calculated when solving the equation of state. Comparisons between the calculated and experimental densities in the single-phase region are shown in Figures 16 and 17.

The limited number of comparisons shown for the ten-component system include some of the best as well as some of the poorest agreement observed between the predicted and experimental results. Generally, the predictions are in better agreement with the experimental data at the higher temperatures. As the system pressure approaches the saturation pressure, the predicted K values do not converge toward unity as rapidly as the experimental K values, particularly at the lower temperatures. This results in a predicted saturation pressure which is greater than the experimental value, and in erroneous liquid-to-vapor ratios at elevated pressures. This appears to be a problem which would be expected to occur in reservoir fluid systems. The predicted results correctly accounted for the effect of the addition of an aromatic component to an otherwise *n*-paraffin system.

Application to Petroleum Reservoir Fluids

Once generalized parameters are available for an equation of state, the major problem is the development of a method to properly characterize the heavy components of the fluid. Good definition of the major nonhydrocarbon components, nitrogen, CO_2, and H_2S, and the light-hydrocarbon components through *n*-pentane is available by normal analyses from many laboratories. The hexanes and heavier (or heptanes and heavier) components historically have been reported as a combined fraction because of the complexity of these components and because no analytical method has been available to provide additional information at a reasonable cost. The combined fraction can contain molecules with as many as 20 to 30 carbon atoms (gas condensates) to as many as 50 to 60 carbon atoms (crude oils). All combined fractions contain various amounts of the different molecular types of hydrocarbons (*n*-paraffins, cycloparaffins or naphthenes, and aromatics) as well as molecules which consist of at least two of the molecular types ("mixed" molecules). The latter type of molecule causes problems in trying to determine the paraffin–naphthene–aromatic (PNA) analysis for characterizing the combined fraction (*110, 111*), but a major problem arises in correctly identifying the molecules which are a mixture of more than one molecular type, particularly for crude oils and some condensates. For the lighter fluids, such as low-molecular-weight lean oils and light-gas condensates, the mixed molecules can be fragmented by a mass spectrometer (MS) to provide a representative PNA analysis. The heavier fluids cannot be analyzed adequately in this manner, and a PNA analysis by chemical methods does not provide representative results.

Jacoby (*112*) illustrated that the molecular weight and specific gravity of pure components can be used to reflect the nonparaffinic nature of the naphthenic, aromatic, and mixed molecules as shown in Figure 18. Since the molecular weight and specific gravity often are measured for the combined fraction of reservoir fluids, these properties are selected as the basic parameters for reflecting the nature of the combined fraction. Since the combined fraction includes a wide range of components, it is also necessary to separate the total fraction in a manner such that the vaporization behavior can be modeled accurately. Historically, this separation or breakdown of the combined fraction has been done using a distillation analysis. More recently, a temperature-programmed chromatographic analysis of the combined fraction is becoming common. The advantages and disadvantages of each analysis for separating the combined fraction and for characterizing each separate subfraction are given below.

	Distillation Analysis	*Chromatographic Analysis*
Advantages	1. Sufficient material is collected to measure molecular weight and specific gravity	1. Good separability—definition of components to C_{50}
	2. Average boiling point available	2. Fast and relatively cheap
		3. Carbon number and average boiling point available via calibration curves—molecular weight known within small deviation
Disadvantages	1. Time consuming and relatively expensive	1. No material collected for specific gravity determination (unless preparative chromatograph is used)
	2. Poor separability—Azeotropes common	
	3. Separations through about C_{20} to C_{25}, even with best techniques	

The chromatograph analysis method is the preferred method to use if the specific gravity for the subfractions (carbon number fractions in this case) can be obtained. The results of specific gravity measurements

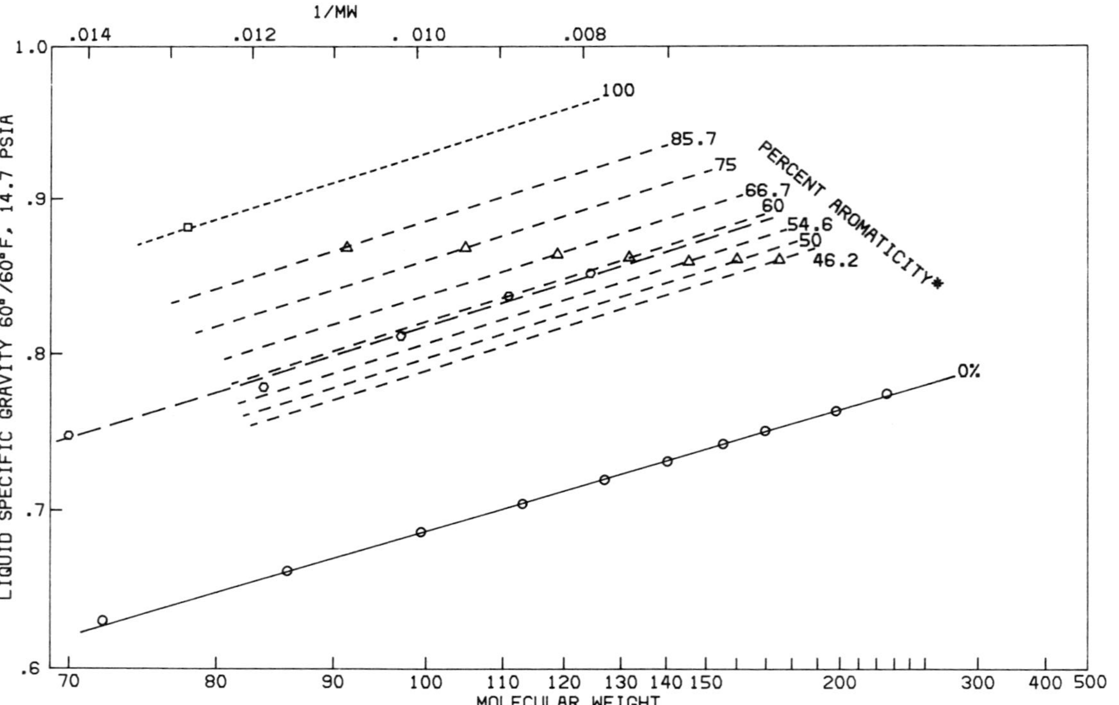

Figure 18. Relation between specific gravity and reciprocal molecular weight for hydrocarbons of various molecular types. Aromaticity value is calculated by the percentage of total carbon atoms in the molecule which are within the benzene ring. (□), Benzene; (⬡), cycloparaffins; (△), n-*alkyl benzenes; (○),* n-*paraffins.*

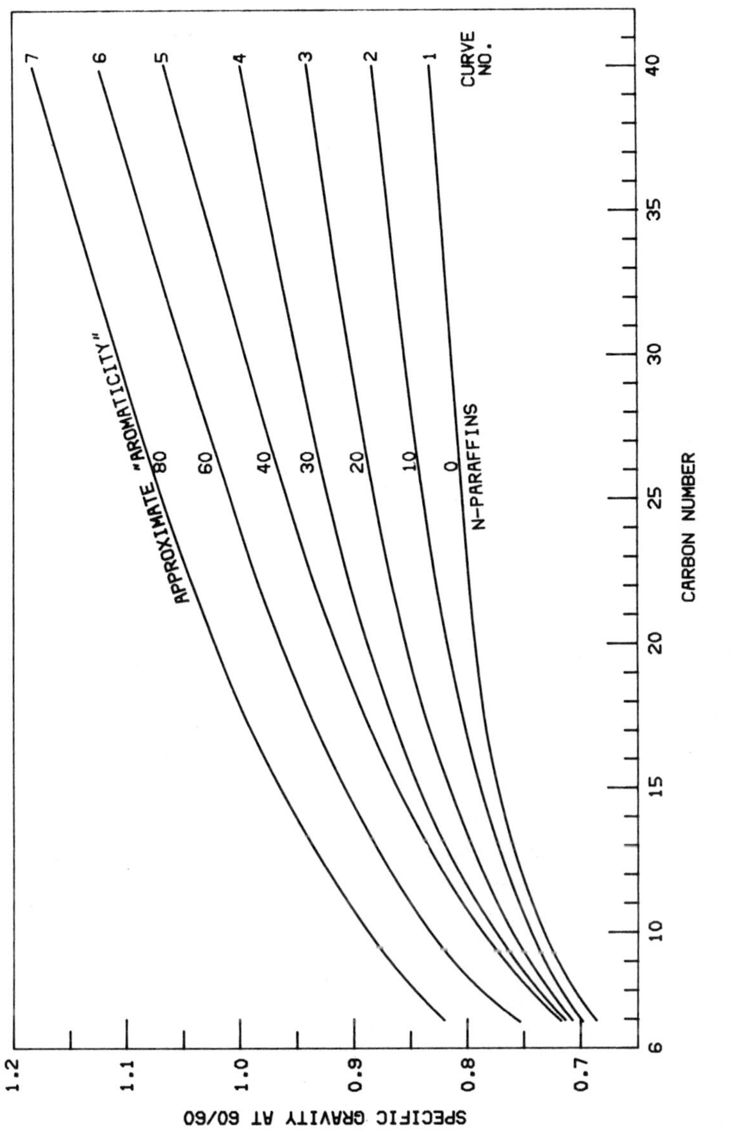

Figure 19. Arbitrary curves relating specific gravity to carbon-number fraction which allow for different molecular types of hydrocarbons

on narrow boiling fractions obtained from several true boiling point distillations of gas condensates and light crude oils are plotted against molecular weight in a manner suggested by Jacoby. Each carbon number is assigned a molecular weight value, and several arbitrary curves are drawn to depict the deviation from *n*-paraffin behavior, as shown in Figure 19. These curves deviate from those of Jacoby to reflect the behavior of the distillation fractions for carbon numbers up to C_{13}; also the heavier subfractions are relatively more naphthenic and aromatic than the lighter subfractions (*113*). Once Figure 19 is established, then for each carbon-number fraction obtained from a chromatographic analysis, one molecular weight and several specific gravity values could be obtained. The carbon-number fractions can be combined to provide a calculated molecular weight and several calculated specific gravities for the combined fraction for comparison with the molecular weight and specific gravity measured on the combined fraction. The curve which properly represents the fluid of interest will provide the physical properties required to estimate the mean boiling point, critical temperature, and critical pressure of each carbon-number fraction (*114*). The acentric factor of each fraction can be estimated then (*115*). Sufficient information has been developed to calculate the equation-of-state parameters. For oils, the acentric factor limitation of 1.6 restricts the carbon-number breakdown to about C_{40}.

Comparisons with Data on Petroleum Reservoir Fluids

Several comparisons are made with published data and some of the results are shown. The first system is data published by Hoffman, Crump, and Hocott (*116*) for a gas condensate system. The composition is reported for carbon-number fractions through C_{22} along with K values at six pressures and a phase-distribution curve. The molecular weight and specific gravity are reported for each fraction. Comparisons of the predicted K values and the measured data are shown in Figures 20 and 21a. Excellent agreement is shown between the predicted and experimental saturation pressures and phase-distribution curve. Relatively good agreement is shown for the methane and ethane K values, but somewhat poor agreement is shown for most of the heavier components. Some of the disagreement could be the result of the smoothing techniques used by Hoffman et al.

The second system is data published by Roland, Smith, and Kaveler (*117*) for a gas condensate system. The heptanes plus fraction is divided using a distillation analysis reported by those authors. The comparisons between the predicted and experimental K values at three temperatures are shown in Figures 21b and 22. The agreement between the predicted and experimental K values is good except for the heptanes plus K value at 120° and 200°F.

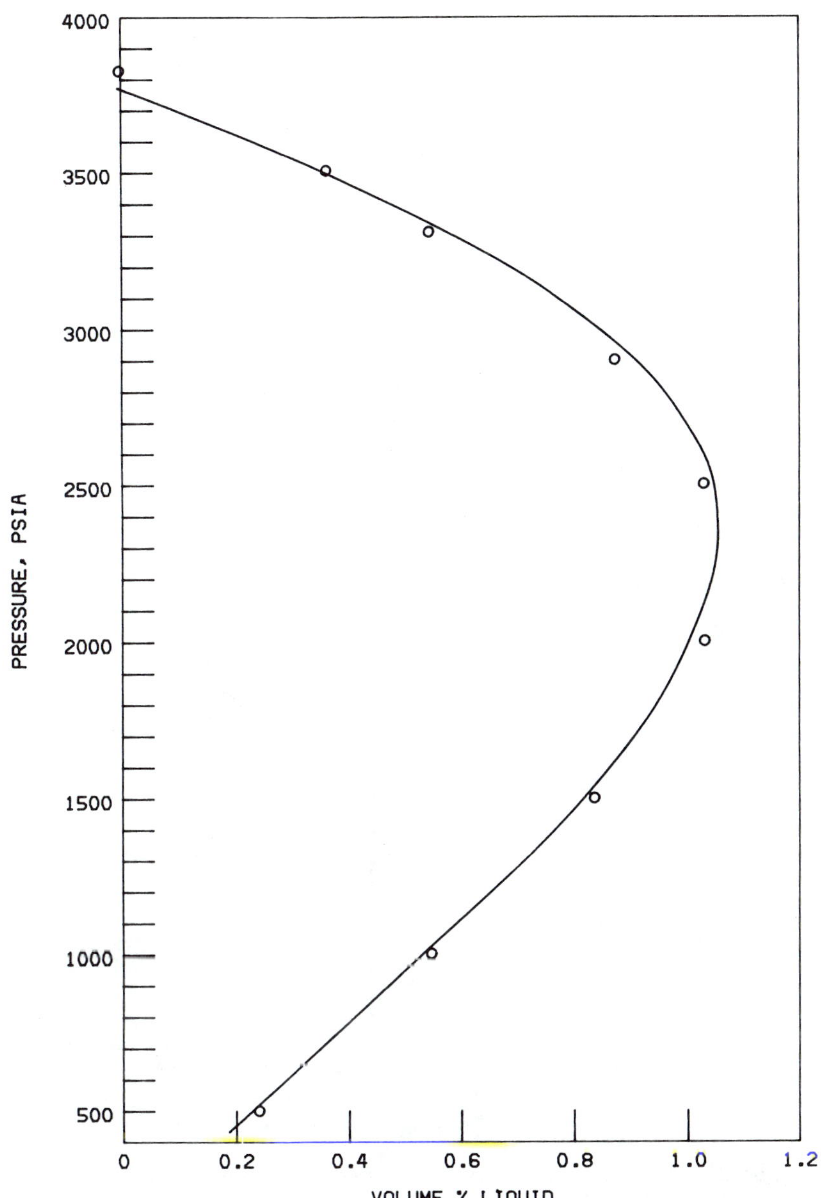

Figure 20. Comparison of experimental (data of Hoffmann, Crump, and Hocott) and predicted dew-point and phase distribution for a lean gas condensate fluid. The temperature is 201°F. (——), Predicted; (○), experimental.

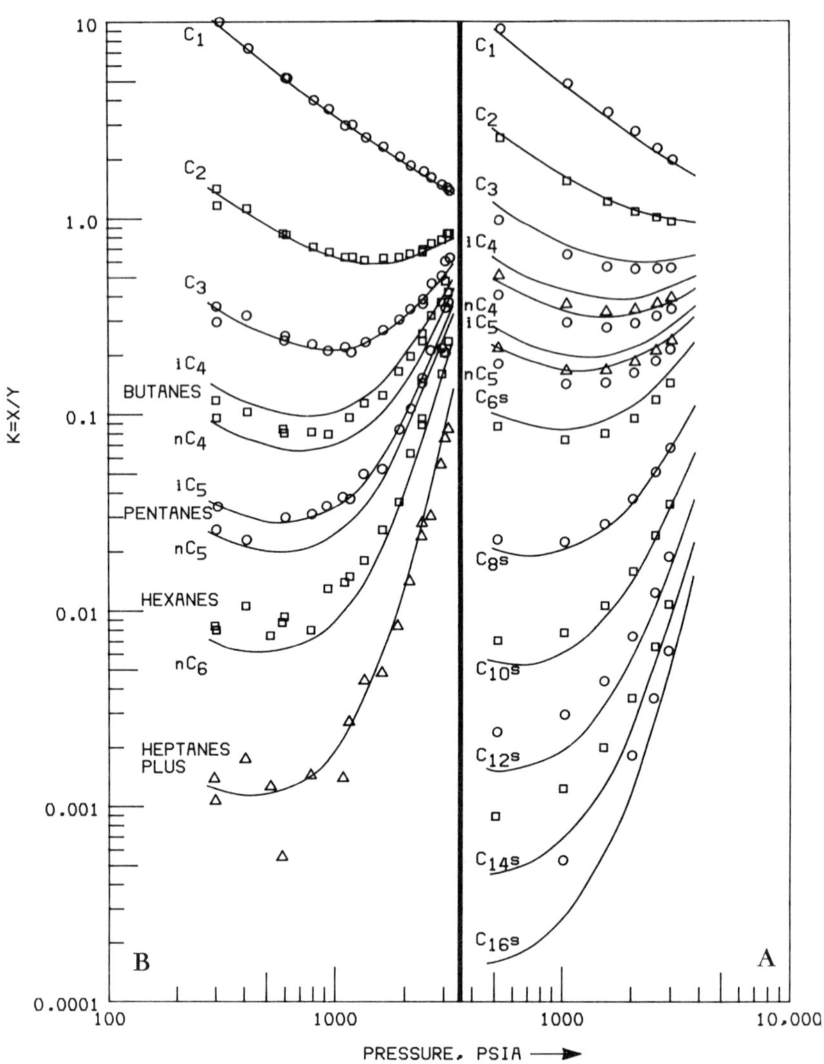

Figure 21. Comparisons of experimental and predicted K *values for two gas condensate fluids: (——), predicted. (A) (○, □, △), Data of Hoffmann, Crump, and Hocott, 201°F; (B) (○, □, △), data of Roland, Smith, and Kaveler, 40°F.*

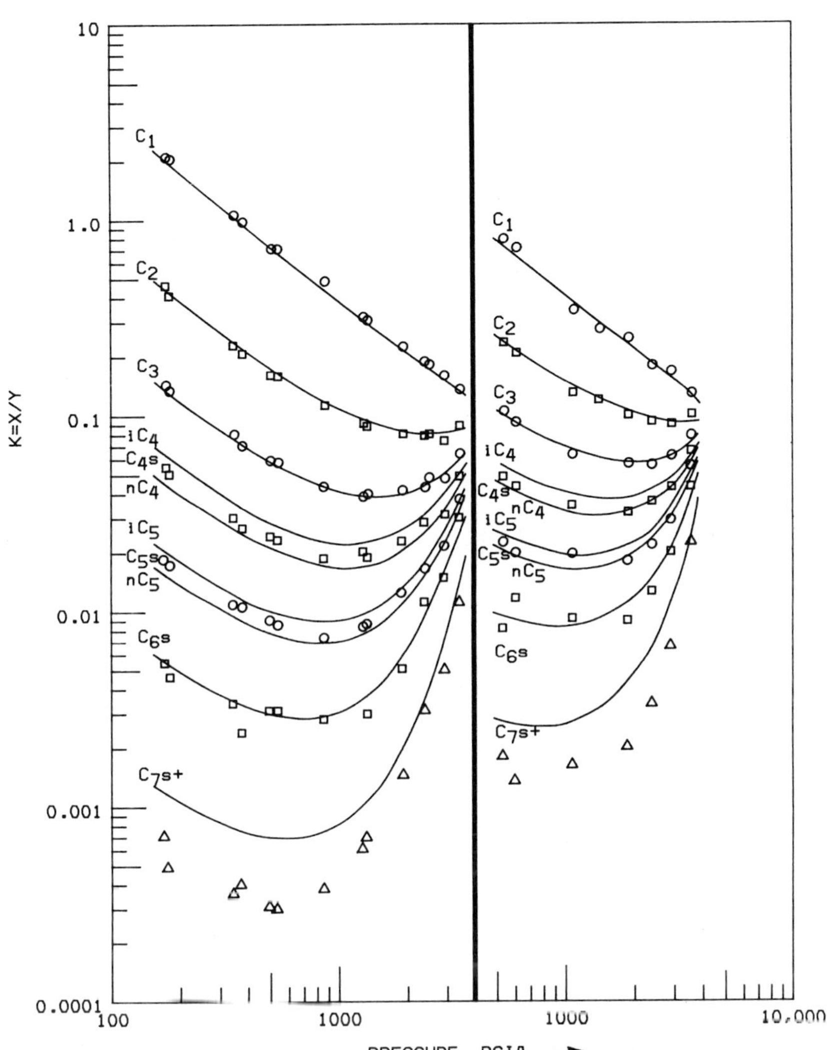

Figure 22. Comparisons of experimental and predicted K values for a gas condensate fluid at 120°F and 200°F. (○, □, △), Data of Roland, Smith, and Kaveler; (——), predicted.

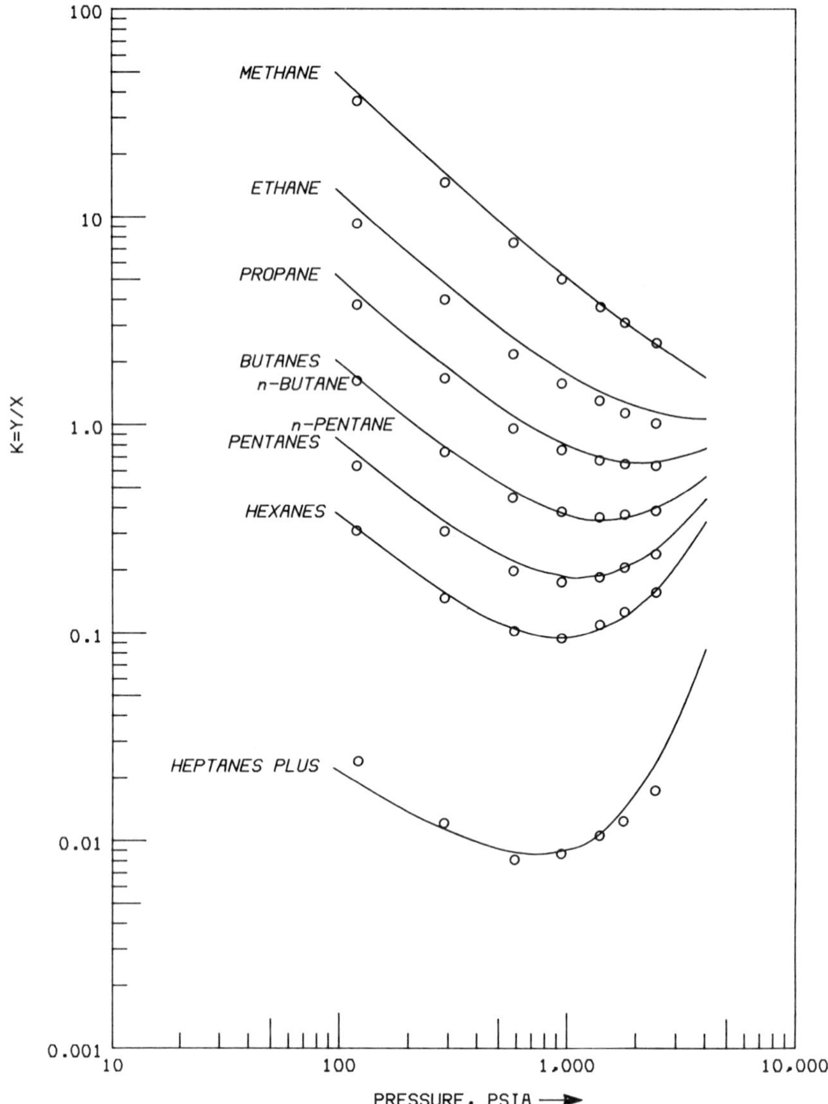

Figure 23. Comparisons of experimental and predicted K values for a black oil at 200°F: (○), data of Katz and Hachmuth; (——), predicted.

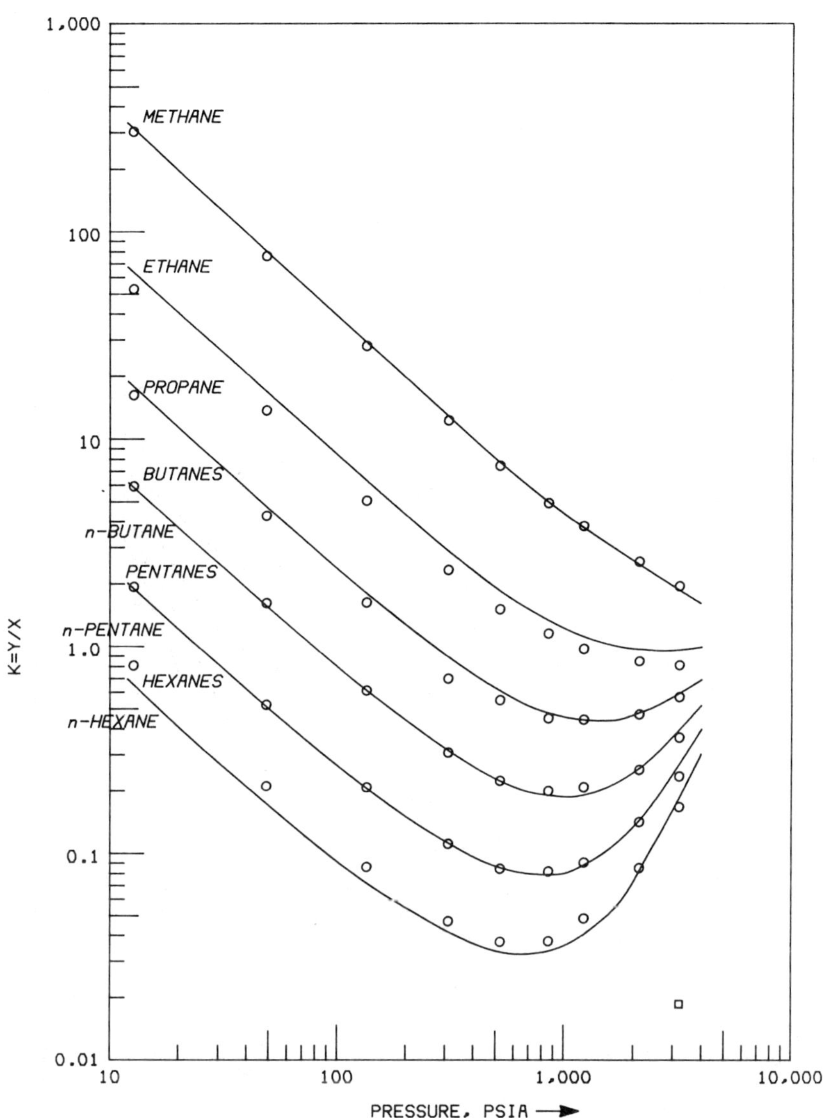

Figure 24. Comparisons of experimental and predicted K *values for a black oil at 120°F: (○), data of Katz and Hachmuth; (——), predicted.*

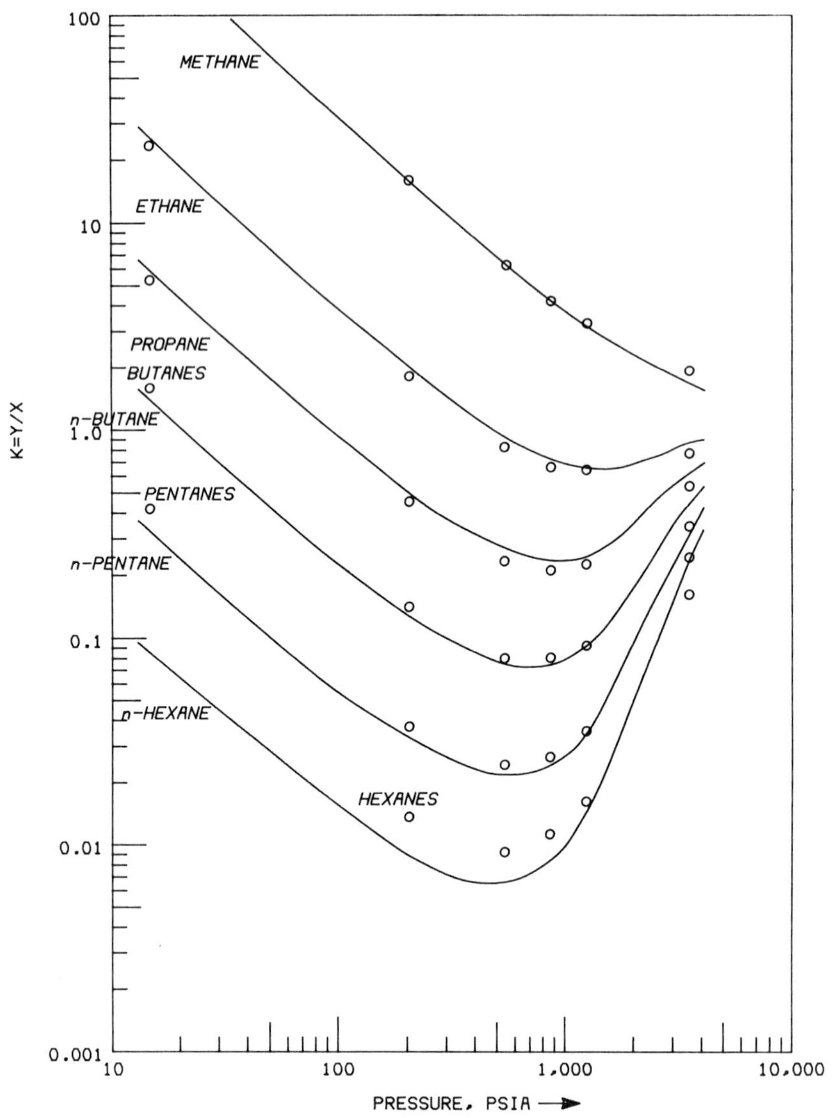

Figure 25. Comparisons of experimental and predicted K *values for a black oil at 40°F: (○), data of Katz and Hachmuth; (——), predicted.*

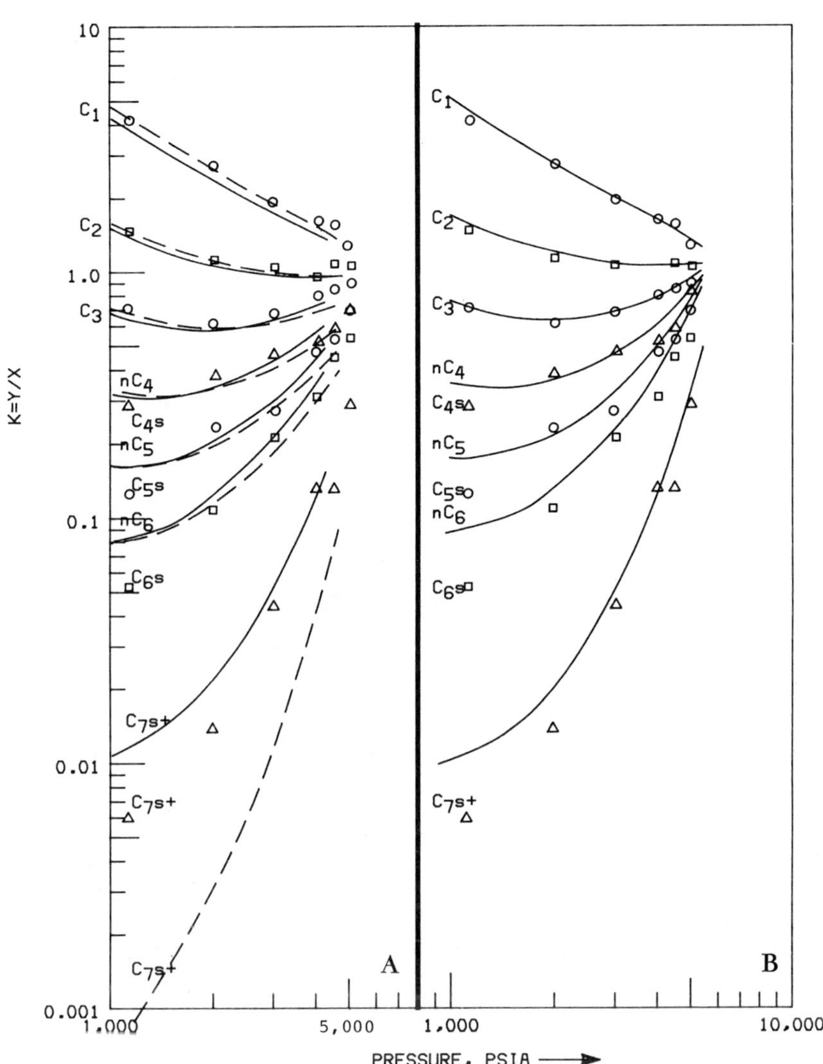

Figure 26. Comparisons of experimental and predicted K *values for a volatile oil at 190°F: (——, – – –), predicted; (○, □, △), data of Evans and Harris. (A) (——), C_7's+ into 14 fractions using the reported USBM distillation; (– – –), lumped C_7's+. (B) (——), C_7's+ into 34 fractions via chromatographic analysis of Bismarck–Fuhrmann unit oil (also 7 fractions).*

The third system is data for a natural gas–crude oil system published by Katz and Hachmuth (*118*) which is typical of a black oil. The heptanes plus is separated and characterized using a chromatographic analysis for an oil which has a molecular weight and specific gravity similar to the oil used by Katz and Hachmuth. Comparisons between the experimental and predicted K values at three temperatures are shown in Figures 23–25. The agreement is good except for ethane at 200° and 120°F.

The fourth system tested is a high-shrinkage-to-volatile-type oil reported by Evans and Harris (*119*). Four methods are used to characterize the heptanes plus fraction for the predictions. One method uses the combined heptanes plus as as one fraction. A second method uses the distillation reported by those authors. Comparisons between the experimental K values and those predicted by the two characterization methods mentioned are shown in Figure 26a. For both methods the calculated methane K value is too low, resulting in the predicted bubble-point pressure being lower than that estimated by Evans and Harris. An interaction parameter between methane and the heavy components can be found which will provide good agreement between the predicted and experimental methane K values and the bubble-point pressure, but after considerable study of this approach, it has been rejected because no consistency in the interaction parameters was noted from one system to another. This conclusion is supported by the results of an independent study (*120*). The third method used to characterize the heptanes plus fraction uses a chromatographic analysis for an oil which has a molecular weight and specific gravity similar to the oil used by Evans and Harris. After the physical properties are obtained for the 34 subfractions (C_7 through C_{40}^+), these fractions are combined to give only 7 subfractions (fourth characterization method). The K values predicted using both the 34 and 7 subfractions are shown in Figure 26b. The component-by-component agreement is generally good, and the predicted saturation pressure is just greater than 5,400 psia as compared with the value of between 5,200 and 5,300 psia estimated by Evans and Harris based on their laboratory data.

The results just discussed indicate that for bubble-point systems, the light-component K values are relatively insensitive to the number of subfractions (between 7 and 34) used for the heptanes plus, especially when the heptanes plus is characterized properly. This behavior has been noted for several oils and is important for two reasons: (1) in a compositional reservoir model study, the number of components must be limited to 14–18 to keep computer time and costs from being prohibitively high; and (2) when vaporization of an oil is being calculated, extremely high saturation pressures and erroneous phase-equilibrium results can be predicted when the heptanes plus is divided through C_{40}^{+}. (An example of the latter case is illustrated in the next system.)

The fifth system tested is that of Roland (*121*). This system is unusual in that Roland combined a natural gas and a crude oil in proportions similar to that produced from a gas condensate reservoir. The resulting mixture will not occur normally in a reservoir but could be similar to a system which would result from gas injection into an oil reservoir at very high pressure. For the predictions, the oil was separated into 34 fractions using a chromatographic analysis for an oil which has a molecular weight and specific gravity similar to the oil used by Roland. The predictions made using the physical properties obtained in the usual way gave light-component K values which were higher than the experimental values. To decrease the light-component K values, the boiling point and criticals for the subfractions were adjusted to reflect more paraffinic behavior than predicted using the specific gravity curve in Figure 19 selected in the normal manner. This additional correction is required for oils from several producing horizons in the world. After the adjustments were made, the predicted K values for the light components show good agreement with the experimental data at pressures up to about 5,000 psia. At pressures greater than 5,000 psia, the light-component K values diverged from unity instead of continuing to converge toward unity. When the 34 fractions were combined into 7 fractions, the predicted K values shown in Figure 27 (two temperatures) were obtained. The measured and predicted K values show excellent agreement for all components at both temperatures. Comparisons between the predicted and measured molecular weights and specific gravities for the combined heptanes plus fractions in both the vapor and liquid phases are shown in Figure 28. The agreement is within experimental error at both temperatures for pressures from 1,000 to 9,400 psia, indicating that the heavy-component K values are correct which agrees with the results shown in Figure 27. The mole percent of the total fluid in the liquid phase was reported by Roland based on material balance calculations for methane and heptanes plus. The results at 200°F are shown in Figure 29 along with the predicted values. The agreement is excellent at pressures up to 6,700 psia, and the predicted dew-point pressure is within 250 psi of the value estimated by Roland.

The sixth system used for comparison is a gas condensate fluid. Four sets of companion separator samples were collected during the initial flow tests on the producing well, and an experimental phase-distribution test was performed using one set of the samples. The compositions of the recombined wellstream based on the producing gas–oil ratio and the

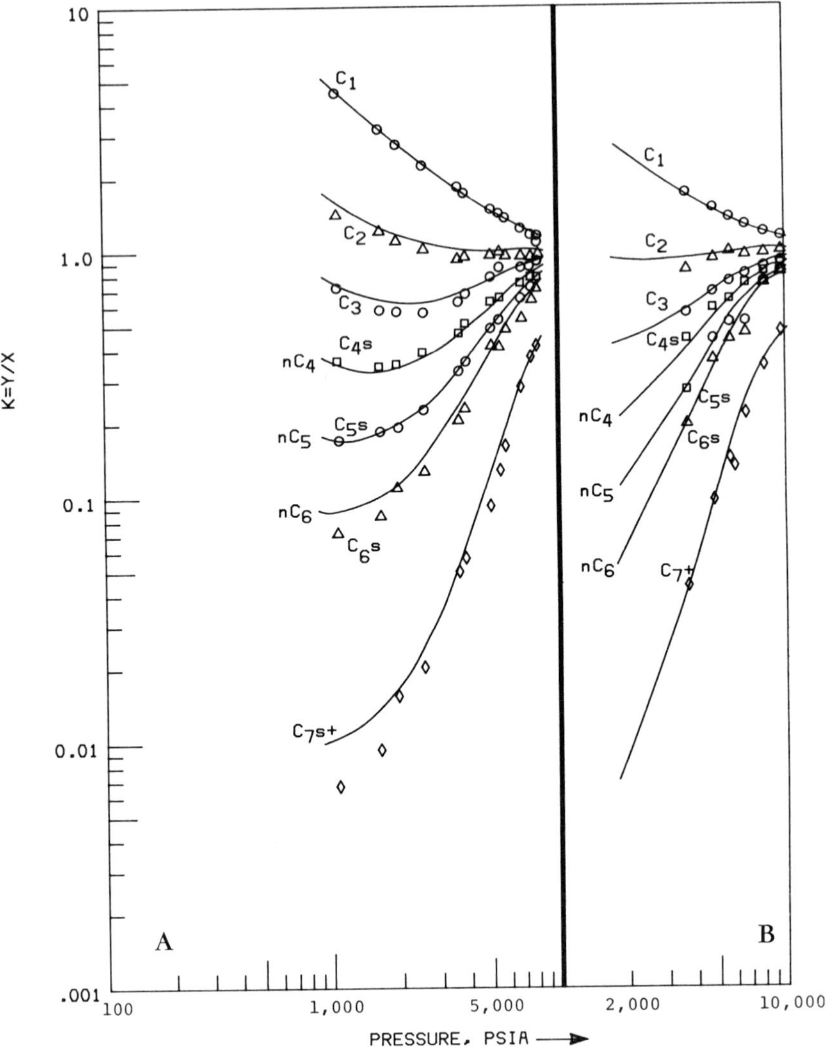

Figure 27. Comparisons of experimental and predicted K values for a natural gas–crude oil mixture at two temperatures: (◇, △, □, ○), data of Roland; (——), predicted. (A) 200°F; (B) 120°F.

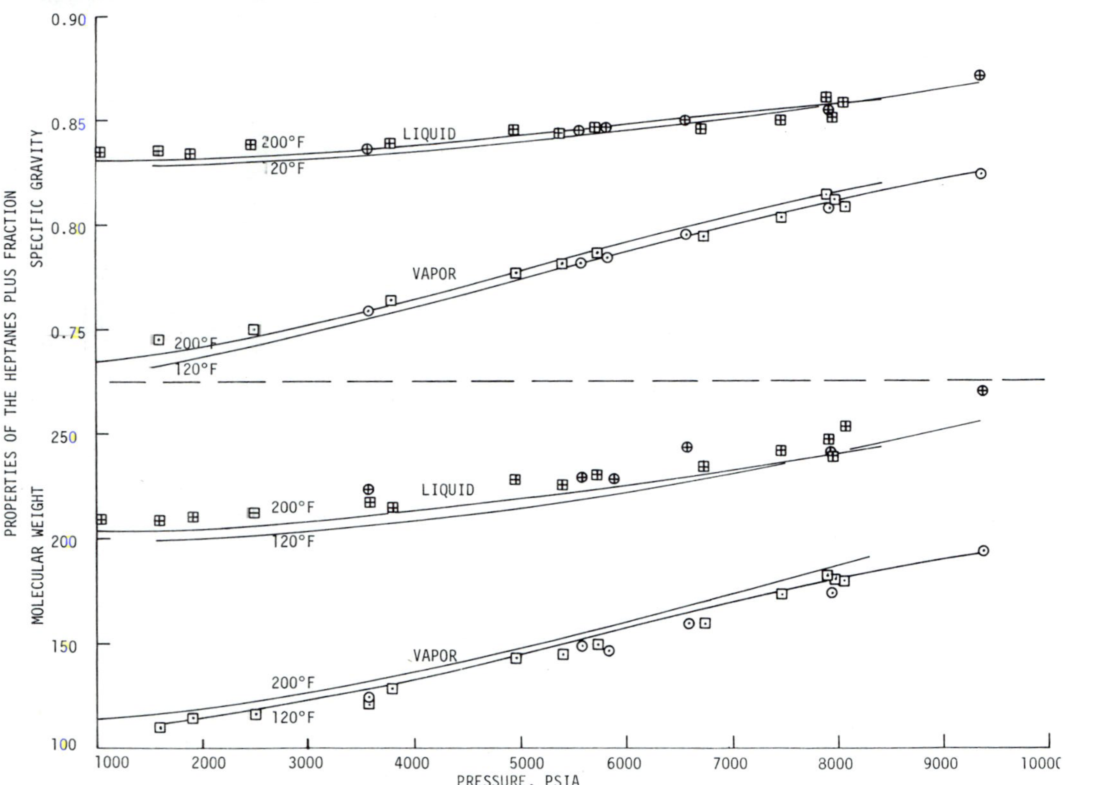

Figure 28. Comparisons of measured and predicted properties of the heptane plus fraction for a natural gas–crude oil mixture at two temperatures: (——), predicted. Data of Roland: (○), 120°F; (□), 200°F; (+), liquid; (•), vapor.

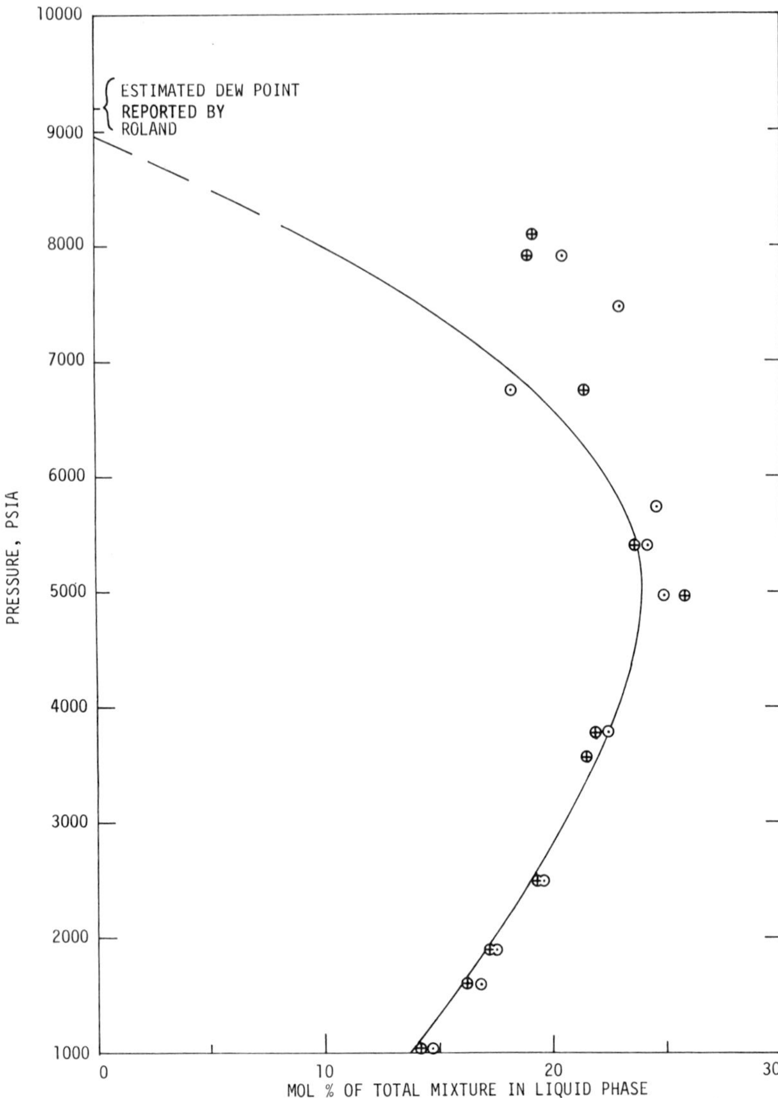

Figure 29. Proportion of total moles in liquid phase calculated from data of Roland and predicted at 200°F: (——), predicted; (⊙), calculated using methane data; (⊕), calculated using heptanes plus data.

separator fluid analyses are given in Table VI. The chromatographic analysis for a sample of gas-free separator liquid is shown in Figure 30. The components heavier than C_{32} were thought to be contaminants from a previous sample in the cylinder, and when the analysis through C_{40}^{+} was used to divide the heptane plus in the wellstreams, the predicted dew-point pressure was much higher than the experimental value. When the chromatographic analysis was terminated at C_{32} (using the solid curve in Figure 30), the results calculated for two wellstream compositions are shown in Figure 31. When the heaviest fraction used was limited to C_{31} and the heptanes plus in the wellstream was adjusted to 7.10 mol %, the prediction was in excellent agreement with the measured results, as shown in Figure 32. This illustrates the need for accurate field measurements, the collection of good samples, and accurate laboratory analyses in order to predict phase behavior with the accuracy required to provide acceptable match to experimental data for this type of fluid. Even with the phase-behavior prediction methods available at this time, often some experimental data are required for reservoir fluids before reservoir simulations are performed, especially for volatile-type fluids (gas condensate and volatile oils).

The final system is a high-shrinkage oil. The recombined wellstream was found experimentally to consist of two phases, gas and oil, at bottom-hole conditions. The equilibrium oil phase was subjected to an experimental test where the solution gas is liberated differentially from the oil phase, and the results are shown in Figures 33 and 34. These results will

Table VI. Recombined Wellstream Compositions for Produced Gas Condensate Fluid

Component	*Sample Set 1*	*Sample Set 2*	*Sample Set 3*	*Sample Set 4*
Nitrogen	0.64	0.63	0.58	0.57
Methane	77.69	77.39	78.01	77.86
Carbon dioxide	5.98	6.06	6.04	6.04
Ethane	4.12	4.15	4.02	4.04
Propane	1.86	1.89	1.78	1.80
iso-Butane	0.56	0.58	0.53	0.53
n-Butane	0.76	0.77	0.70	0.71
iso-Pentane	0.53	0.57	0.51	0.47
n-Pentane	0.36	0.40	0.36	0.32
Hexanes	0.51	0.53	0.60	0.43
Heptanes plus	7.00	7.02	6.87	7.24
Totals	100.00	100.00	100.00	100.00
Reported GOR	5656	5629	5861	5847
Heptanes plus				
Molecular weight	188	185	187	180
Specific gravity	0.8113	0.8123	0.8113	0.8114

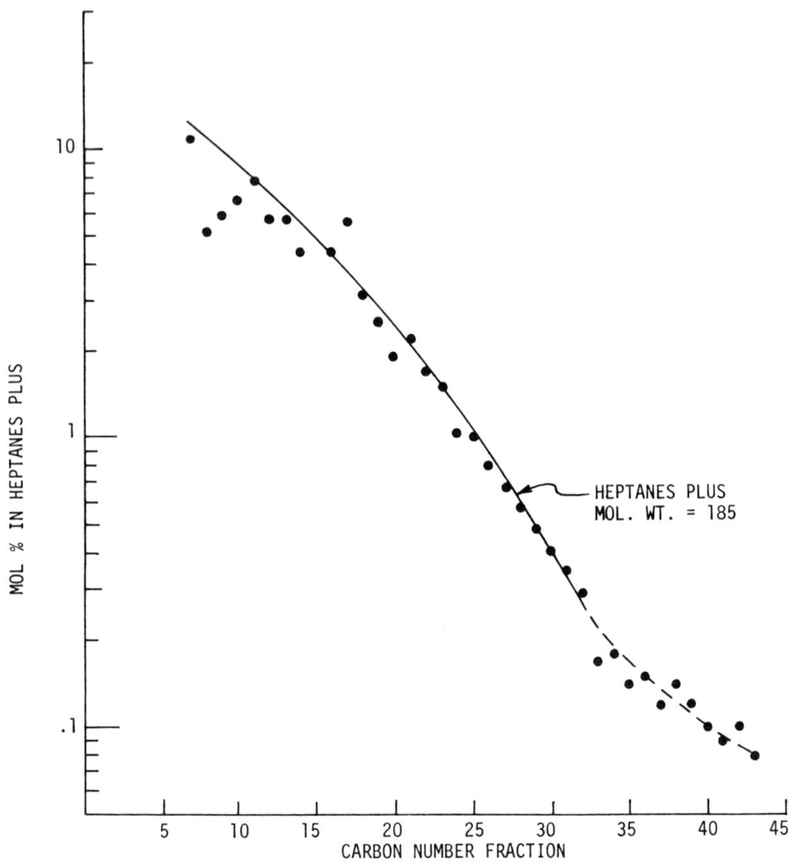

Figure 30. Carbon-number distribution of gas-free condensate

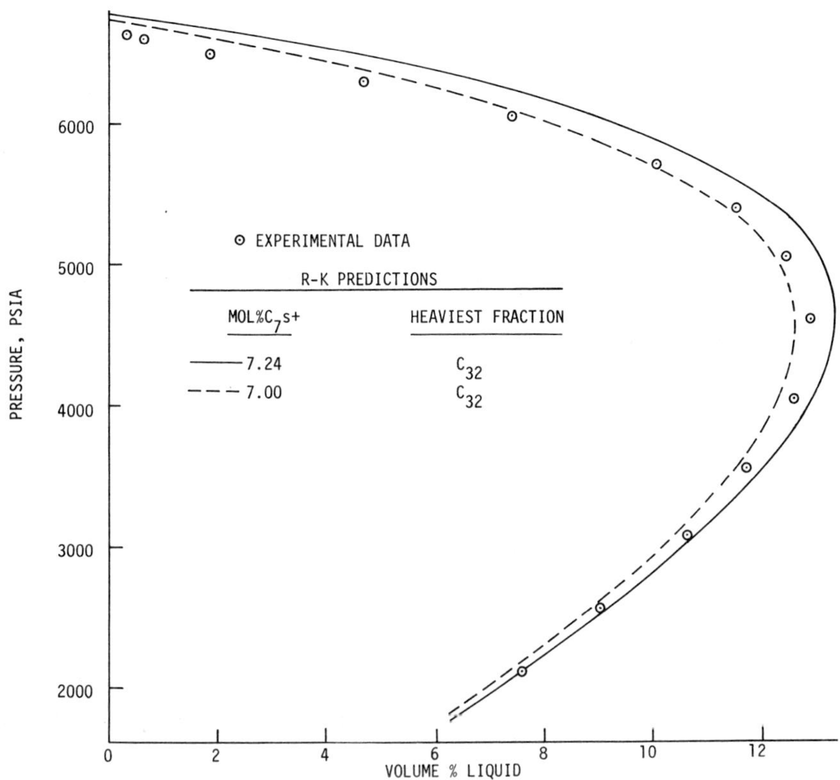

Figure 31. Measured and predicted phase distributions for heptanes plus divided through C_{32}'s. Temperature = 335°F.

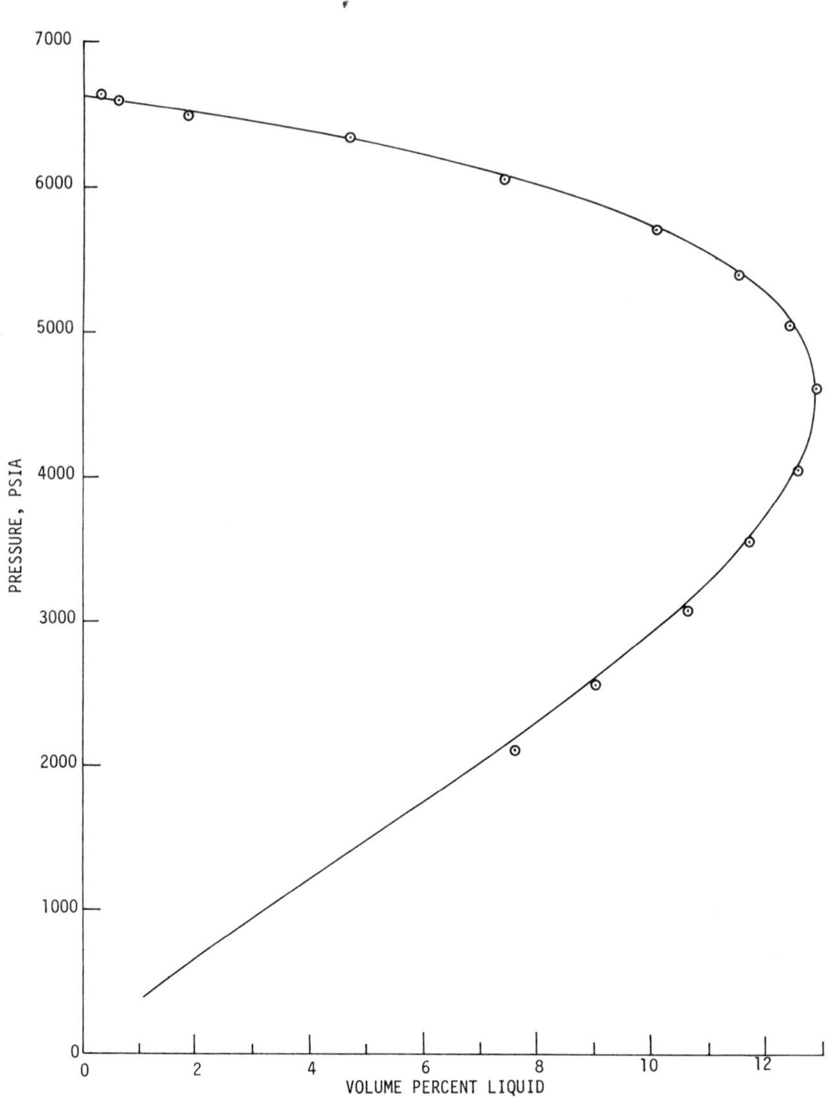

Figure 32. Measured and predicted phase distributions for heptanes plus divided through C_{31}'s: (⊙), experimental data; (——), prediction; 7.10 mol % C_7's+; C_{31} heaviest fraction. Temperature = 335°F.

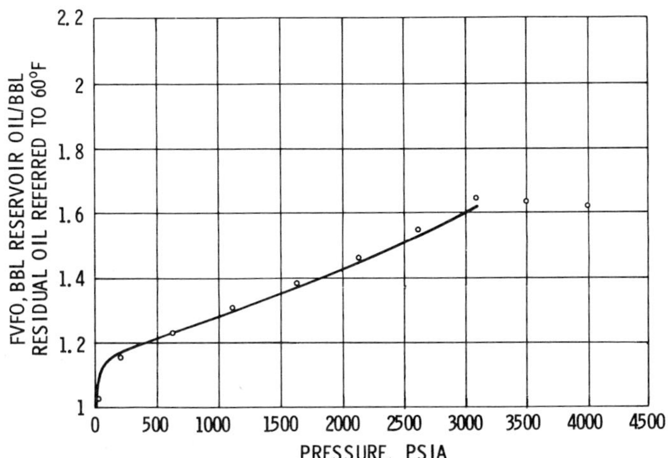

Figure 33. Comparison of experimental and predicted oil-volume factor curve for a high-shrinkage oil: (——), predicted values; (○), analysis by differential vaporization. Recombined oil; test temperature = 114°F.

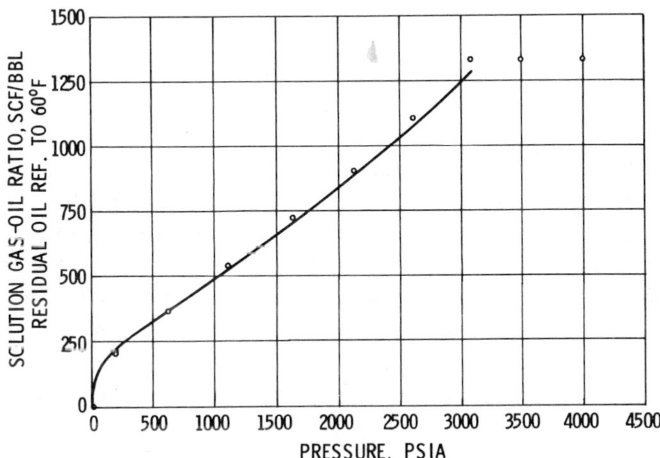

Figure 34. Comparison of experimental and predicted solution gas–oil ratio curve for a high-shrinkage oil: (——), predicted values; (○), analysis by differential vaporization. Recombined oil; test temperature = 114°F.

be used (along with other information) to calculate the fluid in place and the oil recovery by primary depletion. For comparison, the recombined wellstream composition, with a chromatographic analysis used to divide the heptanes plus fraction, was used as a basis for a flash calculation at bottomhole conditions. The predicted equilibrium oil composition was used in the equation of state while a series of flash calculations were performed to simulate the differential liberation test. The predicted results are compared with the experimental results in Figures 33 and 34.

The curves shown in Figures 33 and 34 are not calculated using liquid densities predicted via the equation of state. Testing of the liquid densities predicted for volatile, high-shrinkage, and black oils against experimental data showed that the predicted densities were high by as much as 10 to 15%. This was somewhat surprising since the densities for the ten-component system were predicted generally within 2%. For oil systems (and some gas condensates) the Standing–Katz liquid-density correlation (*122*) is used for liquid densities with modifications to the thermal expansion correction and the method for handling nitrogen and CO_2.

Concluding Remarks

Generalized temperature-dependent parameters have been developed for the RK equation of state which allow calculations to be made for components for which the critical temperature, critical pressure, and the acentric factor are known. The equation of state can be used to predict phase behavior for petroleum reservoir fluids which contain a combined heavy fraction when this fraction is characterized properly. A means for using a gas chromatographic analysis of the combined heavy fraction along with the measured molecular weight and specific gravity of the fraction to obtain a proper characterization is given. This characterization method appears to be better than using a distillation analysis, especially for oils where the distillation does not appear to provide sufficient definition of the distribution of heavy fractions.

A chromatographic analysis for one oil can be used to aid in the characterization of another oil which has similar physical properties. Calculations then can be made for fluids for which a detailed analysis is not available. The characterization procedure and the equation of state presented have been used to: (1) predict the saturation pressure and phase behavior for reservoir fluids; (2) predict properties for estimating the quantity of reservoir fluid in place and the recovery of that fluid by primary depletion; and (3) predict phase behavior during compositional model studies for: primary depletion (*123*); pressure maintenance by gas injection; gas cycling during declining pressure (*124*); miscible displacement by a hydrocarbon solvent (*125, 126*); and miscible displacement by CO_2 injection (*127*).

Although the equation of state can be and has been used to predict phase behavior for petroleum reservoir fluids for which no physical property data are available, it is recommended that some data, at least a saturation pressure, be measured in addition to a detailed component analysis of the fluid. This is particularly recommended when expensive compositional model studies are to be performed.

Glossary of Symbols

a = equation of state parameter
b = equation of state parameter
C_n = hydrocarbon with n carbon atoms
C_{ij} = unlike pair interaction parameter
K = vapor–liquid equilibrium distribution ratio
P = absolute pressure
R = universal gas constant
T = absolute temperature
V = molar volume
y = mole fraction (in the figures; y = mole fraction in vapor, x = mole fraction in liquid)
Ω = parameter defined by Equation 4
ω = acentric factor

Abbreviations

CO_2 = carbon dioxide
H_2S = hydrogen sulfide
N_2 = nitrogen
PNA = paraffinic–naphthenic–aromatic

Subscripts

a = refers to parameter a
b = refers to parameter b
c = critical point value
i = Component i
j = Component j
m = mixture value

Literature Cited

1. Redlich, O.; Kwong, J. N. S. "On the Thermodynamics of Solutions. V. An Equation of State. Fugacities of Gaseous Solutions," *Chem. Rev.* **1949**, *44*, 233–244.

2. Zudkevitch, D.; Joffe, J. "Correlation and Prediction of Vapor-Liquid Equilibria with the Redlich-Kwong Equation of State," *AIChE J.* **1970,** *16,* 112–119.
3. Chang, S.-D.; Lu, B. C.-Y. "Prediction of Partial Molal Volumes of Normal Fluid Mixtures," *Can. J. Chem. Eng.* **1970,** *48,* 261–266.
4. Joffe, J.; Schroeder, G. M.; Zudkevitch, D. "Vapor-Liquid Equilibria with the Redlich-Kwong Equation of State," *AIChE J.* **1970,** *16,* 496–498.
5. Pitzer, K. S.; Lippmann, D. Z.; Curl, R. F., Jr.; Huggins, C. M.; Peterson, D. E. "The Volumetric and Thermodynamic Properties of Fluids. II. Compressibility Factor, Vapor Pressure, and Entropy of Vaporization," *J. Am. Chem. Soc.* **1955,** *77,* 3433–3440.
6. Carruth, G. R. "Determination of the Vapor Pressures of n-Paraffins and Extension of a Corresponding States Correlation to Low Reduced Temperatures," Ph.D. Thesis, Chemical Engineering Department, Rice University, Houston, TX, 1970.
7. Rackett, H. G. "Equation of State for Saturated Liquids," *J. Chem. Eng. Data* **1970,** *15,* 514–517.
8. Lyckman, E. W.; Eckert, C. A.; Prausnitz, J. M. "Generalized Reference Fugacities for Phase Equilibrium Thermodynamics," *Chem. Eng. Sci.* **1965,** *20,* 685–691.
9. Bloomer, O. T.; Parent, J. D. *Chem. Eng. Prog., Symp. Ser.* **1953,** *49*(6), 11.
10. Skripka, V. G.; Nikitina, I. E.; Zhidanovich, L. A.; Sirotin, A. G.; Benyaminovich, O. A. *Gazov. Prom.* **1970,** *12,* 35–36.
11. Ellington, R. T.; Eakin, B. E.; Parent, J. D.; Gami, D. C.; Bloomer, O. T. "Thermodynamic and Transport Properties of Gases, Liquids, and Solids"; American Society of Mechanical Engineers: New York, 1959; p. 180–194.
12. Schindler, D. L.; Swift, G. W.; Kurata, F. *Hydrocarbon Proc.* **1966,** *45*(11), 205–210.
13. Roberts, L. R.; McKetta, J. J. *AIChE J.* **1961,** *7,* 173.
14. Akers, W. W.; Atwell, L. L.; Robinson, I. A. *Ind. Eng. Chem.* **1954,** *44,* 2539.
15. Skripka, V. G.; Barsuk, S. D.; Nikitina, I. E.; Gubkina, G. F.; Benyaminovich, O. A. *Gazov. Prom.* **1969,** *4,* 41–45.
16. Poston, R. S.; McKetta, J. J. *J. Chem. Eng. Data* **1966,** *11,* 364.
17. Akers, W. W.; Kehn, D. M.; Kilgore, C. H. *Ind. Eng. Chem.* **1954,** *46,* 2536.
18. Azarnoosh, A.; McKetta, J. J. *J. Chem. Eng. Data* **1963,** *8,* 494.
19. Miller, P.; Dodge, B. F. *Ind. Eng. Chem.* **1940,** *32,* 434.
20. Robinson, D. B.; Besserer, G. J. NGPA RR-7, June, 1972, NGPA, Tulsa, Oklahoma.
21. Genichi, K.; Toriumi, T. *Kogyo Kagaku Zasshi* **1966,** *69*(2), 175–178.
22. Zenner, G. H.; Dana, L. I. *Chem. Eng. Prog. Symp. Ser.* **1963,** *59*(44), 36–41.
23. Kaminishi, G.-I.; Arai, Y.; Saito, S.; Maeda, S. *J. Chem. Eng. Jpn.* **1968,** *1*(2), 109–116.
24. Donnelly, H. G.; Katz, D. L. *Ind. Eng. Chem.* **1954,** *46,* 511.
25. Khazanova, N. E.; Lesnevskaya, L. S.; Zakharova, A. V. *Khim. Prom.* **1966,** *42*(5), 364–365.
26. Kurata, F.; Swift, G. W. NGPA RR-5, Dec., 1971, NGPA, Tulsa, Oklahoma.
27. Akers, W. W.; Kelley, R. E.; Lipscomb, T. G. *Ind. Eng. Chem.* **1954,** *46,* 2535–2536.
28. Poettman, F. H.; Katz, D. L. *Ind. Eng. Chem.* **1945,** *37,* 847-853.
29. Reamer, H. H.; Sage, B. H.; Lacey, W. N. *Ind. Eng. Chem.* **1951,** *43,* 2515.

30. Olds, R. H.; Reamer, H. H.; Sage, B. H.; Lacey, W. N. *Ind. Eng. Chem.* **1949,** *41,* 475–482.
31. Reamer, H. H.; Sage, B. H. *J. Chem. Eng. Data 1963,* 8, 508.
32. Wan, S.; Dodge, B. F. *Ind. Eng. Chem.* **1940,** *32,* 95.
33. Bierlein, J. A.; Kay, W. B. *Ind. Eng. Chem.* **1953,** *45,* 618.
34. Sobocinski, D. F.; Kurata, F. *AIChE J.* **1959,** *5,* 545–551.
35. Reamer, H. H.; Sage, B. H.; Lacey, W. N. *Ind. Eng. Chem.* **1951,** *43,* 976.
36. Kohn, J. P.; Kurata, F. *AIChE J.* **1958,** *4,* 211–217.
37. Kay, W. B.; Brice, D. B. *Ind. Eng. Chem.* **1953,** *45,* 615.
38. Kay, W. B.; Rambosek, G. M. *Ind. Eng. Chem.* **1953,** *45,* 221.
39. Brewer, J.; Rodewald, N.; Kurata, F. *AIChE J.* **1961,** *7,* 13.
40. Robinson, D. B.; Hughes, R. E.; Sandercock, J. A. W. *Can. J. Chem. Eng.* **1964,** *42,* 143.
41. Reamer, H. H.; Sage, B. H.; Lacey, W. N. *Ind. Eng. Chem.* **1953,** *45,* 1805–1809.
42. Reamer, H. H.; Selleck, F. T.; Sage, B. H.; Lacey, W. N. *Ind. Eng. Chem.* **1953,** *45,* 1810–1812.
43. Levine, M. I.; Ipatieff, V. V.; Postnov, N. I. *Khim. Teknol. Topl. Masel* **1960,** *5*(1), 17–18.
44. Price, A. R.; Kobayashi, R. *J. Chem. Eng. Data* **1959,** *4,* 40.
45. Wichterle, I.; Kobayashi, R. *J. Chem. Eng. Data* **1972,** *17,* 9–12.
46. Reamer, H. H.; Sage, B. H.; Lacey, W. N. *Ind. Eng. Chem.* **1950,** *42,* 534, 1258.
47. Akers, W. W.; Burns, J. F.; Fairchild, W. R. *Ind. Eng. Chem.* **1954,** *46,* 2531.
48. Wichterle, I.; Kobayashi, R. *J. Chem. Eng. Data* **1972,** *17,* 4–9.
49. Olds, R. H.; Sage, B. H.; Lacey, W. N. *Ind. Eng. Chem.* **1942,** *34,* 1008.
50. Sage, B. H.; Hicks, B. L.; Lacey, W. N. *Ind. Eng. Chem.* **1940,** *32,* 1085.
51. Roberts, L. R.; Wang, R. H.; Azarnoosh, A.; McKetta, J. J. *J. Chem. Eng. Data* **1962,** *7,* 484.
52. Amick, E. H.; Johnson, W. B.; Dodge, B. F. *Chem. Eng. Prog. Symp. Ser.* **1952,** *48*(3), 65.
53. Taylor, H. S.; Wald, G. W.; Sage, B. H.; Lacey, W. N. *Oil Gas J.* **1939,** *38*(10), 46.
54. Shim, J.; Kohn, J. P. *J. Chem. Eng. Data* **1962,** *7,* 3–8.
55. Poston, R. S.; McKetta, J. J. *J. Chem. Eng. Data* **1966,** *11,* 362–363.
56. Reamer, H. H.; Sage, B. H.; Lacey, W. N. *J. Chem. Eng. Data* **1956,** *1,* 29.
57. Chang, H. L.; Hurt, L. J.; Kobayashi, R. *AIChE J.* **1966,** *12,* 1212–1216.
58. Kohn, J. P.; Bradish, W. F. *J. Chem. Eng. Data* **1964,** *9,* 5.
59. Shipman, L. M.; Kohn, J. P. *J. Chem. Eng. Data* **1966,** *11,* 176.
60. Reamer, H. H.; Olds, R. H.; Sage, B. H.; Lacey, W. N. *Ind. Eng. Chem.* **1942,** *34,* 1526–1531.
61. Koonce, K. T.; Kobayashi, R. *J. Chem. Eng. Data* **1964,** 9, 490.
62. Beaudoin, J. M.; Kohn, J. P. *J. Chem. Eng. Data* **1967,** *12,* 189–191.
63. Puri, S.; Kohn, J. P. *J. Chem. Eng. Data* **1970,** *15,* 372–374.
64. Reamer, H. H.; Sage, B. H.; Lacey, W. N. *J. Chem. Eng. Data* **1958,** *3,* 240.
65. Chang, H. L.; Kobayashi, R. *J. Chem. Eng. Data* **1967,** *12,* 520–523.
66. Elbishlawi, M.; Spencer, J. R. *Ind. Eng. Chem.* **1951,** *43,* 1811–1815.
67. Chang, H. L.; Kobayashi, R. *J. Chem. Eng. Data* **1967,** *12,* 517–520.
68. Matschke, D. E.; Thodos, G. *J. Chem. Eng. Data* **1962,** *7,* 232.
69. Djordjevich, L.; Budenholzer, R. A. *J. Chem. Eng. Data* **1970,** *15,* 10–12.
70. Kay, W. B. *Ind. Eng. Chem.* **1940,** *32,* 353–357.
71. Mehra, V. S.; Thodos, G. *J. Chem. Eng. Data* **1965,** *10,* 307.
72. Reamer, H. H.; Sage, B. H.; Lacey, W. N. *J. Chem. Eng. Data* **1960,** *5,* 44.

73. Zais, E. J.; Silberberg, I. H. *J. Chem. Eng. Data* **1970,** *15,* 253–256.
74. Mehra, V. S.; Thodos, G. *J. Chem. Eng. Data* **1965,** *10,* 211.
75. Rodrigues, A. B.; McCaffrey, D. S.; Kohn, J. P. *J. Chem. Eng. Data* **1968,** *13,* 164.
76. Reamer, H. H.; Sage, B. H. *J. Chem. Eng. Data* **1962,** *7,* 161.
77. Kay, W. B.; Albert, R. E. *Ind. Eng. Chem.* **1956,** *48,* 422–426.
78. Kay, W. B.; Nevens, T. D. *Chem. Eng. Prog. Symp. Ser.* **1952,** *48*(3), 108.
79. Hipkin, H. *AIChE J.* **1966,** *12,* 484–487.
80. Nysewander, C. N.; Sage, B. H.; Lacey, W. N. *Ind. Eng. Chem.* **1940,** *32,* 118.
81. Vaughan, W. E.; Collins, F. C. *Ind. Eng. Chem.* **1942,** *34,* 885–890.
82. Sage, B. H.; Lacey, W. N. *Ind. Eng. Chem.* **1940,** *32,* 992.
83. Reamer, H. H.; Sage, B. H. *J. Chem. Eng. Data* **1966,** *11,* 17.
84. Glanville, J. W.; Sage, B. H.; Lacey, W. N. *Ind. Eng. Chem.* **1950,** *42,* 508–513.
85. Kay, W. B. *Ind. Eng. Chem.* **1941,** *33,* 590.
86. Reamer, H. H.; Sage, B. H. *J. Chem. Eng. Data* **1964,** 9, 24–28.
87. Cummings, L. W. T.; Stones, F. W.; Volante, M. A. *Ind. Eng. Chem.* **1933,** *25,* 728.
88. Price, A. R.; Kobayashi, R. *J. Chem. Eng. Data* **1959,** *4,* 40.
89. Wichterle, I.; Kobayashi, R. *J. Chem. Eng. Data* **1972,** *17,* 13–18.
90. Wiese, H. C.; Jacobs, J.; Sage, B. H. *J. Chem. Eng. Data* **1970,** *15,* 82–91.
91. Billman, G. W.; Sage, B. H.; Lacey, W. N. *Trans. Am. Inst. Min., Metall., Pet. Eng.* **1948,** *174,* 13–22.
92. Carter, R. T.; Sage, B. H.; Lacey, W. N. *Trans. Am. Inst. Min., Metall., Pet. Eng.* **1941,** *42,* 170–177.
93. Dourson, R. H.; Sage, B. H.; Lacey, W. N. *Trans. Am. Inst. Min., Metall., Pet. Eng.* **1943,** *151,* 206–215.
94. Wiese, H. C.; Reamer, H. H.; Sage, B. H. *J. Chem. Eng. Data* **1970,** *15,* 75–82.
95. Reamer, H. H.; Fiskin, J. M.; Sage, B. H. *Ind. Eng. Chem.* **1949,** *41,* 2871–2875.
96. Reamer, H. H.; Sage, B. H.; Lacey, W. N. *Ind. Eng. Chem.* **1951,** *43,* 1436–1444.
97. Ibid. **1952,** *44,* 1671–1675.
98. Van Horn, L. D.; Kobayashi, R. *J. Chem. Eng. Data* **1967,** *12,* 294–303.
99. Koonce, K. T.; Kobayashi, R. *J. Chem. Eng. Data* **1964,** 9, 494–501.
100. Velikovskii, A. S.; Stepanova, G. S.; Vybornova, Y. I. *Gazov. Prom.* **1965,** *6,* 45–49.
101. Herlihy, J. C.; Thodos, G. *J. Chem. Eng. Data* **1962,** *7,* 348–351.
102. Mehra, V. S.; Thodos, G. *J. Chem. Eng. Data* **1963,** *8,* 1–8.
103. Ibid., **1966,** *11,* 365–372.
104. Ibid., **1968,** *13,* 155–160.
105. Chang, S. D.; Lu, B. C.-Y. *Chem. Eng. Prog. Symp. Ser.* **1967,** *63*(81), 18.
106. Roberts, L. R.; McKetta, J. J. *J. Chem. Eng. Data* **1963,** *8,* 161–163.
107. Azarnoosh, A.; McKetta, J. J. *J. Chem. Eng. Data* **1963,** *8,* 513–519.
108. Saxena, A. C.; Robinson, D. B. *Can. J. Chem. Eng.* **1969,** *47,* 69–75.
109. Yarborough, L. "Vapor-Liquid Equilibrium Data for Multicomponent Mixtures Containing Hydrocarbon and Nonhydrocarbon Components," *J. Chem. Eng. Data* **1972,** *17,* 129–133.
110. Hopke, S. W.; Lin, C. J. "Application of the BWRS Equation to Absorber Systems," *Proc. 53rd Annual Gas Processors Assn. Convention,* 1974, 63–71.
111. Robinson, D. B.; Peng, D.-Y. "The Characterization of the Heptanes and Heavier Fractions for the GPA Peng-Robinson Programs," Gas Processors Assn. Research Report 28, Tulsa, OK, March 1978.

112. Jacoby, R. "NGPA Phase Equilibrium Project," *Proc., Am. Pet. Inst., Sect. 3* **1964,** 288.
113. Nagy, B.; Colombo, U. "Fundamental Aspects of Petroleum Geochemistry"; Elsevier: New York, 1967; Chapters 3 and 4.
114. "Technical Data Book—Petroleum Refining"; American Petroleum Institute: Washington, D.C., 1970; Figures 2B2.1, 4A1.2, and 4B1.2.
115. Edmister, W. C. "Applied Hydrocarbon Thermodynamics"; Gulf Publishing Company: Houston, 1961; p. 28.
116. Hoffmann, A. E.; Crump, J. S.; Hocott, C. R. "Equilibrium Constants for a Gas Condensate System," *Trans. Am. Inst. Min., Metall., Pet. Eng.* **1953,** *198,* 1–10.
117. Roland, C. H.; Smith, D. E.; Kaveler, H. H. "Equilibrium Constants for a Gas-Distillate System," *Oil Gas J.* **1941,** *39*(46), 128–132.
118. Katz, D. L.; Hachmuth, K. H. "Vaporization Equilibrium Constants in a Crude Oil-Natural Gas System," *Ind. Eng. Chem.* **1937,** *29,* 1072–1077.
119. Evans, R. B.; Harris, D. "Equilibrium Vaporization Ratios—Hydrocarbon Mixture Containing Two Concentrations of Heptanes and Heavier Fraction," *J. Chem. Eng. Data* **1956,** *1,* 45–50.
120. Firoozabadi, A.; Katz, D. L. "Predicting Phase Behavior of Condensate—Crude Oil Systems Using Methane Interaction Coefficients," *Soc. Pet. Eng. Fall Meeting, Denver, Colorado, October 9–12, 1977.*
121. Roland, C. H. "Vapor-Liquid Equilibria for Natural Gas-Crude Oil Mixtures," *Ind. Eng. Chem.* **1945,** *27,* 930–936.
122. Standing, M. B.; Katz, D. L. "Density of Crude Oils Saturated with Natural Gas," *Trans. Am. Inst. Min., Metall., Pet. Eng.* **1942,** *146,* 159–165.
123. Fussell, D. D. "Single Well Performance Predictions for Gas Condensate Reservoirs," *J. Pet. Technol.* **1973,** 860–870.
124. Fussell, D. D.; Yarborough, L. "The Effect of Phase Data on Liquids Recovery During Cycling of a Gas Condensate Reservoir," *Soc. Pet. Eng. J.* **1972,** 96–102.
125. Metcalfe, R. S.; Fussell, D. D.; Shelton, J. L. "A Multicell Equilibrium Separation Model for the Study of Multiple Contact Miscibility in Rich Gas Drives," *Soc. Pet. Eng. J.* **1973,** 147–155.
126. Fussell, D. D.; Shelton, J. L.; Griffith, J. D. "Effect of Rich Gas Composition on Multiple Contact Miscible Displacement—A Cell-to-Cell Flash Model Study," *Soc. Pet. Eng. J.* **1976,** 310–316.
127. Metcalfe, R. S.; Yarborough, L. "Effect of Phase Equilbria on the CO_2 Displacement Mechanism," *Soc. Pet. Eng. Fifth Symposium on Improved Methods for Oil Recovery,* Tulsa, Oklahoma, April 16–19, 1978, paper 7061.

RECEIVED October 5, 1978.

INDEX

INDEX

F

G

J

K

L

Q

R

S

W

Jacket design by Carol Conway.
Editing and production by Candace A. Deren.

The book was composed by Service Composition Co., Baltimore, MD, printed and bound by The Maple Press Co., York, PA.